FADED FOOTPRINTS

Other Lost Treasure and Ghost Town Books
by George A. Thompson

Some Dreams Die: Utah's Ghost Towns and Lost Treasures
$19.95. New edition forthcoming in 1997. ISBN 0-942688-22-8

Treasure Mountain Home: Park City Revisited
$16.95. ISBN 0-942688-89-9
100 copy hardbound, slipcased signed limited edition: $100.00. ISBN 0-942688-90-2

Throw Down the Box! Treasure Tales from Gilmer and Salibury The Western Stagecoach King
$15.95. ISBN 0-942688-73-2

George Thompson's books are available in stores or directly from the publisher.
Please add $5.00 per order for shipping. Utah residents add 6⅛% sales tax.

FADED FOOTPRINTS

THE LOST RHOADES MINES
and Other Hidden Treasures of the Uintahs

GEORGE A. THOMPSON

This book is a completely rewritten and expanded version of an earlier work by the author.

The new work was finished by the author prior to his untimely death in May of 1995.
It is over 80 pages longer than the old edition and features almost 200
photographs, over sixty of which have never before been published.

© 1991 George A. Thompson
Revised edition © 1996 by the estate of George A. Thompson and Dream Garden Press

All Rights Reserved

ISBN 0-942688-37-6

Library of Congress Catalog Number 96-70968

All photographs are courtesy of the author or
the Utah State Historical Society unless otherwise noted.
Front cover photograph of Spanish cannon courtesy of Mr. Rubin Guillen.

Cover design: Craig Geertsen
Book design: Richard Firmage
Typesetting: Patricia Parkinson

Dream Garden Press
P.O. Box 27076
Salt Lake City, Utah 84127

Printed in the United States of America

PREFACE

This book is about lost mines and treasures in the Uinta Mountains of northern Utah; a few of them found, but many more still lost. It is also about Spanish silver and gold, for those who have spent much of their lives exploring and prospecting the Uinta Mountains know that most of the lost mines and caches hidden there are of Spanish origin. But mostly this book is about the biggest treasure of them all, the place which prospectors, rockhounds, treasure hunters and armchair historians all search for; the fabled Lost Rhoads Mines, for Rhoads gold and Spanish gold are one and the same. Except for the author's *Some Dreams Die* and *Lost Treasures On The Old Spanish Trail* and *Footprints In The Wilderness,* written by Kerry Ross Boren and the late Gale Rhoades, there is little on record regarding Spanish mines and treasure in Utah's northern mountains.

Nearly every school child in Utah knows that Spanish conquistadors explored much of their state long before arrival of the first Mormon pioneers, and in a vague sort of way, most Utahns accept the fact that their state was once Spanish territory, and that sometime in the distant past there were Spanish mines, missions and settlements at places with names like San Rafael, Montezuma, Spanish Fork and Escalante. The first Mountain Men and fur trappers to venture into those then unknown mountains discovered ancient ruins, mine shafts and smelter sites; old and long abandoned when they first saw them. Mormon explorers encountered Mexican pack trains carrying heavy loads of gold and silver bullion from the mountains high above the Kamas and Heber valleys, and later from the wild and forbidding retreats located north of the Uinta Basin.

Mormon settlers were noted for keeping detailed diaries and family journals, and in many of those accounts there are references to artifacts of European manufacture being found: swords, rusted pieces of armor and antiquated mining tools. They also recorded finding names, dates and strange signs and symbols cut into the bark of patriarch pines, or inscribed deeply into sandstone ledges. When those first Mormon pioneers arrived in the valley of the Great Salt Lake, they were destitute, among the poorest people in the land, yet only a year later they established a mint to coin gold pieces. For a time, few people seemed to know where the gold used to mint those coins came from, but it wasn't long until the name of Thomas Rhoads became as well known as that of Brigham Young. The legend of the Lost Rhoads Mine had been born. It wasn't long until some of those pioneer settlers began finding Spanish mines, from Kamas Valley where the Wasatch Range butts up against the Uinta Mountains, to the Green River along the Colorado border.

Until now, information on those old Spanish mines and hidden caches has been the closely guarded secret of a small brotherhood, a closely-knit and close mouthed fraternity. But over the years most of that elite club have crossed the Great Divide, while those few remaining belong to the over-the-hill gang. It's time to let a new generation of treasure hunters in on the fun! For the first time many of those jealously-guarded waybills and mysterious maps will be made public. Tales, tips and a thousand tidbits of treasure lore acquired through more than forty years of search, research and interviews with hundreds of old-timers will be told. But I'm not going to make finding those lost treasures easy, for I know from experience that half the fun of finding treasure is putting the pieces of the puzzle together. Since the first edition of *Faded Footprints,* several treasure hunters have thanked me for giving them the clues necessary to find treasure, while others have said the information they gleaned from that volume furnished the missing pieces they needed to solve their own puzzle. the clues to many other treasures still unfound are all here, but you might have to read between the lines to find them!

In order to properly set the stage for readers not already familiar with early Spanish exploration and mining in the Utah mountains, part of Chapter One will provide a brief description of the Old Spanish Trail, which led from Santa Fe to the northern mountains. Those who have read the author's earlier works will be familiar with that ancient track, but even they will find more specific details of how that trail relates to mines and mining in the Uinta Mountains; while updated information about treasures mentioned in earlier books will be found in this new volume. Don't skip quickly over references to the Old Spanish Trail, for many vital treasure clues can be found there. The reader may also find some parallels with Footprints In The Wilderness, since an understanding of the life and times of Thomas Rhoads and his son, Caleb Rhoads, is necessary in order to weave all of the loose threads of the tale together. For many readers, that book raised more questions than it answered. This new edition will attempt to answer some of those questions. Much of the legend of the Lost Rhoads Mines is so closely intertwined with other lost Spanish mines and caches that occasional reference to people and places mentioned in that book will be made.

In several places in following chapters, references will be made to mining terms and technology, as well as to geology, metal detectors and electronic treasure finders; the latter sometimes called "doodlebugs." There are good reasons for including that information, not the least of which has been frequent requests that I do so. I have often been asked: Is it possible for minerals to occur in the rock formations one encounters in the Uinta Mountains? The answer is there would be no reason for me to encourage prospecting in those mountains if I thought they were barren of valuable minerals. To corroborate that opinion, several references to geological studies conducted in the Uinta Mountains will be cited. A Spanish mining terms glossary will be found in Appendix One. There are many maps associated with the Lost Rhoads Mines, the most significant of which will be described at length in the text and shown in detail in a Map Index found in Appendix Two.

A parallel theme of this book is a strange phenomenon which I will refer to as the "Killer Mountains." One cannot search the remote places of the Uinta Mountains, or research the history of lost mines and treasure caches hidden there, without encountering tales of mysterious murders and unexplained disappearances of others who earlier sought those same treasures. Prospectors and treasure hunters have been found with their necks broken, others have been shot, while a few have been hanged with their hands tied behind their backs. A retired mortician who once served that area stated that he had participated in the investigation of at least two dozen such murders over a period of forty years! A retired sheriff attributes many of those killings to a few "crazies," but many prospectors believe they are the work of a small but dedicated band of Indians whose job it is keep treasure hunters away from the Spanish caches they regard as being sacred. But whoever guards the hidden treasures of the Killer Mountains, don't under-estimate them. Some of those who did are now in lonely and forgotten graves all across those blood-stained mountains!

Those who have contributed to this volume will be recognized, except in a few instances where they have requested anonymity, or to ensure their personal safety. Their stories are true, which subsequent events and circumstances have verified, even though in a few instances their tales may be based upon legend or folklore. But remember, even legends and folklore are based upon actual facts and events. In the case of Indian legends, oral history is all we have, since the Utes had no written language. The exact location of some Spanish names, dates and symbols will not be given to protect them from vandals. Too many old treasure signs and trail markers have already been damaged or destroyed by thoughtless persons.

Several people have contributed photos or research material, and deserve special credit. I am indebted to Wayne Handy, who shared information acquired from a lifetime of treasure hunting in the Utah mountains. I also want to thank Kerry Ross Boren, who allowed the author use of the extremely rare "Happy Jack Journals." I am especially grateful to Bill Bleazard for sharing his memories of early homestead and mining days on the old Ute Indian Reservation. Thanks also is due Dan Tucker, a companion of many long, hard days on the treasure trail, and also to Eric Barnes, whose enthusiasm kept out hopes high when our spirits were low. And appreciation is due Ken Sanders, whose publishing expertise made this and other books by the author possible.

CONTENTS

George A. Thompson
January 27, 1930–May 3, 1995

The late author, George Thompson,
at a mine somewhere in the Uintas.

More than forty years of in-the-field search and research has established Mr. Thompson as the foremost authority on Utah's lost mines and treasures. He is not a coffee-shop prospector or armchair historian, but a man who has traveled the dim trails he describes and has been to the places he writes about. He is a great-nephew of the fabled Ute Indian prospector white men called Pick Murdock, and counts among his associates such outstanding mining men as Bill Bleazard and renowned treasure hunter Mel Fisher, discoverer of the treasure ship Atocha. Join him and his many friends in this intriguing tale of treasure!

CHAPTER 1

THE OLD SPANISH TRAIL
Footpath in the Wilderness

The Old Spanish Trail in Utah was an obscure sort of path through the wilderness, an indefinite and sometimes dim trail, its course changed by the whim of every exploring and mining party that followed it north from Santa Fe to Utah Valley. Each band of prospectors, miners and slave traders who pursued its tortuous zig-zag miles across red sandstone desert or through dark forests of pine and aspen chose their own route, each believing his way was the best or most direct. Each party picked whichever river fork they chose to follow, or decided which mountain pass seemed easiest to climb. No wonder that old trail is such a difficult track to retrace today! Using modern place-names for convenience, the trail from Santa Fe led nearly due north past the frontier outposts of Abique and Chama, en-route to the headwaters of the San Juan River. It followed that blood-red stream nearly straight west to Davis Canyon and Horse Mountain, in the four-corners wilderness of southeastern Utah. There it turned to the northwest, following the trail of the Ancient Ones, the Anasazi, into the Needles Country of present-day Canyonlands Park. It wound its way through that moon-like sandstone desert to Red Lake Canyon, where it descended to the Spanish Bottoms on the Colorado River.

From the Colorado, the Rio Tizon of Spanish days, the trail climbed steeply through a jagged gash in the cliffs to the canyon rim high above, where it entered the Land of Standing Rocks. That now forgotten track, in places worn deeply into the soft sandstone rock of the desert by the hooves of countless horses and pack animals, turned westward again, to pass through the weird world of the Maze and Land's End. It snaked its way up through green cedars into Sunset Pass, and then dropped down the north side of that narrow defile into North Hatch Canyon and to its mouth at the turbid waters of the Dirty Devil River. Forgotten names and dates as early as 1642 have been found along the trail in that still wild and seldom-seen country.

The Old Spanish Trail continued to the northwest, following the course of the Dirty Devil, that roily-red river named by Major John Wesley Powell's expedition of 1869. It passed by the steeple-spires of Temple Mountain and into the mysterious realms of the Sinbad Desert, along what some have called Camino del Diablo, the Devil's Road, and on to the San Rafael Knob and Eagle Canyon. It pursued the twists and turns of Coal Wash down to Salt Wash, a wild and broken gorge leading to the towering mesas and seemingly bottomless canyons of the San Rafael River. Historian Bert Silliman discovered an eerie place of unmarked graves along that stretch of trail, where a Spanish pack train heavily-laded with bars of gold bullion was attacked by waiting Ute Indians. In his notes, on file at the Utah Historical Society, Silliman describes that burial site, where he found pieces of rotted leather pack sacks, rusted iron buckles and brass uniform buttons in the sandy desert wastes. He wrote of that old trail, worn deep into the gray shale rock from years of use. There he found an inscription cut into the canyon wall, dated 1777, a Catholic cross, an illegible name nearly worn away by the sand-laden winds of time and some twenty old grave sites.

Ute Indians still tell of a battle between Utes and Spanish miners in the long-ago of their grandfather's time, a battle in which a large number of Spaniards were killed, only a few escaping the trap set for them. The Indians captured the miner's pack animals, which were loaded down with heavy bars of silver and gold. According to Ute tradition, that bullion was cached not far from those grave mounds. That white men once trav-

It took centuries for pack animals to wear steps in stone along the Old Spanish Trail.

eled those desolate waste-lands and may have engaged in such a battle is given further credence by the discovery of a Catholic rosary and crucifix, found along the San Rafael by sheepherder George Wareham, as described in a history of Emery County. If one examines a map of that region, it becomes clear why those travelers of long ago chose to follow the San Rafael and its life-saving waters to the plateau country which forms a natural pass from east to west, in a land where such passes are few and far between.

From Salt Wash, the trail turned westward, following the twists and turns of Hornsilver Gulch to Castle Valley. In a seldom visited side gulch along the way, there is a row of seven Catholic crosses on a weathered stone ledge, while a little higher in a nearby canyon there is a lone cross with the date 1667 inscribed below it - - - taunting symbols from a forgotten past which mock our attempts to interpret their lost message. From Castle Valley, that now dim trail followed the Price River north to a deep canyon which climbs up through brush covered hills to a mountain pass at Soldier Summit, at the head of Spanish Fork Canyon. Down that long, narrow gulch on the north side of the mountain, the trail continued ever northward until it reached the canyon mouth where in the distance a great lake of clear blue water beckoned to weary travelers. It was the Spaniard's Laguna de Copalla in the Valle de Teguayo, today's Utah Lake, the place Father Silvestre Escalante would later call "the most pleasing, beautiful and fertile spot in all New Spain."

The Old Spanish Trail had many variations. Side tracks frequently led from the main route to mines and mission sites in the distant deserts and mountains. In the red-rock desert near the Head of Sinbad, a fork turned eastward across Spanish Valley to join the Escalante Trail near present-day Moab. That trail also connected Santa Fe with Utah Valley and the northern mountains, but along a route far to the east of the main traveled trail. The Dominguez-Escalante Trail of 1776 headed almost straight north from the San Juan River, roughly paralleling the present Utah-Colorado state line. It followed a route blazed two decades earlier by the adventurous Father Juan Maria Rivera. There were so many variations and side routes along the old trail that one early text describes it as the trail which never was, for there were many old Spanish trails.

Those who doubt that Fathers Dominguez and Escalante followed a trail already well known before their epic journey need only study Escalante's journal, which contains many entries such as the following: "On August 22 we went down from the mesa by a rugged but short slope, the one which Don Maria Rivera describes in his journal as being very difficult." Such entries indicate that they were in fact following a trail used earlier by Rivera. At the junction of the Uncompahgre and Gunnison rivers, the two explorers "found carved on a white poplar tree" a Catholic cross, Rivera's name and the words "Viva Jesus." It is well documented that included as guides and interpreters with the Dominguez-Escalante Party were two brothers, Lucrecio and Andres Muniz, both of whom had accompanied the earlier Rivera expedition. Andres Muniz spoke to Escalante of the "ancient friendship" he had with the Indians they encountered along the way, and at one camping place he told Father Dominguez of "veins of metallic rock in the nearby sierra," which he had seen while with Rivera.

That eastern section of trail passed by many rich silver mines in the La Plata Mountains, and further north skirted the famous Hesperous Mine, to wend its way to the crossing of the Buenaventura (Green) River, northeast of Vernal in Uintah County. That Dominguez and Escalante already knew of the Hesperous Mine and the La Plata Mountains is evident from reading their journal entries, one of which states: "The La Plata River flows through a canyon in which there are veins and outcroppings of metal. According to some citizens of this kingdom, there are mines of silver, which caused the mountains to be called Sierra de La Plata (Silver Mountains)." At the crossing of the Green River north of Vernal, Don Joaquin Lain carved his name, two

Catholic crosses and the date 1776 on a "black poplar" tree, probably a cottonwood. Some of the first American prospectors in that area saw that ancient message and noted it in their journals.

Note that many important physical features of the country through which Dominguez and Escalante passed were already well known to the Muniz brothers, with names which apparently had been in common usage even before their travels; names like La Plata and Hesperous. It is also obvious that the La Plata Mountains had been prospected long before the Dominguez-Escalante expedition, and had in fact been named for the mines of silver (La Plata) ore found there. It is also on record that in 1766, the Marquis de Rubi made a tour of inspection for a distance of one-hundred leagues (250 miles) north of New Mexico, which suggests that even Father Rivera was following a trail already well-worn. At the Gunnison River, Escalante recorded that Father Rivera had reached that same crossing in the year 1761.

From the crossing of the Green River, the Dominguez-Escalante Trail followed the Rio de San Cosme (Duchesne River) westward along the south slope of the Uinta Mountains to the Rio Santa Catarina de Sena (Strawberry River) at the site of present-day Duchesne City. Note that although they called the Uinta Mountains, Sierra Blanca de los Lagunas, translated literally as the snow-capped peaks, the name Uinta is actually a Ute Indian word: Country at the foot of long-leafed timber pines, clear stream flowing. The explorer's trail continued westward along the Strawberry River to its confluence with Currant Creek, which it followed to Deep Creek Canyon, which led the explorers to Strawberry valley. They turned south from that mountain meadow and entered Diamond Fork Canyon.

Their path wound down that steep, brush-choked gulch to Spanish Fork Canyon where it met the more traveled section of the Old Spanish Trail, which they followed downstream to the canyon mouth. Some believe that Spanish Fork Canyon was named by Mormon pioneers, but its name was on the maps of Mountain Men long before the first Americans saw that steep gulch. The Indian name for the canyon was Pick-Quanah-We-Woods. Today a monument commemorates the place where Dominguez and Escalante first looked out over the valley below. Escalante named that valley, Tinpanogatz, his name for the Indian band living there. There is no doubt that Escalante knew of that name even before his arrival in the valley, for Father Rivera had used that name to describe the valley and its residents. In his journal, Escalante wrote that the Tinpanogatz Indians had earlier been known as

Who left this ancient date as evidence of his passing along the old trail to the northern mines?

Lagunas, because they lived on the shore of a great fresh water lake.

The name Tinpanogatz is an Aztec word meaning "fish eaters," an appropriate description of the Ute Indians who lived along the lake shore. The late Utah historian, George Stewart, agreed that the name is of Aztec origin, but he told the author that it also translates to "stone person," no doubt a reference to the summit ridge-line of Mt. Timpanogos, which outline has the appearance of a reclining person, some say an Indian princess. It is also interesting that the Aztec "Tinpanogatz" is very similar to the Ute's "Timpanogo," which means "Rock River," the name they gave to today's American Fork Canyon Creek, a stream which flows into Utah Lake. But whichever route was followed by Spanish explorers, mission priests, soldiers and miners, the Old Spanish Trail north from Santa Fe ultimately led to Utah Valley and the now lost mines and treasure caches hidden in the Uinta Mountains beyond, the land Spaniards called El Dorado.

The few records so far uncovered at the Archives de la Nacion at Mexico City or at the Archives of the Indies at Seville, Spain, do not tell us who the first Spanish conquistador was to cross the Sinbad Desert to make his way down Spanish Fork Canyon to the valley of the Utes. It may have been Father Geronimo Salmeron in 1621. His description of the great lake and the Indians who lived by its shore is one of the earliest known. He called the Utes who lived by that lake, "Guasvatas," which was his understanding of their name for the high mountains near where they lived; but in translation, his "Guasvatas" was pronounced phonetically to become "Wasatch," the name by which those mountains are still known. More than a century later, Father Escalante would write of trading with the Guasvatas Indians along the Tizon (Colorado) River. In his journal, Father Salmeron reckoned the location of Lake Copalla (Utah Lake) as being two-hundred leagues (500 miles) north of Santa Fe. He wrote of the riches of the Guasvatas (Wasatch) Mountains:

> There are mines of silver, copper, lead and lodestone (magnetic iron). In all the ranges there are mineral deposits, where I discovered many, and which I claimed for His Majesty. I took eighteen arrobas (450 pounds) of ore, pieces of which I distributed at places I passed, so that all might see and recognize them.

Another who wrote of the rich mines of the north was Father Alonso Benavidas, who wrote that he found "a great treasure of mines, very rich in both gold and silver, as well as large deposits of very fine garnets." Nearly every Spanish expedition into the northern mountains mentioned finds of precious metals. The chronicles of the Humana-Bonilla expedition recorded that the Spaniards found so much gold they were worn out from the handling of it, and as they rested, Indians set fire to the grass and forest around them, so that many in their party died. Many years later another Spanish mining party found "some things of iron, pieces of leather boots and the bones of men and horses" where those earlier Spanish explorers had died, but the great hoard of gold the Humana-Bonilla party was said to have been carrying has never been found.

Almost all accounts of expeditions north of Santa Fe were recorded by Franciscan or Jesuit priests, who were required by royal decree to accompany every party licensed by the king to pass beyond the Spanish borderlands. Many small parties of traders and miners covertly penetrated into those unknown and uncharted lands north of New Mexico, but because they did so illegally, no priest accompanied them, therefore no official record of their travels were recorded. Each adventurer followed his own trail, trying to find a route where water and horse feed was available, while at the same time eluding wandering bands of Indians, all of whom prized Spanish scalps.

That there were many illegal parties venturing into the northern frontier in search of gold and silver without paying the Royal Fifth, or twenty percent tax levied by the king, is well known. Faded court documents in the archives at Santa Fe tell of many groups seeking mines in the north, or trading with the Utes. Some traders were known to go among the Indians and remain with them for three or four months at a time, "for the purpose of gathering peltry or to see what they could see." Spanish authorities were well aware of those illegal parties, but the frontier was so broad and poorly guarded that it was easy to evade the king's tax collectors. One history text states: "Many entradas were made for the purpose of capturing slaves to be used in mining." Another reported that a number of unauthorized

What tales of treasure do these ancient crosses in stone have to tell?

groups went into the northern mountains "in search of slaves and booty."

When gold deposits were first discovered in New Mexico, the Spanish king demanded fifty percent of all riches found, since it was thought that he alone owned the earth and everything on it. But when his excessive taxation led to illegal mining parties paying no tax at all, a twenty percent or one-fifth tax was decided on, which became known as the Royal Fifty or Quinto. Bars of gold or silver on which the Royal Fifth had been paid were marked or stamped with the letter "V" to show the tax had been paid. But not all of the illegal treasures being brought from the north were intercepted, and whether legal or illegal, not every pack train of gold and silver reached the gulf coast where treasure galleons waited to carry that great wealth to Spain. Most illegal parties were small and poorly armed, and were often unable to defend themselves from Indian attacks. Often a mining party under siege was forced to cache their heavy packs of gold and silver bars in order to save their lives. Many never lived to return to claim those cached hoards.

Most of those soldiers of fortune were illiterate, but even so, today a cryptic waybill or intriguing map hastily sketched by a now forgotten expedition leader is found in some dusty archive or musty ledger. They challenge even the most skilled cryptographer to decipher their mysterious message. Among the most enigmatic of all Spanish waybills is the puzzling set of figures cut into a large boulder in Clear Creek Canyon, along the western section of the Old Spanish Trail. They were discovered by a deer hunter before 1900, and have challenged experts ever since. Cryptographers, university professors, map dowsers and other authorities have been baffled by that strange inscription. One scholar even wrote a twenty-page thesis to explain its riddle. Perhaps you can decipher that ancient message. It could lead you to treasure!

Even the most doubtful skeptic can't question the truth of a waybill written by Father Alonzo de Posada in the year 1686, as part of his official report to the Council Of The Indies. In that report, Father Posada described the trail from Santa Fe north to Provincia de Teguayo and Laguna de Copalla, today's Utah Valley. Some background information may be in order to explain the reason for Father Posada's report. In 1660, Don Diego de Penalosa, a descendant of the earliest conquistadors, was appointed Governor and Captain-General of New Mexico's northern provinces. Upon his return from a long and hazardous journey of discovery into the far north, Penalosa told of seeing a great fresh water lake, and of rich mines in the mountains beyond. In his records he made the following statement: "From those mines of Moctesuma, there can be drawn every year from twenty to twenty-five millions in silver." But Governor Penalosa had the misfortune of placing more importance on locating mines of precious metals than he did in doing the bidding of the Catholic clergy. He made the mistake of ordering his soldiers to violate the law of sanctuary by seizing a prisoner from the mission chapel, for which he was placed under arrest by the clergy.

On charges brought by the priests that Penalosa used "unrestrained language against priests, and also spoke absurdities which border on blasphemy, as well as dressing in black velvet robes and other fine dress," he was brought before the Inquisition, was arrested and taken to Mexico City in chains, where he was incarcerated for thirty-two months. All of his property was seized, he was forbidden to ever hold public office again and was expelled from New Spain. In addition, he was ordered to dress in penitent's garb and go bare-foot through the streets carrying a lighted candle. After years of appeals and waiting, Penalosa was finally granted an audience with the Royal Court of St. James

Jesuit priests followed the trail of Father Posada northward beyond the Spanish borderlands.

Father Escalante's view of the Mineral Mountains from Spanish Fork Canyon.

at London, where he told King Charles of his discoveries, of the great Lake of Copalla, and the regions beyond where there were mountains "abounding in mines of gold, silver and other precious metals." But the English king dismissed his story, so Penalosa went to Paris where Louis X1V took great interest in his tale.

Penalosa petitioned the French king to grant him leadership of an expedition to seize the lands north of New Mexico from the Spanish. He assured the king that the Indians would welcome the French over their Spanish task-masters and would willingly work in mines for them. He described how the great wealth from those mines could easily be taken overland to the Mississippi River and thence to the Gulf of Mexico, avoiding the long and perilous trail across Ute and Apache country. Penalosa then promised the king that if everything was not as he promised, the king could order him to be hanged from the first tree the expedition came to.

The French monarch was so impressed by Penalosa's tale that in 1684 he outfitted an expedition, but he placed it under the leadership of Cavelier Sieur de la Salle. Unfortunately, La Salle's mercenaries became lost in the broad Mississippi Basin, only a very few ever making their way back to the Texas sea-coast, where La Salle died, in 1687, the same year that Penalosa died in poverty at Paris. When the Spanish king learned that the French were investigating Penalosa's story, he quickly reconsidered and sent Father Posada to investigate those claims. Father Posada had been a missionary in the northern borderlands for many years, so today most scholars agree that his "Informe" on the lands explored by Penalosa probably contain the best description of Utah and the Great Basin made during the mid-1600s, and as such is a valuable historical document. Paraphrased somewhat due to its length and difficulty of translating exactly its now archaic wording, Father Posada's "Informe" in essence relates the following description of the Old Spanish Trail:

> The trail to the north passes by several Indian rancherias before reaching the village of Chama and the Rio San Juan, which river runs nearly straight west for seventy leagues, passing through the lands of the Navajo Nation. Near where the San Juan joins the Rio Tizon (Colorado), the trail passes into the land of the Yutahs, who are skilled horsemen and brave warriors. Turning to the northwest, the trail continues for another sixty leagues, all the time passing through a desolate land of stone and sand (today's Canyonlands), which country provides very little feed or water for either man or beast. Beyond that awful desert are some brush covered hills, and further north can be seen mountains which are always white with snow. Crossing over a high pass between those peaks, the trail follows down a long and rocky defile for fifty leagues, moreor less, to enter the Valle of Teguayo, which place the Indians call Tinpanogatz. According to their traditions, it is the place where all Indians originated, even the Aztecs and Incas.

One doesn't have to be a skilled mathematician to recognize that Father Posada's journey of 110 leagues "more or less" northwest from the confluence of the San Juan and Colorado rivers, converted from leagues to miles at a ratio of 2.52 leagues to the mile, would place a traveler almost exactly at today's Utah Valley. It is the only place in the western country which has a large fresh water lake surrounded by high mountains. No other place answers his description.

Father Posada, like Father Salmeron sixty years earlier, placed Lake Copalla (Utah Lake) some 200 leagues north of the New Mexico settlements. He also wrote of mines and precious metals: "The lands of those kingdoms are very rich in silver and gold, plentiful in foodstuffs and are well populated. When Indians saw me eating from a silver plate, they told me there was much of that same metal in their country. They also said there were mines of quicksilver." But the riches of Tinpanogatz notwithstanding, Father Posada advised the king that he did not recommend establishing either a military province or a church mission so far north of Santa Fe:

> Such an isolated settlement, so far from the lands peopled by the Spaniards, could not expect aid or assistance against their enemies. Nor could carettas be taken there because of the roughness of the land, cut by dangerous river canyons and rocky mountains, as well as the length of the trail, more than two-hundred leagues, all of it through a des-

olate land peopled by barbarous and infidel enemy Indians.

Even though the good padre hoped to discourage settlement, he actually encouraged adventurers who also read the following in his report: "The Indians at Teguayo clean gold from the sands of the rivers just as they do in New Mexico. They wear ear-rings and have bracelets of gold, which they always wear on their left arm."

It is well known that Father Escalante studied the Posada report just as he had read Father Rivera's journal, for he referred to it several times in his own daily record, on one occasion when he came to the Green River: "This is the same river which Father Posada gave as dividing the Ute Nation from that of the Commanches." The years following publication of Father Posada's "Informe" saw even more illegal mining parties slipping out of Santa Fe during the dark of the moon, each hoping to be the first to discover the treasures to be found beyond the northern borderlands. Those now unknown and forgotten argonauts of three centuries ago explored nearly every desert range and forested mountain peak long before the first American fur trappers or Mormon pioneers saw those same places. They followed every river and stream and penetrated into the heartland of even the most inaccessible places, all at a time when British subjects were first gaining a feeble foothold on the Atlantic Coast, two-thousand miles to the east.

Utah Valley became an important Spanish outpost in the northern mountains shortly after it was discovered, and certainly by the time of Father Posada. Like the Indians before them and the Mountain Men who came afterwards, Spanish adventurers made their camps in that valley, where mile-high mountains protected them from winter storms and where deer, elk, bison and fresh-water fish were plentiful. By the early 1700s the name Teguayo and Lake Copalla were replaced by the Indian's Tinpanogatz, that name spelled several ways, but eventually to become Timpanogos, by which name the valley and its most prominent peak are still known. As described earlier, Father Posada did not recommend that the region be designated as a Spanish province because of its isolation, but no doubt to the surprise of some armchair historians as well as some skeptical historical societies, contrary to his recommendation, the valley and mountains of Timpanogos were designated as an official province of New Spain's internal regions.

Because of continued Indian problems in the northern borderlands, the Spanish king decided to establish a new region which would be called the Internal Provinces. In 1782, Manuel Agustin Mascaro had combined the separate maps made by Don Bernardo Mierra, cartographer for the Dominguez-Escalante expedition and the map of Father Pedro Font, the latter having charted a route connecting Santa Fe and the Moqui lands with Monterey in 1777, but the detail of Mascaro's map was incomplete and a more accurate chart was needed. We now know that such a map was made, for research has uncovered that ancient chart, dated 1783 and titled Mapa de los Provincias Internas Septentrioal. (See Map A) That map bears the dedication, Al Exmo Senor Principl de la Paz. Literally translated, that map and title reads: Map of The Internal Provinces of North America, Dedicated to The Very Distinguished Prince of La Paz. Of great interest to lost mine aficionados and treasure hunters is an inscription on that map adjacent to Laguna de los Timpanogos which states: Provincia de Timpanogos!

Do these strange inscriptions at Clear Creek Canyon lead the way to treasure?

That the new Internal Provinces were soon well known was noted in a report prepared by Lt. Jose Cortez, an officer in the Royal Engineers, who used the La Paz Map when he wrote of the population of those lands: "The long line of presidios and the poor quality of our troops which garrison them, and also considering the nature of the savage Indians against whom we wage war, and the need for a system of pacifying them, make doubly great the dangers facing the Spanish in those parts." In addition to the Utes who lived in the lands adjacent to New Mexico, which peoples Lt. Cortez called the Noaches and Payuchis, was still another band of Utes of whom he wrote: "Far to the northwest, more than 200 leagues, are tribes known as Zaguaganas, whose language is Timpanocuitzis, and whose numbers are quite considerable."

Considering how notoriously slow the Spanish bureaucracy worked, added to the fact that designating Timpanogos as a province required approval by the king, with all correspondence between Mexico City and

Seville sent by sailing ship, it is not unreasonable to believe that the province shown on the 1783 La Paz Map had the king's approval a decade or more before that date. Provincia de Timpanogos became the Spaniard's refuge in the mountains as well as the gateway to the treasures of the Uinta Mountains beyond. The La Paz Map shows the Old Spanish Trail leading north from Santa Fe to Provincia de Nabajoa and the land of the Yutah Payuchas, thence west of the Sierra de la Sal, now the La Sal Mountains, east of Moab in Grand County. That trail continued northwest across Spanish Valley to the Rio San Rafael and thence north to Vegas de Puerto, the pass leading into Spanish Fork Canyon and Lake Timpanogos in the valley beyond.

Several early explorers described the mineral wealth of the Timpanogos Mountains, but none excited the miner's imagination more than Don Bernardo Mierra, cartographer for the Dominguez-Escalante expedition. He was among the first to use the name Timpanogos on a map. On that famous map, he named the range east of Utah Valley, the Mineral Mountains. In his derrotero (Diary or record) he wrote: "The Timpanogos Indians say that tribes living in those sierras make the tips of their arrows and lances and also their macanos (war clubs) of a yellow metal." (See Map B)

Don Mierra, unlike Fathers Dominguez and Escalante, was greatly interested in the mineral resources of the lands he helped explore, which fact Escalante often noted in his record of the party: "On the side of the canyon there is exposed a vein of metallic ore. Don Mierra said it is of the sort which miners call tepustete, and that it is an indication of gold." On another occasion Escalante recorded: "There are many deposits of gypsum, and also some of metallic ore." In his own report to the king, Mierra noted: "There still remain ruins of many large and ancient settlements of Indians, and furnaces where apparently they smelted metals." Mierra later petitioned the king for permission to return to those lands, and made a pledge: "God giving me life and health, I promise that in the term of three years to have at least three settlements established, and at least one mine of gold, silver and copper located." An entry not in the Escalante journal, but noted in Father Juan Agustin's Descripcion Geografica del Nuevo Mexico of 1782, there is described an incident in which "a ball of silver the form and size of an egg was found mixed with quicksilver." That quicksilver was separated from the silver in the presence of Father Escalante, "Much to his surprise, since it had been thought that no one had used quicksilver in that kingdom."

One of the sources of the yellow metal mentioned by Don Mierra was later discovered near the mouth of

Spanish explorers and miners of old left a record of their journeys in stone.

Spanish Fork Canyon, just before the old Dominguez Escalante Trail emerged from Diamond Fork Canyon. Today a dim path leads to the northeast up a steep side-gulch to an old mine working. Just under the top of a ridge trailing down from 10,00-foot Spanish Fork Peak are two inclined shafts, located only a few yards apart. They were both sunk on the same ore vein. Their real depth has never been determined, since both were partially filled with caved rock and debris when discovered. It appeared that tree limbs had been purposely thrown into the shafts, perhaps by Indians who wanted to conceal their location after they drove Spanish miners away from them.

In 1956 a deer hunter named Clark Rhoades came upon the two old shafts. He later returned to explore them and found one to be about 90-feet in depth while the other was about 110-feet deep. From the depths of the deeper pit he recovered two old wooden shoulder yokes, of the type used by Spanish miners to carry sacks of ore. They were made of native cedar wood and were about three feet in width, having a curved cut in the center to fir the bearer's neck. They have notches cut into each end from which "zurrons," rawhide ore sacks, hung. The miner, or his Indian slave, would climb out of the depths of the mine shaft using shallow steps or footholds cut into its side walls, where a fall would have meant certain death.

Rhoades planned to dig into the ore vein found at the bottom of the deeper shaft, but an unexpected discovery discouraged any attempt to do so. As he began to excavate loose rock which had fallen into the shaft during a century or more of time, he found that the floor of the shaft was not its actual bottom, but only a platform of log timbers which had been cleverly fitted

The mysterious Brass Plates of Laban. What message do their strange symbols conceal?

into notches cut into the shaft wall. Those ancient timbers were nearly rotted away, so that pieces of rock dislodged by digging fell into empty space between them, ringing as they bounced from side to side into unknown depths. Rhoades feared that the entire log platform might collapse, sending him crashing into that bottomless void. But he was able to see where the center of a log which had been wedged across the shaft had a deep groove worn into it, where a rope or rawhide riata had cut into it as heavy sacks of ore were lifted from that black pit below.

His duties as a law officer kept Rhoades busy seven days a week, so that he had little time to spend solving the mystery of that ancient shaft. Not long afterwards those old diggings were located as mining claims by Wayne Handy, but legal problems arose concerning his partnership with others, and a lack of funds to purchase necessary tools and machinery to reopen the shaft kept any work from being done. Over the years since, those old diggings have reverted to ruin again. It is a riddle which may never be solved.

Just across a mountain ridge from the Rhoades shaft, another mystery remains unsolved in the Salem Hills. In Water Canyon, at the foot of an unstable and dangerous rock slide, a large square-cut mine tunnel was found in 1894. Although obviously dug long before arrival of the Mormon pioneers, it was definitely dug by miners, their tool marks being evident on its walls. Close by that tunnel, strange and still undeciphered "hieroglyphics" were found cut into the canyon wall, while just a short distance down canyon there was a slag pile where evidence indicated that gold ore had once been refined. Those "hieroglyphics" show a line of strange looking pack animals which resemble South American llamas, being led by a man dressed in what looks very much like ancient armor. At the time of its discovery, a rock slide partially covered the tunnel entrance, but even then the opening was large enough for a man on horseback to enter. Henry Armstrong, a pioneer who lived in the valley below that tunnel, knew well the value of ore found there, for he saw it, and later said that samples from that Water Canyon mine contained more gold than rock! He was so impressed by that gold ore that for eight years he invested every penny he could spare, financing what would later become part of the famous Koyle Dream Mine.

Several years after the mystery tunnel was discovered, John H. Koyle began working at his now well known Dream Mine, located east of the settlement of Salem, and just north of Water Canyon. Koyle's Dream Mine was situated on Knob Hill, where he filed mining claims, including one on the old mine in Water Canyon. Koyle never attempted to reopen that tunnel, instead devoting all of his time and energy on his Dream Mine. For more than fifty years Koyle pursued his dreams of finding riches on Knob Hill, but the pay-streak always eluded him. After a lifetime of sinking shafts and burrowing deep into the mountain, work slowed to a near-stop. Over the years that rock slide in Water Canyon slowly covered the old tunnel until in time it was forgotten. During the 1930s hikers in the canyon felt cold air coming from a small hole at the edge of the slide and tried to remove enough rock so that they could crawl back into what looked like a large room, but for every stone they removed, another slid down the mountain until the hole was covered again. No one has tried to open it since.

Today some wonder just where that old tunnel was located. It shouldn't be hard to find, but it may be located on claims owned by the Koyle estate. In 1959, the Koyle Dream Mine was re-incorporated as the Relief Mining Company, with offices at Provo, Utah. Stock in the Relief Mine is still in demand by those who still have faith in Koyle's dream; usually at about one dollar per share. There are some who purchase every share they can find, confident that their stock will be worth a thousand dollars or more when Koyle's great bonanza ore body is found. But your best bet might be that ancient tunnel in Water Canyon. If you would like to try your

Mormon Selman. He saw Spanish pack trains heavily laden with bars of bullion.

luck, first locate the "hieroglyphs." They are still there on the canyon wall.

The Timpanogos Valley of the eighteenth century was a place of beauty, as yet not visited by the Mountain Men or settled by the Mormons. From places where Spanish miners had camped, trails led off to distant regions like spokes in a wheel. One of those trails led northward along the foothills of Mt. Timpanogos. In time the near-vertical slopes of that 12,000-foot giant proved the accuracy of Father Escalante's report of 1776, which stated in part: "The sierras to the east of this valley have many rivers, springs and timber, including royal pines, and the veins in the sierras appear to have mineral in them." After arrival of the Mormons, mining camps blossomed in the shadow of the mountain. Both gold and silver were found in the depths of American Fork Canyon, at places like Forest City, Mary Ellen Gulch and Sultanna Smelter. Even today weekend recreational gold prospectors using small portable dredges recover grains of gold from the canyon. Other mining camps grew up just over the mountain's shoulder, at Alta and Brighton. So it was of little surprise to early settlers of Mountainville (now Alpine) when Ute Indians brought chunks of shiny silver ore to trade at the white man's store. They would only say that it came from an old Spanish mine located somewhere deep in the recesses of the mountain canyons.

An Indian who often traded pieces of silver ore for the white man's trinkets, once told a store owner who had befriended him, "You will only find the silver if a big fire comes, or if a flood washes down the mountain." Was he trying to tell his friend that the mine was covered with brush and timber so that only a fire would reveal it, or was the entry hidden behind earth and rocks which only a flash flood might carry away? Those who watched the Indian coming and going all agreed on one thing; he always left Mountainville late in the afternoon, probably so that he could not be followed, and that he would return by noon the following day, with as much of the heavy silver rock as he could carry. The Indian's mine can't be very far from the mouth of American Fork Canyon, perhaps hidden in one of the many caves found there. A more recent clue may be of significance. Gold dredgers who wash the sands of Mary Ellen Gulch, located just west of Graveyard Flat and close to old Forest City, often find pieces of native silver in their sluice boxes. They might be closer than they think to the Indian's mine on Mt. Timpanogos.

A treasure hunter whose employment recently required him to move out of state discovered some interesting Spanish signs on Mt. Timpanogos, which might lead to the Indian's mine. Near Camp Timpanokee, along the Aspen Loop Road, he found the letter "R" and a distance marker indicating 33 varas (about 82 feet) carved into the bark of an old aspen tree. Walking in a circle at that distance from the tree sign, he came upon another symbol, that of a man holding a sombrero over his head. Continuing on a line along those marker trees, he came to other symbols, including an "X" and an arrow pointing the way to where a dim trail led further up the mountain. Along that trail he came to still more marks, among them the numbers 70 and 73, as well as a coiled snake whose head pointed up-canyon. The finder hopes that someone will follow up on his discovery. Who knows what you might find?

Some say there is a treasure more valuable than either gold or silver hidden near the foot of Mt. Timpanogos. Frank Peay was one of the first settlers at Mountainville, and it wasn't long before his neighbors noticed that he was spending more time digging for treasure than he was at farming. Written family journals from those days state that Peay uncovered the entrance to a "sealed tunnel which has the structure of a cavern." Those who attempted to enter Peay's sealed cavern claimed that they "were hit with a hot wind," which spoke to them with a voice directing them to leave at once. Several people claimed that a ghostly voice filled them with such an overpowering fear that they fled from the cavern and were afraid to return.

His neighbors watched the progress of Peay's digging, most of which was done by hired hands. One close acquaintance overheard Peay tell his workers that they should dig straight ahead for two more feet, at which point they should turn the tunnel's course to the right, "where they would hit it." Within several days, Peay's workers uncovered an ancient appearing stone box, inside of which there were seven brass plates, obviously of great antiquity. Those who saw the plates said they were the Brass Plates of Laban, which pertained to the sacred Sword of Laban, referred to in the Book of Mormon. Church authorities at Salt Lake City were quickly notified and soon arrived at Peay's property. The mysterious plates were carefully examined, photographed and then spirited away from Mountainville, some said to a secret vault hidden deep beneath the Mormon Temple.

Some years after Peay's death, Wayne Handy conducted an interview with Alfred Devey, then the oldest resident of Mountainville. Devey recalled that during the 1880s, people used to come to Peay's homestead and would camp there, some helping to dig his tunnel for a week or more at a time. There were those who said that the Brass Plates of Laban revealed the location of still other plates which were made of solid gold, hidden somewhere in the foothills near Peay's home. During that interview, Devey stated: "What was found there was worth more than all of Mountainville!" An account written when the Brass Plates were first found and photographed gives the location of Peay's tunnel: "near Peay's homestead there is a red house facing west, which house was built by Johnny Main. The walls of Peay's treasure chamber are in perfect alignment with the west wall of Main's house." If you want to try your luck searching for those golden plates, just ask of the old-timers who live there for directions. All you have to do is find the house which Johnny Main built. Don't forget to take your metal detector!

Since the first edition of *Faded Footprints* was published, many persons in the Brass Plates of Laban have contacted the author, among them Mr. Bob Ater, probably the country's outstanding dowser and renowned researcher of ancient writings and inscriptions. Ater stated that the inscriptions on the Brass Plates are very similar to those on ancient Numidian plates, and appear to tell the story of ancient peoples buried near where the plates were found. They also contain information about where those peoples came from and what they were doing at Utah Valley so long ago. Ater says that the message inscribed on the plates also tell of the maker's hope for resurrection, or life after death.

Even more exciting was information received from Dr. Lyman Pratt, Doctor of Archaeology and Spanish History in America. Dr. Pratt is a scholar who is compiling a thirty-seven volume record of Spanish explorations in the Americas. He reported that the Brass Plates of Laban are not unique to Utah Valley, for similar plates have been found in nearby San Pete Valley, while still others were recovered from ancient tunnels uncovered by Mormon pioneers while constructing a foundation for the Church's Manti, Utah, Temple. The Deseret News of April 12, 1871, reported an even stranger discovery:

> A curious relic of the past has been ploughed up near the mounds in the vicinity of Spanish Fork. It is a plate of metal, seemingly copper, on which is engraved the figure of a man in priestly robes, with a censer (an incense burner used in religious rituals) in his hand, and an alter on which a fire is kindled. There is also a tree, probably emblematic of the Tree Of Life, and above the tree, a representation of the sun. After being carefully cleaned, the engraving is as distinct as if recently done.

The first Mountain Men and Mormon pioneers discovered old mines and smelters; proof of early Spanish miners.

Still another intriguing find was made only recently in a another canyon a few miles north of old Mountainville. A plate approximately three by six inches in size, but only one-eighth of an inch thick, was discovered when a hiker noticed a face of granite rock which appeared to have a cleverly made recess or box cut into it. A thin slab of rock had been precisely fitted into that recess, so that the rock face appeared to be solid. The finder stated that if the sunlight and shadows had not been exactly as they were, or had been at even a slightly different angle, he would not have noticed the repository. That plate, which appears to be made of a silver colored platinum-like metal, has a number of rows of strange characters inscribed on it. They are plainly visible but are totally indecipherable, bearing a strong resemblance to hieroglyphics or the "reformed Egyptian" figures found in the Book of Mormon. A university professor, skilled at making interpretations of similar figures, is at a loss to explain their meaning. But one can't help but notice that they are very much like the characters inscribed on the mysterious Brass Plates of Laban.

Of all the trails leading into the Mineral Mountains above Utah Valley, none was more traveled or more important than the one which followed the Rio San Antonio de Padua, today's Provo River. That stream led northeast to Valle Redondo, present Heber Valley. The earliest settlers there knew that Spanish miners, by that time usually called Mexican miners, still worked mines of gold and silver in the mountains high above their valley home. Who first worked those old mines is now unknown, but in 1805 Manuel Meastes was commended by Governor Alencaster of New Mexico for more than fifty years of faithful service bringing pack trains of ore from those northern mountains. Simple arithmetic reveals that Meastes was engaged in mining in those mountains from about the year 1755. He was also credited with recovering a band of horses which had been stolen from Spanish miners by Ute Indians living near Heber Valley.

Early settlers at Heber Valley often witnessed pack trains of metal bars being brought down from the mountains and taken across their valley to the Provo River Canyon. After crossing Utah Valley, those pack trains turned up Spanish Fork Canyon. In 1852, Marshal William H. Kimball, who was responsible for enforcing territorial laws banning Indian slavery, arrested a band of Mexican miners in Spanish Fork Canyon for having Indian slaves with them. Their leader, an old man named Pedro Leon, readily admitted that he had been engaged in mining and slaving in those mountains for a half-century. Heavy fines were levied against the party and Leon was ordered not to return without a special permit from the Alcalde at Santa Fe and written permission from Governor Brigham Young. It is of interest what Leon's miners and other parties were carrying on their pack animals.

Daniel Jones was a Mormon frontiersman who traveled extensively throughout the Utah Territory. In a book written during the sunset years of his life, titled Forty Years Among The Indians, Jones described how "descendants of Coronado from Santa Fe" secretly worked mines of gold and silver in the Utah mountains. Without obtaining permission from the Spanish Alcalde, they covertly made their way north to where they worked at mining, keeping the gold and silver for themselves, without paying the king's Royal Fifth. They treated the Indians shamefully at their mines, working them until the miners had acquired a fortune in precious metals. Jones told how the ore was smelted into bars of bullion, and how frequently there was so much treasure to be taken south on their pack animals that some of the heavy bars had to be left behind, cached at or near their mines.

Mormon V. Selman was one of the earliest settlers at Utah Valley, his father having a homestead at the mouth of Provo Canyon. In his journal, Selman described one of those Mexican pack trains which brought gold down from the mountains:

> I remember a time in the early days when a pack train came down Provo Canyon and camped near our place for a few days to rest their animals. Those mules were loaded with heavy packs which did not seem to be very large, yet it was all they could carry. The miners kept an armed guard at their camp and we were not allowed near, as if whatever was in those packs was very valuable.

Neither Selman or his neighbors had time to investigate those pack trains which crossed the valley each fall, for their every waking minute was dedicated to building homes and wresting a living from the raw land. Still, they could wonder, and perhaps dream a little.

In *Lost Treasures On The Old Spanish Trail,* the author described how an aged Mexican came to Utah Valley during the early 1920s, where he worked for a cement contractor while searching for a mine and a concealed cache in the mountains above the valley. As a small boy he had been to that mine with his father and a party of miners, and was later one of those who had camped at Mormon Selman's farm. After weeks of searching, the old man finally had to admit that he could not find the mine, even though he had his father's map, a rough sketch drawn on a piece of animal skin. His employer was allowed a brief glance at that map, and being familiar with the nearby mountains, he easily recognized the landmarks on it. He tried to convince the old man that he had been searching in the wrong area, but the aged Mexican wouldn't listen to him. After the old man left the valley at winter's first snow, the contractor began his own search for the lost mine.

The landmarks shown on the Mexican's map were few and readily apparent to the contractor. The map showed a lake where the fish-eaters lived, today's Utah Lake, and a river flowing into it, the Provo River. The old mine was located near the headwaters of the first fork of that river, just below two high peaks. It was reported in the author's *Old Spanish Trail* that the mine had been located, southeast of Wallsburg in Wasatch County, on a point of rocky ridge above Little Valley. Everything seemed to match the landmarks on the Mexican's map. The mine appeared to be where the map indicated it was, but since that time further investigation has revealed that is not the true location. That suspected site is indeed an old diggings, with shafts and tunnels supported by underground timbers, as well as relics left by unknown persons sometime in the long ago, but it is not the mine shown on the Mexican's map.

The real lost mine has since been found, not far from the originally reported location. The finder has had little time to inspect its old workings since his discovery, not long before winter's first snow fell, but the coming spring should reveal its hidden secrets. At that newly discovered location, there is a tunnel driven through solid rock for some thirty-feet, where it makes a turn to the left for thirty-feet more. At that point, a vertical shaft descends an additional thirty-feet. At the shaft bottom, a tunnel can be followed for twenty-feet to where it is caved. All of the diggings are of the old-time Spanish style, the tunnels being arched and entirely without supporting timber.

Those miners of old appear to have followed a narrow but rich ore vein, where very little waste rock was removed, in the fashion of Spanish mines. The floor of the lower tunnel has a hollow sound when it is walked on, as if there might be another vertical shaft covered over with timbers and rock. Perhaps it was covered to conceal the ore vein below. The finder does not want the location revealed until examination and ore sampling is completed, but if you study this account carefully, and compare its information with a good map of the area, you should discover where the error was made in describing what was once thought to be the old Mexican's mine, and you will know in what area to look for the real mine. Good Luck!

Early Mormon settlers and the Mountain Men and fur trappers before them discovered unquestionable proof that Spanish miners had inhabited the Utah mountains long before, and had camped in the valleys before the first Americans crossed the Rockies. Ancient trail markers and treasure symbols inscribed on rock ledges or cut into the bark of centuries-old trees was proof of that. Those settlers found well-worn trails leading off into unknown canyons where mines of gold and silver had been worked by Spanish miners and their Indian slaves. Long abandoned tunnels and shafts gave even further proof that they were not the first to visit those places.

Treasure hunters who seek the lost mines and secret caches of those early Spanish adventurers know that treasures are still being found. They know that is true, even though you might not read about those finds in your daily newspaper. Of course a few recoveries do make the headlines, such as the fantastic hoard of gold and silver recovered from the treasure ship Atocha by Mel Fisher, or the astonishing quest for the Doc Noss treasure at Victoria Peak. But understandably, most treasure hunters are quite reluctant to make their finds known, for who wants the IRS knocking at their door? Be advised: treasure is considered as income at the time of its recovery, whether or not it is exchanged for cash, then or later. If you don't have a competent attorney, it is best to keep your own counsel. There are plenty of Philadelphia lawyers out there who are slicker than bear grease. If the IRS doesn't get you, those guys in silk suits and patent-leather shoes will!

But even though you make your search in a seldom-seen canyon or far off the beaten path in some mountain wilderness, you are still following the Old Spanish Trail blazed by Coronado's children in the long ago. Northward from the snow-capped crest of Mt. Timpanogos, that old trail led up the narrow, twisting Provo River Canyon to Heber Valley, Rhoades Valley and the Spaniard's El Dorado in the Uinta Mountains beyond. It was in those mysterious mountains that Thomas Rhoads discovered El Dorado, the place prospectors and treasure hunters now call the Lost Rhoads Mine.

CHAPTER 2

THOMAS RHOADES
The Wealthiest Mormon

Tales of Spanish gold in the Uinta Mountains and the story of Thomas Rhoads are one and the same. They are inseparable. To understand one, you must know the other. This is the Lost Rhoads Mine story.

Thomas Rhoads was born July 13, 1794, at Boone's Fort, Kentucky. As a young man he served his country during the War of 1812, after which he moved to Missouri and later to Illinois. By trade he was a surveyor. He married Elizibeth Foster, their marriage in time producing twenty children, seven of whom died as infants. Their fifteenth child, Caleb Baldwin Rhoads, was born April 4, 1836. Young Caleb was named for Mormon Church Elder Caleb Baldwin, who had baptized Thomas Rhoads into the Latter Day Saints (Mormon) Church only a year earlier. That youngest son, Caleb, would one day become the West's most mysterious gold miner; more about him later. Thomas Rhoads rose rapidly in the esteem of his church leaders, soon becoming an Elder. Later he was a Lieutenant in the Nauvoo Legion, the church's military arm. By 1845 he was described by one of his peers as being an honorable man.

But one of the things we don't know about Thomas Rhoads is the correct spelling of his name. As was commonplace for frontiersmen of that period, he was barely literate, and spelled his name in various ways: usually Rhoads, but sometimes Rhods, Rhodz or even Roads. His German ancestors spelled their name Roesch. Some of his descendants spell their name Rhoades. To avoid confusion, we will use the more common Rhoads spelling.

Thomas Rhoads cast his lot with the Mormons, and suffered with them during their trials in Missouri and Illinois. When Brigham Young and his followers were driven out of Illinois by angry mobs, most thought that their prophet would lead them to California, but the sheer logistics of out-fitting and feeding such a multitude of people forced him to make a starvation winter camp on the west bank of the Mississippi River. The Rhoads family with several of their Missouri neighbors were far better equipped and prepared for the anticipated journey west. With Young's blessing, they started out to find a wintering place somewhere further west where there would be adequate food and shelter.

When the Rhoads group reached the Missouri River on May 5, 1846, they encountered several other wagon trains ready to make the long trip to California and Oregon. Rhoads decided to attach his followers to that large but loosely knit group of about three-hundred emigrants. Most of that party was made up of people with the Harlen-Young and Donner-Reed wagon trains, led by George Harlen, who had brought fifty wagons from Michigan. All of the members agreed to travel within sight of each other for mutual protection, but without any designated leader over the entire group. One party might move out first one morning and lead the way that day, while at other times they might make a late start and fall behind. At times one party might be several days ahead of the others. The Rhoads party, being small and well organized, often led the way.

The Rhoads Party consisted of Thomas Rhoads and his wife, Elizabeth; their son John and his wife Matilda with their six children; their son Daniel and his wife Amanda; and their daughter Polly and her husband, Turner Elder, with one son. There were also eleven unmarried children: Issac, Thomas, Elizabeth, Sarah, William, Catherine, Foster, Foster Jr., George, Henry, Lucinda and Caleb Baldwin. Also attached to the party were John Patterson and his wife Christine with one

Thomas Rhoads. He brought the sacred gold of Carre-Shin-Ob to Brigham Young.

child; Mathew and Robert Fanning; Joseph and Issac House; Mr. Whitman and the Esprey boys.

The combined wagon trains traveled more or less together as far as Fort Laramie, which place they passed in early June. They continued on to Fort Bridger, encountering only minor Indian problems along the way. They arrived at Fort Bridger on July 17, where dissension flared between the different groups regarding which trail to follow. Part of the travelers decided to turn northwest towards Oregon, but most turned to the southwest, towards the Great Salt Lake. Their trails split again along the Weber River at the mouth of Echo Canyon. The Harlan-Young faction followed the river downstream to Devil's Gate, where they had to winch their wagons up over steep cliffs and were eventually forced to disassemble the wagons and carry them piece by piece through a narrow gorge. Based on information given to them by Lansford Hastings, the Donner-Reed band chose to cut a trail through heavy brush up and over the summit of Big Mountain, a choice which took them twenty-one days to advance only thirty-six miles! It is generally agreed by historians that Thomas Rhoads guided his followers up river to Silver Creek and Parley's Park, the site of present Snyderville. They then crossed the Golden Pass, now Parley's Summit, and followed down the steep western slope of the mountains into the Salt Lake Valley.

Upon their arrival in the valley, by then nearly a month ahead of the Harlen-Young and Donner-Reed groups, Rhoads decided that his people had ample time to reach California before winter. He led his party around the Great Salt Lake to where they met the Fort Hall to Humboldt River trail, which led them straight west to the Sierras, but their exact route is unknown. In his notes, Rhoads wrote that they went around the south end of the Great Salt Lake. Heinrich Lienhard, one of the Harlen-Young group, kept a detailed journal, in which he wrote that they followed the Rhoads wagon tracks southward along the eastern side of the lake. Historian Dale Morgan described their route as passing through the north end of the valley and around the north end of the lake to the Humboldt Trail. Whatever their route, somehow they avoided the worst part of the Great Salt Lake Desert. The Donner-Reed wagons followed the Hastings Cutoff which led them west into the worst part of the mud flats. They spent five days bogged down in bottomless mud and were forced to abandon many of their wagons, including Reed's fancy two-story Palace Wagon. They discarded extra wheels and axles, heavy tools and farm equipment, trunks full of clothing and personal items, firearms and according to those who knew, a heavy iron chest owned by George Donner, which contained $15,000 in gold coins. That money, with which Donner planned to purchase a ranch in California, has never been found. The first snow of winter already covered Pilot Peak before the badly demoralized party finally struggled through to solid ground at the west edge of the desert. While the Donners were inching their way across the salt flats, the Rhoads group was rapidly crossing the Nevada desert.

In his journal, Daniel Rhoads described the trail which the Rhoads group followed over the Sierras:

> We crossed the Truckee River twenty-seven times, on rocks from the size of a washboard to that of a kettle, so close together that neither oxen or wagon ever touched bottom. The nearer we got to California, the worse was the road. Our oxen could be trailed from bottom to top by the blood from their hooves. We all arrived safely in California, except for John Patterson, who died along the way and was buried on the plains of California.

The Rhoads wagons crossed the crest of the Sierras during the last days of September and arrived at Sutter's Fort on October 4, 1846. On that same day the Donner-Reed Party was camped at Gravelly Ford, near present Battle Mountain, far to the east of the Sierras. All of the Harlen-Young Party, well ahead of the Donners, crossed the mountains before winter snow closed the high passes.

Shortly after their arrival in California, three of Rhoads' daughters married: Elizabeth to Sebastian Keyser, a trapper; Sarah to William Daylor; and Catherine to Jarad Shelton. Daylor and Shelton were co-owners of a large Mexican land grant of 22,000 acres, stretching seventeen miles along the Cosumnes River. Thomas Rhoads and the rest of his family settled on Dry Creek, a fork of the Cosumnes, in Calaveras County, at a place later shown on maps as Rhoads Settlement.

Almost immediately, Rhoads and his family discovered gold at their new home. Even the Rhoads women found gold. Sarah struck it rich when she pulled up a clump of bunch-grass by her cabin door and saw flakes of yellow gold clinging to the roots. She later recalled that she picked up enough gold there to have two fine gold watch cases made.

Thomas Rhoads and his family soon made other gold discoveries. Rhoads worked two rich placers at a small place known as Rhoads Diggings, in El Dorado County near Placerville. He also mined gold at still another Rhoads Diggings in Sacramento County near Folsom; and at a place called Alder Creek, also in Sacramento County. Apparently all of the adults and grown children engaged in mining, for the family quickly accumulated a small fortune in placer gold. Regarding those early days of discovery, nearly two years before James Marshall officially discovered gold at Sutter's Mill on January 24, 1848, William Grimshaw wrote: "Before the advent of white strangers, the surface was worked with knifes, spoons and pointed sticks. Where the dirt had to be washed, we used water-tight baskets made by Indian women."

Rhoads' son-in-law, William Daylor, owned a large store which sold prospecting supplies at a place called Weberville, a rich placer area worked almost exclusively by Mormons. Several pioneer journals tell of the fortunes found at those diggings long before the first 49ers arrived in California. In his records, William Daylor wrote: "The ravines were so rich that at the end of one week, three men divided $51,000. Nuggets could be picked from the dirt with knifes. The daily yield was up to five pounds of gold per man." It was during those heady days that Rhoads' wife died, on September 25, 1847, while she was enroute to San Francisco to seek medical help. She was buried along the way near present Benicia, at a place where Rhoads paid a local man to erect a tombstone But no stone was ever erected, so that when Rhoads returned, he couldn't find his wife's grave site.

Within weeks of the Rhoads party's arrival at Sutter's Mill, it became apparent that the Donner-Reed Party had not crossed the Sierras. It took several weeks more to organize a relief party, which both John and Daniel Rhoads volunteered to accompany. Thomas Rhoads was considered to be too old to participate in the search. Daniel Rhoads later wrote of that search: "It was two weeks before any person would consent to go, but John and I concluded that we would go or die trying, for not to assist them would be a disgrace as long as time lasted." They were among fourteen men who dug their way through the deep snow of February to lead part of the stranded emigrants to safety. John made three trips into those terrible mountains, and both he and Daniel were ruined in health because of their efforts. Of the eighty-seven members in the Donner-Reed group, only forty-five lived to reach California. The others starved or froze to death in the Sierras.

Early in 1849, Brigham Young contacted the California Mormons and instructed them to return to the Salt Lake Valley, where the center stake of the church had been established. Most of the Rhoads family refused to exchange their new homes at the gold diggings for the dreary desert wastes of Utah, but by mid summer, Thomas Rhoads decided to heed the call and began preparations for the hard journey back across the Sierras and the Nevada deserts to the Salt Lake Valley. On August 10, 1849, Rhoads headed eastward, accompanied by only three daughters and his youngest son, Caleb Baldwin. Also returning with Rhoads were the families of William Glover, Samuel Ladd, Harvey Green and Levi Riter.

West-bound parties of gold seekers were amazed to meet the Rhoads wagon train traveling eastward. The trail west was almost a steady stream of 49ers, and probably a dozen of those groups made journal entries about meeting the Rhoads Party. Several noted that the "aged patriarch of the group, Mr. Thomas Rhoads," openly displayed fine specimens of gold, and readily admitted that his well-armed group had $40,000 in their possession. He freely offered advice on where the richest diggings might be found. In retrospect, it seems strange that Rhoads was so open and forthright about his gold, for in later years he became very reclusive and secretive about his wealth. According to the Journal History of the Church, the Rhoads Party arrived at Salt Lake City on September 28, 1849, having traveled the northern route via the City Of Rocks and the north end of the Great Salt Lake.

While Thomas Rhoads was still in California, Salt Lake City had become the cross-roads of the West. Every major trail to the gold fields passed through the Mormon capital. Many of the 49ers who arrived there were anxious to trade jaded livestock for fresh animals, or to purchase farm goods and other supplies, but there was no medium of exchange to conduct business with. As governor, Brigham Young had the authority to establish a mint, and during the summer of 1848 he made the decision to mint gold coins. All Mormons in the valley were told to surrender their gold to the mint, whether it be in the form of wedding rings, heirloom watch cases, prized lockets or whatever they might have. The first crude ten-dollar gold pieces were minted in December, 1848, their actual gold content being $10.50. Only fifty-six ten-dollar coins were minted before the crudely made dies broke. New dies were ordered from England but did not arrive at Salt Lake City until September 12, 1849, when manufacture of coins resumed. Those new coins were finely made. U.S. Marshal Joseph Heywood took several of the new issue to the Philadelphia Mint, where they were judged to be of high quality. It was a happy coincidence that those new dies and the Rhoads wagon train arrived in the valley at almost the same time.

It is well known that Thomas Rhoads brought a considerable quantity of gold to Salt Lake City. That fact was even verified in a speech given by Brigham Young at the church tabernacle, in which he stated: "Old Father Rhoads is the wealthiest man to come from the mines, bringing with him $17,000 in gold." In the Journal History of the church there is a little known historical record titled "The Gold Account," a rare document which was discovered by Dr. J. Kenneth Davies, Professor of Economics at Brigham Young University, Provo, Utah. The Gold Account is a sort of banking statement recording the amount of gold each Mormon pioneer surrendered to the church for the purpose of mining gold coins. All of the contributors except Thomas Rhoads had their relatively small deposits listed under the name of Brigham Young. Most gave only a few ounces of gold dust or perhaps some cherished jewelry keepsake, such as a locket, bracelet or wedding ring.

William Glover was one of the Mormons who went to California on the sailing ship Brooklyn in 1846 with Samuel Brannan, one of the church elders. He worked at the gold diggings with Thomas Rhoads, accumulating a considerable quantity of gold dust and nuggets. In Glover's writings there is a notation that the Rhoads Party surrendered one-hundred pounds of gold to the mint. Glover returned to Utah with the Rhoads Party, and upon his arrival, Brigham Young directed him to give all of his gold to the church. According to Glover's journal, he gave Young "a peck of gold," keeping back nothing for himself or family. A "peck" is the equivalent of one-quarter of a bushel, or eight quarts. Desperately poor after giving Young everything he had, Glover was presented with the opportunity to purchase a small farm north of Salt Lake City, but he had not a penny with which to do so. To his great surprise, his wife revealed that she had five-hundred dollars in gold dust which she had panned while at California, so they were able to purchase the land they wanted.

In his "Gold Discovery Journal," Azariah Smith told of the gold he brought from California, but he wasn't as generous towards the church as Glover had been. Referring to the diggings near Sutter's Mill, Smith wrote: "Mr. Marshall gave us the privilege of picking up gold, and I gathered up considerable." On his return to Salt Lake City, Smith added: "I gave $500 to Brigham Young, some to the poor and a dollar each to the twelve apostles." Levi Riter, who also returned to Utah with Rhoads, told of bringing a sack of gold so heavy he could not lift it. It required both he and Harvey Green to unload it from his wagon and carry it into the church mint. But Thomas Rhoads was the only Mormon to contribute such substantial amounts that a separate account was kept in his name. On October 9, 1849, the first entry credited him with a deposit of $10,826 in raw placer gold, an amount considered to be a fortune at that time. That amount was part of the $17,000 Young had referred to in his tabernacle speech.

From 1848 to 1862 gold coins were minted at the Mormon mint. A portion of the gold used to make those coins came from gold items surrendered by church members, and after arrival of the Rhoads wagon train, placer gold from California went into their manufacture. But after that gold was consumed, gold used to continue minting coins came from a secret source known only to Brigham Young and Thomas Rhoads, as will be described in following chapters. It will never be known how much gold passed through the Mormon mint, for many transactions were never recorded, others were lost and if any records still remain, they are secure from prying eyes. For example: The Gold Account reveals that on November 7, 1849, $6,000 in gold was withdrawn from the mint and was traded for silver. A month later, on December 7, Brigham Young withdrew $1,000 in gold which he gave to Isaac Higbee. None of those transactions are recorded in mint records. The Journal History mentions amounts of gold contributed which mint records do not reflect, such as $25,000 in 1850 and $71,000 in 1851. Records do show deposits to Brigham Young's account in excess of $80,000 between 1848 and 1851, but they do not explain the origin of those amounts.

Business at the Mormon mint was conducted at a brisk rate. Examination of the scant records still available clearly reveal that gold coins were minted and issued at a much faster rate than all of the deposits of gold made by valley residents and California miners combined could account for. In addition to the amounts already referred to, it is known that Brigham Young paid fur trapper Miles Goodyear $1,950 in gold coins for his Spanish land grant, now the site of Ogden City. $8,000 in gold was paid to Jim Bridger and Louis Vasquez for still another Spanish land grant, the site of Fort Bridger. Church records reveal that an additional $12,000 was spent in rebuilding and renovating that fort. Still another Spanish land grant is referred to in church records: "Upon arrival of the Mormons, they found that the Mexican government had given a grant to a man who was living near (present day) Riverton." The church purchased this man's holdings: "This gave the church exclusive control of this section." It is unknown what was paid for that grant. That there was

Dies used by John Kay to mint Mormon gold coins from Rhoads gold. (Courtesy: Deseret News)

The crude 1849 ten-dollar Mormon coin compared to the beautifully made 1860 design. (Courtesy: Deseret News)

no shortage of gold to mint coins is shown in the case of Cary Peebles, a California merchant who demanded gold coin for a wagon-load of merchandise. It is recorded that he stood by at the mint while $4,000 in coin was minted for him.

The church owned Deseret News also reported on activities at the mint, such as the following: "Considerable gold being brought to the mint is keeping John Kay and Willard Richards busily engaged making bars." Another article reported: "John Kay and Thomas Bullock are engaged in melting gold at the mint." That melted gold was rolled into long strips of varied width and thickness, depending upon the size and value of the coins to be made from it. Occasional entries in Bullock's journal noted gold being made into bars, 183 bars during a two week period. Kay's journal mentions that he sometimes took small gold bars to his home where his children would play with them like toy building blocks. Most Saints called the mint, "Bullock's Money Mill," a structure located located on the site of the later Hotel Utah.

An entry in the Journal History on October 22, 1850, reported that Brigham Young had stated: "More gold than has already come to the mint has just been brought in, as well as a box of silver so heavy that it required three men to lift it from the wagon." Several documents mention large quantities of silver being brought to the mint, but none reveal what happened to that silver. There were no silver coins manufactured at the mint, and only a very small amount was added to later issue coins for hardness, so both its source and its disposition remains a mystery. John Kay was said to be busy stamping coins with a hand press, while another report stated "Considerable gold is on hand." For more than a dozen years, large quantities of coins were minted, in denominations of 2, 5, 10 and 20 dollars; the first twenty-dollar gold coins minted in the United States.

Most Mormon coins were designed by Robert Campbell, while the dies and other paraphernalia were made by John Kay and Alfred Lambson, except for the 1861 ten-dollar gold piece, which was designed by James M. Barlow. The Barlow coin featured a crouched lion centered below three mountain peaks on the obverse and an eagle and beehive on the reverse. Only 472 Barlow coins were minted, making it one of the rarest of all Mormon coins. At a recent New York City coin auction, the opening bid for a Barlow Mormon coin was $155,000! Common Mormon coins are found in at least seven different designs, but it is possible that there may have been at least one more. In 1852, William Kelly, author of *Across The Rocky Mountains And A Visit To The Celebrated Mormons* described a design unknown today:

> The authorities have a mint, from which they issue gold coin only. The coins are plain, but massive, and without any alloy. I saw two issues, five and ten dollar pieces, with the amount on one side and the date of issue on the other, without any emblem or device whatsoever.

Brigham Young often admonished his followers not to engage in mining. Typical of his speeches at the tabernacle is the following: "If I knew where there was a gold mine, I would not tell you. I do not want you to find one. If you do, it shall be over my faith. We have enough gold, and it is the Lord's!" Because of his frequent tirades against mining, there are some who believe he would not allow Mormons to engage in mining, counseling members not to go to the gold fields, or chastising them for catching "yellow fever." Those persons have not closely examined Young's real attitude toward mining. If mining was of benefit to Young or the church, his attitude was quite the opposite of what it appeared to be. He actually promoted mining, so long as it was under his direction and control. Selected members were frequently encouraged to participate in mining. The Journal History reveals that certain trusted members were sent on "missions" to California and Nevada to work at the gold mines. They were instructed that upon their return, they were to bring gold dust or nuggets instead of coins, "since our own mint is now in operation."

In April, 1849, Young sent Amasa Lyman to the California gold fields with a letter for Sam Brannan, the church's ranking member there. Brannan was not in good standing with the church hierarchy at that time, having been somewhat remiss in sending tithes he had collected to Salt Lake City. Included in Young's letter was a request for $20,000, "which is but a trifle to you when gold is so plentiful." In sterner language, Brannan was then directed to send Young $100,000, which amount he had allegedly been paid in tithing by California Mormons. But when Lyman told Brannan that he had come for the Lord's gold, Brannan replied: "Give me a receipt signed by the Lord and you can have the gold!" Brigham Young didn't win them all!

Thomas Rhoads led an exciting life and he had acquired a small fortune in gold at the California diggings, so he found it difficult to settle down in the Salt Lake Valley and be content as a farmer. After his return to the valley, he married plural wives Mary White, Eliza Johnson and Jacobennie Johnson, and in time increased the number of his children to thirty-two, by some accounts thirty-six. For a time he was busy having a new home built only a block south of the temple site; the finest home in the valley at that time. But his neighbors couldn't help but notice that he began spending less time at his home and gardens and more time going on mysterious trips into the mountains. Those in high places in the church hierarchy nodded knowingly, and during the summer of 1852, talk was heard on the streets that a secret agreement had been made between Brigham Young, Thomas Rhoads and Chief Wakara of the Utes. Only two years earlier, Young had arranged for Wakara to be baptized as a Mormon, by Bishop Isaac Morley, on March 13, 1850. Wakara was appointed as an Elder in June, 1851, after-which Brigham Young adopted him as a son. In private counsel, Young directed Rhoads that he was to go on a church mission with Chief Wakara, to a place the Utes called Carre-Shin-Ob, at which place he would find plenty of gold. To the Utes, Carre-Shin-Ob was a sacred or holy place, that name meaning, "There dwells the Great Spirit."

Thomas Rhoads was a logical choice to be sent on that gold mission, since he had spent considerable time since his return from California exploring and prospecting the northern mountains of the territory. He was also an interpreter of the Ute language, had been trained as a surveyor and was without doubt the most experienced gold miner among the Mormons. Rhoads learned that although white men called the Ute Chief Wakara, his given name at birth had been In-Carre Winker. Wakara was born about 1808 along the Spanish Fork River, in the Tumpanawach Band of Utes, a faction of the Tinpanogatz or fish-eaters of Utah Valley. Wakara said that while still a youth, the Ute God Towats had spoken to him in a vision, revealing a place in the Uinta Mountains where a great hoard of gold was cached. In that vision, Wakara had been told by Towats that the yellow metal belonged to the "high hats" who would one day come into his country, and that he was to give the gold to them. After that time his name became Yah Keera, "Keeper of the yellow metal." Howard R. Driggs, a noted Utah historian, wrote that Mexican miners gave Wakara ornaments of gold, which he wore with great pride. His Ute brothers called him Oker, which in their language means "yellow," and because of the gold ornaments he wore, he became known among his own people as Chief Yellow Metal. Most white men simply called him Walker or Wakara.

Wakara agreed to show Rhoads the sacred Carre-Shin-Ob repository only after both he and Brigham Young gave their word that no other man would ever know its location. After that first trip with Wakara, Rhoads was allowed to visit Carre-Shin-Ob with only a Ute guide, and later by himself. During those trips into the mountains, Rhoads discovered several old Spanish mines, and although he later brought gold from them for his own as well as for church use, none of his personal gold ever came from the sacred cache first shown to him by Chief Wakara. Very likely some of the Ute guides who accompanied him to the place where the sacred gold was hidden also showed him places where old Spanish mines were located.

Through his close friendship with Brigham Young, Rhoads obtained a large tract of mountain land where some of those mines were located. Until now, few treasure hunters have known exactly where that land grant was located, some believing it was on Indian reservation land in present Duchesne County. The recent discovery of an 1858 Mormon Church map now reveals the boundaries of that grant. (See Map C) Even more revealing is a detailed description of a trip Rhoads made into the mountains with William Wines Phelps, a long time friend from early days in Illinois. That trip was made to ascertain the exact boundaries of the lands he was seeking title to, in order to keep others from locating mines there. It is fortunate that Phelps kept a detailed record of their explorations in his journal. His account follows, condensed somewhat due to its length, but abbreviated only in those areas not pertinent to its boundaries.

> Mr. Rhoads and I left the city on Saturday the 29th (1855) by way of Little Mountain and Lamb's Mill, and camped at the foot of Parley's Park, about three miles from Snyder's Mill. On the 30th we passed east, where we came to a fine prospect for grazing on Silver Creek. From that place we made the best way we could, north-easterly among hills and canyons to the Weber River; country very rough. We struck the Weber and after going up it some four or five miles to the junction of the east and south forks (now the fork of the Weber with Beaver Creek) we came to another site for grazing. We stopped for the night where the South Fork enters the Grand Pass (now the mouth of Beaver Creek Canyon). We saw what we at first thought were two Indian ponies, but upon getting nearer found they were an excellent span of grizzly bears.
>
> We shaped our course over the mountains to the west, a mile or two north of the Provo River. We ascended the mountains and upon looking back to the east, we saw plenty of timbered mountains before us, and about twenty miles in the distance, the headwaters of the Weber and Provo rivers. We passed on seven or eight miles into a small valley full of buffalo bones, where we turned north and passed over the divide into Silver Creek Valley. We took the trail down Big Canyon on our return. The distance to the Grand Pass is about forty miles (actually closer to 70 miles); the way down Weber Canyon would be more.

Upon returning from their secret mountain journey, Rhoads petitioned the Territorial Legislature, then convened at Fillmore, the Territorial Capital, for a land grant of the area described - - - the valley and mountains to the headwaters of the Weber and Provo rivers. On December 23, 1855, the Deseret News reported the following:

> Affairs at the Capital, National Hall, Fillmore:
>
> Several petitions have been presented for large tracts of land to be used for herding and other purposes. That portion within the Territory known as Rhoads Prairie has been granted to Brigham Young and Thomas Rhoads.

That tract of land, when plotted on a modern map, begins at the mouth of the Weber Canyon, near present day Oakley, and follows the Weber River upstream to its

Brigham Young. He recorded the quantities of gold brought from Carre-Shin-Ob in the Church Gold Account.

Ute chiefs Wakara and Arropene. Thomas and Caleb Rhoads promised them they would never reveal the secret of Carre-Shin-Ob.

Towats, the Great Spirit, instructed Chief Wakara to give the sacred gold to the "High Hats."

Smith & Morehouse Fork; then follows up that branch to the divide which is also the headwaters of the North Fork of the Provo River. It continues down the North Fork past its junction with the Main Fork at Pine Valley, and to its confluence with the South Fork. From there it continues downstream to the West Hills, where it continues along the west side of Kamas Valley to the Weber River at its point of beginning.

On May 20, 1857, Phelps reported that Rhoads had established a ranch at Rhoads Prairie (Kamas Valley), and that he (Phelps) had surveyed the site of Fort Peoa on the Weber River, several miles north of the Rhoads ranch. The origin of that name is of interest. Phelps discovered a squared log with the letters PEOHA carved on it. He speculated that it may have been carved by some forgotten fur trapper, or perhaps by some Spanish miner. In addition to Rhoads Valley, several other landmarks would soon be named for Thomas Rhoads, including Rhoads Plateau, a high mountain-mesa area located between the Provo and Duchesne rivers; Rhoads Canyon, a drainage from Soapstone Basin into Wolf Creek; Rhoads Peak, high on Lightning Ridge; and Rhoads Lake, a small mountain lake located on Gardners Fork of the Weber River, close to where one of his old Spanish mines was discovered.

Thomas Rhoads spent much of his time at his Kamas Valley ranch, exploring the mountains which rise from that valley to tower as some of the highest peaks in the state. Much of that high mountain wilderness was included as part of his land grant. That he discovered several old Spanish mines in those mountains is certain; how he found them is not as well known. Spanish (later called Mexican) miners took great care to conceal their mines to keep others from finding them, covering shafts and sliding dirt and rocks down over tunnel portals. They often went to elaborate lengths, leaving enigmatic signs and symbols carved into trees or cut into ledge rock to guide them back to those diggings after long absences; signs which only they would recognize or know the meaning of. Rhoads undoubtedly discovered many of those secret signs and symbols, but it is unlikely that he located many hidden mines by deciphering those cryptic messages.

Many treasure hunters have puzzled over the hidden meaning of those same mysterious signs left by Spanish miners of yesteryear. Most professional treasure seekers agree that there are a few universally used and recognized symbols, and as a matter of fact, certain designated signs were directed by Spanish authorities. But over the years I have come to one conclusion with which I am comfortable: most of those strange signs and symbols mean exactly what they meant to the person who made them. A turtle might very well refer to treasure whether it is found in the Utah hills or somewhere else, but an arrow, a cross, or a sun-sign might mean different things to people at different times and places. So if Rhoads never discovered the old mines he found by interpreting those obscure signs, how did he find them? Part of the answer to that riddle can be found in an investigation into the deaths of a party of Mexican miners who were taking a pack train of gold from the mountains above Kamas Valley south to Santa Fe.

In 1854, as a Lieutenant in the Nauvoo Legion, the Territorial Militia, Thomas Rhoads was designated to lead Company "B" to determine who had attacked and killed the Mexican miners with one of those pack trains when it reached Chicken Creek near present day Levan in Juab County. Rhoads concluded that the miners had been killed by Ute Indians, and that according to settlers who lived nearby, those Indians had recovered a considerable quantity of gold at the massacre site, which they told the settlers would be returned to Mother Earth, from whom the miners had stolen it. What Rhoads did not report was that two old Spanish maps had been found with one of the bodies, and that he kept those maps for his own use.

One of those maps had been sketched on a small

Keep a close watch on your back trail; some found only death in the Killer Mountains!

piece of parchment by Antonio de Reinaldo, and was dated 1851. (See Map D-1) That map showed the location of three mines close by several mountain lakes, near the headwaters of a stream which Rhoads believed was the North Fork of the Provo River. Of interest to treasure hunters is a pine tree with the date 1856 which is shown on some copies of that map. Obviously, that pine tree and date had to be added to that map sometime after Rhoads acquired it. Also, note that the 1851 Reinaldo Map is almost identical to the 1856 Pine Mine Map (See Map H, to be described later) which some persons believe depicts an area along Rock Creek on the Ute Indian Reservation. The legend on that map leaves no doubt that it pertains to the Provo River area, for that legend, which does not appear on copies many lost mine seekers have used, (after translation) reads:

> Explanation: These mines of gold and silver are on the high plateau of the Timpanogos Canyon, and are hidden among many large lakes. Here is much gold and silver in the form of rock crystals.
>
> /s/ Antonio de Reinaldo, 1851

The second map recovered at the massacre site by Rhoads was larger in size and portrayed a much wider area. It too was a Reinaldo map, but with the date 1851-1853, and is titled Paramo de los Fantasmo, literally translated as "The high, cold fantastic region," but also read by some authorities as "The barren plain of ghosts." (See Map D-2) Some who have studied that map carefully believe that "plain of ghosts" is today's Towanta Flat, located north of and on a bench land above lower Rock Creek, northwest of Duchesne City. That larger map depicts a "real" of mines, located among a group of lakes on the upper plateau. There is some question among treasure hunters regarding exactly what constitutes a "real" of mines. In the Rudo Ensayo, Society of Jesus (Jesuit) Father Juan Nentvig defines a "real," which is pronounced Ree-Al, as a "royal mining settlement," also written as "Real de Minas." A small garrison of the king's troops might be stationed at a real de minas for protection of the site, but not necessarily because the mines were the property of the king. That definition might do very well for mines in Old Mexico, but in actual practice, the reals de minas of the northern borderlands were often administered by a priest of the church, not by soldiers. That priest was usually a Franciscan or Jesuit, depending upon the time period in question. That did not, however, preclude the possibility that a small detachment of the king's troops may have accompanied a mining expedition, or that they remained near the mines to protect the clergy and miners from Indian attacks.

Those mines of the upper plateau shown on the Reinaldo Map are located at the headwaters of four rivers having their source at those lakes. Those rivers are designated as the Rio de Timpanogos (Provo), Rio de St. Ana (Weber), Rio de Damian (Duchesne) and the Rio de los Cumanchi (Bear). That large Reinaldo Map also shows the Rock Creek located in Duchesne County, but it is not named. Along Rock Creek there is depicted a gold placer and a "Fundicion," or smelter, which is located just downstream from a branch which appears to be the South Fork of Rock Creek. That map also shows thirteen mines or prospects where there is gold, each marked by an "X," but there is good reason to believe some of those "Xs" may have been added to that map long after Thomas Rhoads acquired it, no doubt to confuse and mislead searchers.

On the original maps as obtained by Rhoads, the mines were all named. Most copies seen today do not bear those names, which included Mina des Damian (Duchesne River Mine), Mina de las Puerta (Mountain Pass Mine), Mina de Alambre (Wire Gold Mine) and Placeres de Oro (gold placers. The 1851-1853 Reinaldo Map is so well drawn that several of the mines shown on it have been found during recent years, and were no doubt just as easily found by Rhoads. The smaller 1851 Reinaldo Map, which shows the mines of the high plateau in the mountains above Kamas valley, was doubtless also deciphered by Rhoads, at least in part, so it wasn't by chance that the Provo and Weber highlands shown on those maps was exactly the same area he sought for his land grant!

Old-timers then living at Kamas Valley pretty well knew the trail Rhoads followed into the mountains, but few chose to follow him, and those who did often regretted having done so, that is if they lived to tell about it. Rhoads warned the valley settlers not to follow him, and as the ranking Mormon and militia leader, his warning was accepted as gospel. Ephraim Lambert, one of those pioneer settlers, later told his son and grandson about a man who followed Rhoads. That story was tape-recorded by the grandson, Glen Lambert. Excerpts from the author's copy of that tape relates the following:

> Now Rhoads was a good rifleman, and he let it be known that no one was to follow him. He wasn't above shooting someone if he found them trying to do so. Of course, a few tried, but they soon lost his trail and came back. But one fellow, a Spaniard, came to Kamas and asked about Tom Rhoads, where he was at or if anyone knew how to find him. At that time Rhoads was away on one of his trips to his gold mine, so this fellow lit out on his trail, going east up Beaver Creek. No one ever learned how far he went or what happened, but a few days later his horse came back without him. No one ever knew what became of that Spaniard, nor did anyone ever ask. Did the Killer Mountains claim another victim?

At that time almost everyone knew that Rhoads went east following Beaver Creek, but somewhere along the way he turned off into rough country where his tracks would be lost on hard ground or while crossing a rock slide area. Some thought he left Beaver Creek to turn up Shingle Creek, while others said he went up the North Fork of the Provo. Aaron Daniels, the early settler at Heber Valley, recorded the following in his journal:

> About 1860 I moved from Heber Valley to Wanship (a tiny settlement located north of Kamas Valley). It was in the mountains near there that I first encountered Thomas Rhoads. He was packing out gold from a mine he had somewhere just over the summit of the mountains above Kamas Valley.

At that time there was a well-worn Indian trail between Kamas Valley and the Uinta Basin. From the headwaters of the Provo River, a rider could pass by Mirror Lake and follow the Duchesne River downstream to the Basin. He could also cross the high country between Mirror Lake and Granddaddy Lake to the headwaters of Rock Creek, and follow that stream down to the Basin. W.P. Mecham, an early-day rancher at Kamas Valley, told of finding a chunk of quartz rock heavy with gold where he crossed that high divide at the headwaters of the Provo. That piece of ore was obviously out of place, as if it had been dropped or lost where Mecham found it; perhaps by a Spanish pack train. When he had that quartz specimen analyzed by an assayer, it was found to carry a fantastic 800 ounces of gold to the ton! That was along the same trail taken by Thomas Rhoads.

It is well known that Rhoads explored as far east as the Uinta Basin, for in September, 1860, he was sent on a mission there by Brigham Young, to locate possible settlement sites. Accompanying him were Jesse Fox and J.W. Cummings. That they were discouraged by farming prospects in the Uinta Basin is evidenced by their report, which stated in part: "Our opinion is that the inducement to locate a settlement for farming in the Uinta Basin is not at all inviting." But less than a year later, Brigham Young became concerned that non-Mormons might claim the Uinta Basin and he had second thoughts regarding the opinion expressed by Rhoads. On August 25, 1861, Young directed thirty trusted Mormon Elders to move their families to the Basin. In doing so, Young stated that if the church did not send settlers there, "The gentiles (non-Mormons) will take possession of that valley, and I do not want them to have it."

The church owned Deseret News reported that settlers could expect to find "fertile fields, extensive meadows and wide pristine ranges." If Young had given more credence to the counsel of Thomas Rhoads, he could have saved those emigrants to the Uinta Basin a long, hard trip for naught. On September 25, only a month later, the Deseret News announced that the Basin settlers had returned, and that they had reported finding "a vast contiguity of waste, measurably valueless, excepting for nomadic purposes, hunting grounds for Indians, or to hold the rest of the world together!" If Young had asked Rhoads his opinion of the mining potential of the Uinta Basin, he might have received a much different answer.

A pioneer settler named Mitchell had a cabin near where Shingle Creek flows into Beaver Creek, and on occasion Rhoads would stop there to visit. One day he left an extra pack horse with Mitchell, and upon his return from the mountains several days later, he gave him a small piece of gold in payment for the care and feeding of his horse. Those who saw the gold ore Rhoads brought from the mountains all agreed that it was gold in quartz rock, not placer gold. Several described it as "sugar quartz, with wires and bits of gold all through it." Cal McCormick, who once helped him load sacks of ore onto a pack animal said that ore was yellow as bees-wax. "When it was melted down, the molten metal looked like honey, but if it was spilled, it would splatter like molten lead. That ore was so heavy that both of us would grunt with the lifting of it."

Another who remembered that ore was an old-timer who spent his last years at Fort Kamas. No one now seems to remember his name, but a few recall that he was small and frail, with long white hair and a beard to match. Wilford Binggeli remembers talking to him more than sixty years ago. "You bet I remember Rhoads," the old fellow said. "He would take six or seven pack horses and go up Beaver Creek, and would be gone for five or six days. When he returned, all of those pack horses would be loaded down with gold ore. I saw a lot of it. You could see the gold right there in that rock. Yes Sir! I remember old Rhoads!"

That Thomas Rhoads took gold from the mountains above Kamas Valley is further evidenced by an interview the author conducted with Mr. David Horne, based on information gleaned from Horne's great-grandfather's journal.

> My great-grandfather was George A. Smith, the famed southern Utah colonizer, and he was the father of George Albert Smith, later President of the Mormon Church. Soon after the Salt Lake Valley was settled, Brigham Young sent my great-grandfather on a mission to Nauvoo, Illinois, from whence the Mormons had come, to meet with members of the Joseph Smith family. He took with him a bag of gold for each member of the Smith family, as a token of friendship from the church.

It is interesting to note that Brigham Young had acquired enough gold to send a present to each of the Smith family, for he and the Mormons were poor as church-mice when they arrived at the Salt Lake Valley only a short time before. The only place where Young could have obtained so much gold so soon after he arrived in the valley was from that which was brought to him by Thomas Rhoads. Note that Smith took each member of the family "a bag of gold." I wonder how much gold there is in a bag of gold?

Ever since the Mormon Temple was built at Salt Lake City, there have been those who say that the gold used to plate the Angel Moroni figure atop the temple was Rhoads gold. Inside the temple there is an enormous baptismal font, a large vessel holding consecrated water, so large that it rests upon the backs of twelve life-size oxen. All of that gigantic structure is gold plated; some say it is solid gold. The late Gale Rhoades claimed to have definite proof that Thomas Rhoads did in fact contribute that gold, and that hidden away deep in vaults under the temple there are still boxes filled with Rhoads gold. Rhoades told the author that when he asked Church Patriarch Joseph Anderson if those stories were true, Anderson made a statement which did not preclude that possibility: "The Church will neither confirm or deny the existence of the so-called Brigham Young Mine, nor will it confirm or deny that the Church has any evidence or artifacts in its possession pertaining to that mine."

Mormon Church records, as well as individual pioneer journals, all agree that when the "Saints" first arrived at the Salt Lake Valley, they had less than one-hundred dollars among them, yet within only a few months Brigham Young paid hard gold coin for the sites of Ogden City and Fort Bridger. Church Historian Andrew Jensen recorded that the payment in gold for Fort Bridger was made in the presence of newly appointed Governor Alfred Cumming, who was a non Mormon. Jensen also noted that shortly after that purchase, Young invested $12,000 in improvements to the fort. Those first Mormon settlers may have been hard-pressed for just about every type of goods and merchandise, but there certainly wasn't any shortage of gold in Zion.

Also of interest is the fact that the church paid the fare from Europe to Utah Territory for approximately 60,000 converts, through their Perpetual Emigration Fund. Historian Leonard Arrington has researched the administration of those moneys and told David Horne that it cost approximately three ounces of gold, the only medium of exchange the church had in those days, to pay the fare for each convert brought to Utah. To bring those 60,000 converts from Europe required approximately 180,000 ounces of gold. If the price of gold at that time was set conservatively at ten-dollars per ounce, the cost to the church would have been more than two-million dollars! No doubt part of that amount was paid with member's tithing, while some came from gold jewelry and ornaments given to the Gold Account, but a large part had to come from some other source. It is only reasonable to believe that source was Rhoads gold, brought to the Mormon Mint from Spanish mines and caches in the mountains above the Heber and Kamas valleys and later from the Uinta Mountains.

Many prospectors have searched for those lost mines and hidden hoards all across the Wasatch and Uinta mountains, from Kamas and the Granddaddy Lake country to Rock Creek and Ashley Valley beyond. There is no doubt that Thomas Rhoads brought gold to the Mormon Mint from the sacred cache at Carre-Shin-Ob and the Spanish mines shown to him by Chief Wakara. The Spanish mines he discovered through use of the Reinaldo maps were located on the Rhoads Land Grant, and in the mountains adjoining it, in what are now property called the Wasatch Mountains, not the Uintas where the sacred cache is located. That fact has been known to serious researchers for some time, and has no doubt been suspected by others who have tried in vain to match those maps to the terrain of the Uinta Mountains.

Until recently, there has been little documentation to suggest where those Wasatch Mountain mines were located, but research has uncovered new information, which will be revealed in following pages. A statement made by one of the first settlers of Kamas Valley adds greatly to that information. Glen Lambert, mentioned earlier, died in 1982 at the age of ninety years, but before his death he recorded his own reminiscences as well as valuable factual material learned first-hand from his father and grandfather, the latter being the same Ephraim Lambert who told of the Spaniard who tried to follow Rhoads into the mountains. Glen Lambert's truthfulness is unquestioned, for he was a man of stature in his community, having served as a United States Forest Ranger for thirty years. Ephraim Lambert was a neighbor and close friend of Thomas Rhoads for many years. The Lambert reminiscences are in part as follows:

> This man Rhoads came to Kamas Valley during the early days as the first settler, and located a ranch just north of where the town is now located. He spent a lot of time trapping and exploring. Both my father and grandfather were well acquainted with him. He prowled around a lot in the mountains, and it was generally well known that he had discovered a rich gold mine somewhere east of Kamas. Nobody knew just where it was, but he would make regular trips into the mountains and would come back with gold. Now aside from what he needed for his own use, he gave that gold to the Mormon Church. He gave it to Brigham Young, and it was used to build the Salt Lake Temple. A lot of material had to be freighted in, and other costs were high. That's where a lot of that gold went, to pay those expenses.
>
> Rhoads made frequent trips into the mountains, and he brought gold into Kamas. A lot of people saw it. My father saw it. He saw one sack full that Rhoads brought to my grandfather Lambert to have it weighed, so he would know how much tithing to pay on it. Grandfather had the only scales in town. Now, Rhoads had nearly sixty pounds of gold in that sack, in chunks and pieces. My father was standing there looking at it. He was about eighteen years old at the time, and Rhoads looked at him and said, "That's a lot more than you can carry, young man!" Later on some people said Rhoads brought that gold from California, but my father knew him better than most did, and he told me that none of that gold came from California. That was quartz gold, not placer. Father heard him say that while on his prospecting trips into the mountains, that he found a likely looking place, so he dug down into it a little ways and he found gold, much richer than it was in California. So from what he said, I have no doubt that his mine is somewhere in the mountains east of Kamas.
>
> When Rhoads took that gold to Brigham Young, he was told never to make his mines public. The people then were poor farmers, and farming was their only source of food, so a gold mine like that would cause a stampede into the mountains and probably a lot of people would starve to death. Young told Rhoads that if he didn't make it public, that he nor any of his posterity would ever want for the necessities of life. All of them, Tom Rhoads, his kids and grand-kids have always been pretty well to do. They never worked any harder than anyone else, but it seemed like they were always able to acquire land, ranches and cattle wherever they went, until it wasn't long until they were the best-off people around.

Caleb Rhoads. You can still follow his faded footprints to the hidden treasures of the Uintas.

The weighing of that sixty-pound sack of gold occurred soon after the first settlers arrived at Kamas Valley, about 1860. Volume 9 of *Our Pioneer Heritage,* published by The Daughters Of The Utah Pioneers, makes reference to that sack of gold which Rhoads took to Salt Lake City. Col. Joseph M. Lock was quoted as follows: "Father Rhoads brought in several sacks of gold, and among them was a sixty-pound sack. That sack of gold was the chief topic of conversation for people in the valley for quite some time!" In 1862, Thomas Rhoads was elected to the Territorial Legislature from Summit County, a position which required much of his time. Sometime before that date, about 1855 according to Lambert, during a period when he was ill and could not go out into the mountains, Thomas Rhoads disclosed the locations of his secret mines to his son, Caleb. But even before that time Caleb had accompanied his father to Chief Wakara's Carre-Shin Ob cache on the Ute Reservation.

Caleb Baldwin Rhoads took the same oath which

his father had taken in the presence of Brigham Young and Chief Wakara, however, Caleb's oath was given in the presence of Young and Seigerouch, known to white men as Chief Arropene, who had become Chief of the Utes on Wakara's death. That oath was given on September 7, 1855, when Arropene was at Salt Lake City to sign a treaty. Caleb Rhoads was then twenty-five years of age. He had recently married Malinda Powell, a daughter of James Powell, a new settler at Kamas valley. Although they were later divorced, Caleb and the Powell family always remained close friends. Caleb later married Sidsie Adams, his wife until death did them part. A physical description of Caleb Rhoads at that time stated that he was large in body size, heavily built and accustomed to hardships. Like his father, he was at home in the mountains, as a trapper, hunter and prospector. With Caleb Rhoads bringing gold from his father's Spanish mines, Thomas Rhoads was free to accept a church mission call from Apostle Erastus Snow in 1864, to develop some church owned mines located southwest of Minersville, in the Pahranagat Mountains. He was also expected to resolve a dispute concerning some mining claims allegedly stolen from Brigham Young by General Patrick E. Connor. He made his home at Minersville and never returned to his Kamas Valley ranch except for occasional visits.

Two of his wives operated a small general store at Minersville, but everyone knew that small store couldn't turn enough of a profit to support those two wives and their large families. A grand-daughter, Ethel Pehrson, later recalled that much of Thomas Rhoads' time was spent in the mountains around Minersville, prospecting and mining for gold. He would periodically visit his other wives at Kamas Valley and Salt Lake City, always leaving them enough gold to sustain them and their children until he came again. Thomas Rhoads died at Minersville on February 20, 1869, just short of his seventy-fifth birthday. His body was taken to Provo Valley for internment. True to his promise to Brigham Young and Chief Wakara, he never revealed the location of the great cache of gold at Carre Shin-Ob, nor any of the old Spanish mines he found, except to his son, Caleb.

CHAPTER 3

WHEN OLD TRAILS WERE NEW
Prospects along the Provo River

Beyond Timpanogos Valley and the Rio San Antonio de Padua (Provo) Canyon, the Old Spanish Trail entered a high mountain valley nearly round in shape, hence its name, Valle Redondo, or Round Valley; today's Heber Valley. Like a giant compass, four trails came together there from the four cardinal points. Using modern place names, the old trail from Utah Valley came from the south, a second led eastwards to Daniels Canyon, another went northward over Lake Creek Summit to the West and North forks of the Duchesne River and the Ute Indian Reservation beyond, while the fourth turned west to Kamas Valley and the Provo and Weber river canyons.

Heber Valley had been a favorite camping place for Ute Indians long before the coming of either the Spanish-Mexican miners or the Mormon pioneers. Two traders who yearly made their way north from Santa Fe to trade with the Utes were Mauracio Arze and Lagos Garcia, even though they often traded without benefit of a license. Their names are still preserved in the shelter of an overhanging cliff along the Old Spanish Trail. In 1813 both were arrested for engaging in illegal trade with the Utes. Court records at Santa Fe reveal that Arze and Garcia traded with the Indians from the Sevier Valley of central Utah north to Lake Timpanogos and beyond. It is interesting to note that trial records do not detail their route as earlier accounts often did, indicating that by then the trail north from Santa Fe was so well known that court scribes apparently felt no description was needed.

Just beyond where that well-traveled trail entered Heber Valley from the south, it crossed a small stream later called Daniels Creek, named for Aaron Daniels, one of the first settlers in the valley. Remember that name, for Daniels was one of the West's first treasure hunters. Daniels was born August 1, 1822, at Dryden, New York, and arrived with the Mormons at the Salt Lake Valley in 1847. Like many Mormons, he kept a detailed daily journal, one which has become a Bible for modern-day gold seekers. His adventures will be documented at length in following pages. Daniels Canyon, the dark recess from which that small creek came, was named for him. An early entry in his journal reads as follows:

> In 1858 I took a herd of cattle to Heber Valley and started a ranch on the Provo River, about one mile north of where Daniels Creek enters that river. I trapped and explored during the winter, and in the spring I discovered a Spanish mine near the summit of the ridge of Daniels Canyon.

In Volume Seven of the Mormon Church's *Enduring Legacy* history series, it is written: "Daniels was a rancher, trapper and prospector. He had a ranch on the Provo River in the Uinta Mountains. He trapped in Daniels Canyon and later found a gold mine somewhere in those mountains. When he was told that he should give ten percent of his gold to the church, he refused and apostatized." Note that Daniels "found a gold mine," a mine already in existence, not some mineral discovery which he found and then developed into a mine.

Daniels didn't elaborate on the exact location of the mine he found; but note that he said it was near the summit of "the ridge of Daniels Canyon," not the summit of some lesser ridge within the canyon as some have thought. The summit ridge he referred to is much higher in elevation than any of the canyon's lesser ridges. Apparently, he didn't record any written claim to that mine, leaving a paper trail for today's prospectors

Mauracio Arze left his name and 1812 date along the trail where he packed gold nearly 200 years ago.

to follow, so that when he moved to nearby Wanship, it was soon forgotten and became a lost Spanish mine again. Also, note that the mine, old even when Daniels found it, could not have been dug by anyone other than Spanish miners, since there had been no Americans in those mountains before Daniels, except perhaps a few Mountain Men, none of whom were noted for digging in mines.

A tale which used to be told by old men around Heber Valley claimed that the Daniels Mine was located in the seventh canyon which branches off from Daniels Canyon. If that tale was true, the mine might be found somewhere in Thornton Hollow, which is the seventh canyon, located seven miles from the mouth of the canyon. That might be worth investigating, for other Spanish mines have been found in that same area. In 1896, Bill Bethers discovered a strange stone not far from the mouth of the canyon. That rock was inscribed with rows of peculiar "hieroglyphic" characters. Records reveal that Bethers took a man "accustomed to deciphering such signs" to interpret that puzzling inscription. Figures on that stone showed a man with his hands thrown up, as if he had suddenly been startled, a burro or similar pack animal and a crescent moon. Those signs were interpreted to be a guide for miners returning to the mountains after a long absence. Some time later a very similar "hieroglyphic" stone was found on the point of a ridge near the mouth of Provo Canyon; its message directing Spanish miners to the stone discovered by Bethers. When the present highway was built through Provo Canyon, that stone was removed by archaeologists from Brigham Young University and was taken to that campus where it still remains.

With Henry Boren as a partner, Bethers began searching Daniels Canyon for some sign of a mine, and atop a high ridge nor far from the mouth of the canyon he found three tunnels close together, driven into the mountain side at different angles. That discovery caused such great excitement at Heber Valley that on February 12, 1897, the editor of the Wasatch Wave published an account of the find:

> All early settlers of this valley have heard that some years before the first pioneers arrived here, there existed valuable mines in the surrounding mountains which were worked by Mexicans. Those miners would return to their homes each fall, taking a long pack train of burros heavily laden with golden riches, which they had dug from mines which all of the first settlers said were located above this valley. One old Mexican said the mines were located about thirty miles from a large lake, and told how miners there had been driven away by Indians. Many miners were killed, only a few escaping to tell the tale of treasure left hidden in those mines. Now comes the claim that those mines have been found by Messers Bethers and Boren.

That report also included an account written by the reporter who accompanied Bethers and Boren to personally determine the authenticity of their find. His inspection was reported as follows:

> The writer took a long and tedious march up a ridge at what seemed to be an angle of forty-five degrees. After about an hour and a half we came to where a tunnel had been driven into solid rock. It followed a vein of ore from the surface, which vein is reported to return very good assays in gold. Cleaning out the debris in the old workings has left a hole through solid rock some three feet wide and five feet high, as far as the tunnel has been explored. Northwest from that prospect is another tunnel which has only recently been discovered, and there is still another about seventy-five yards to the southeast. All of the tunnels have been completely filled in with loose dirt and rock. The work of cleaning them out has exposed chisel marks in the rock which plainly reveals that the implements of man were used by the original miners.

Remember, those were old mines discovered by the earliest settlers. They were not, as some now claim, dug by those pioneers. The mark of wrought iron or brass tools found in the tunnels, as well as the fact that they had purposely been concealed with rock and debris to cover their portals, identify them as being of Spanish origin; while the "hieroglyphic" rocks confirm that fact. The Wasatch Wave reporter also saw and described the inscription stone which first led Bethers to the mines.

> A large granite rock which stands perpendicular to the ground about a half-mile down slope from the mine is covered with peculiar looking hieroglyphics. It is of a wedge shape, about a foot thick on one side and tapering to about two inches on the other. It stands about five feet high and is about four feet wide. There are a number of characters cut into the rock, which though perfectly visible, I was unable to decipher. Whether Bethers and Boren have found the lost mine described on that rock is a matter which will yet unravel, but that they have found tunnels made by human hands is an absolute certainty. We hope they will open up a bonanza!

Bethers and Boren opened one of the tunnels for seventy-five feet before they came to a place too badly caved and dangerous to proceed further. Starting anew, they open an adjoining tunnel for twenty-five feet before winter snow drove them from the mountains. Both men intended to return in the spring, but Bethers moved from the Heber Valley while Boren was called on a church mission to settle new townsites in another part of the territory. Apparently no one else in those olden days had the time or interest to continue digging at those old mines, so once more they were abandoned. Over the years since then, several parties have staked their claim to the tunnels, but none have reopened those caved workings beyond where Bethers and Boren quit work. Few people today know the exact location of all three tunnels, and apparently none of them are interested in digging for Spanish gold.

Across Daniels Canyon from the Bethers and Boren Mine, in Boomer Canyon, Robert Giles discovered several intriguing Spanish tree signs. Cut into a patriarch pine is a Catholic cross, with the letter "U" atop its right arm. Only a short way further up canyon, hidden at the top of a rockslide, is a mine tunnel, no doubt as ancient as that old cross on the pine. Two elk hunters recently came upon it entirely by accident. According to their account, they crawled back beyond its caved entry, using only matches for light. But they made a hasty exit when they came upon two skeletons, one of a man and the other of a mule or burro. The human skull had a hole through it, as if made by a bullet, or perhaps an arrow. They claim taking that skull to the local sheriff's office, but apparently little investigation was made, for records disclose no follow-up. The hunters, more interested in elk than skeletons, never returned to Boomer Canyon.

Aaron Daniels, the pioneer settler at Heber Valley, probably knew of that old mine, for in the history of Wasatch County it is written that he built a small cabin at the mouth of Boomer Canyon. One can't help but wonder why he would build his home in such a gloomy and secluded place, when he had the entire valley below in which to locate his cabin site. Remember, in his journal Daniels wrote of finding an old Spanish mine in Daniels Canyon. Did he build his cabin at the mouth of Boomer Canyon to be close to that mine?

If you can decipher this stone tablet found in the Uinta Mountains, it could lead you to treasure!

There could be a lost ledge of gold near the top of Daniels Canyon. Wilford Binggeli operated a gravel pit close to the mouth of the canyon, but some say he spent more time prospecting than he did selling gravel. Binggeli was an experienced miner, having worked at mines all across Utah and Nevada, so when he confided in Gordon Taylor, a banker friend at Kamas Valley, that he had found a place near the head of Daniels Canyon where sometime in the long ago Spanish miners had dug for gold, Taylor paid close attention to him. Binggeli told Taylor that he had come upon a stringer of gold ore in an exposed rock outcrop, where miners of old had built fires against the ledge to heat it, and then threw water on the hot rock to crack and break it loose. Times were hard then, so Taylor wasn't able to loan Binggeli the money he needed to develop his find, and as the years passed Binggeli had to devote his time to raising a family. But he still maintains mining claims on that area, but with modern environmental restrictions and regulations, he has about given up on working them.

Binggeli's gold ledge in Daniels Canyon might very well be part of a gold vein discovered by William Foreman in 1891. In his journal, Foreman wrote of a find he made in May of that year, but he wrote that it was then part of the Ute Indian Reserve, where white men were not allowed to trespass. But apparently, Foreman worked his find in secret, despite it being on Indian land, for an entry made in his journal the following July reads: "Commenced to run a tunnel, no timber needed." Two years later, In August, 1893, another entry states: "I camped all winter while working at the mine. I am in 250 feet now." Nothing further about his mine appears in his journal that year, but no prudent man would dig a tunnel for 250 feet through solid rock where "no timber is needed" unless he had something darn good; nor was it unusual that he kept quiet about mining on the Indian Reserve.

Shortly afterwards Foreman wrote of a tragedy which befell him: "My son, while fooling with a pistol, shot himself in the head. This has been a sore trial, for it leaves me without my best helper." No doubt the loss of his son contributed to the end of Foreman's mining, as did yet another problem he faced. He was kept busy dodging Federal Marshal Bauchman, who sought his arrest on polygamy charges. In fact, Foreman was arrested soon after his son's death and was sent to the territorial prison for having multiple wives. We can only wonder whether the discoveries made by Foreman and Binggeli were part of the same ore vein. They could be, for only recently a Heber Valley prospector discovered an old shaft in that same area. He has been showing pieces of beautiful blue-green peacock ore, and he says an assay report indicates that ore carries values of several ounces of gold plus substantial values in silver. Daniels Canyon might be a good place to do a little prospecting, especially along the summit ridge described by Aaron Daniels; perhaps somewhere close to Foreman's Spring. That area isn't on the Indian Reserve now.

Nearly ninety years after Foreman dug his tunnel near the top of Daniels Canyon, the Killer Mountains claimed another victim almost in sight of his mine. In September, 1987, local newspapers carried the following headline: "Mysterious Death Of Gold Prospector! Murder In The Mountains Above Heber!" That gruesome account described the killing of Joseph Sheets, age 28, whose body was found "East of Heber City, about three miles from the Daniels Canyon summit." A few days later a follow-up article reported that Sheets had been a prospector and was enroute to his diggings when he was killed. That report added that he had been shot twice in the back of the head. His tragic death was only the latest in a long list of miners and prospectors mysteriously killed in those mountains. To this date no clue has been found to explain his untimely death.

In addition to the lost mines of Daniels Canyon, there is still another lead at Heber Valley which you might want to consider. Some may call it fanciful or wistful dreaming, but there does appear to be good reason to investigate it more thoroughly. There are two old

A centuries-old Indian petroglyph records the coming of the Black Robes to the Uinta Mountains.

Few men today follow the dim trail to the lost Foreman Mine in Daniels Canyon.

mine dumps on the mountain side above Midway. You can easily see them from town. The earliest settlers at Midway heard Ute Indians tell that in the long ago, Spanish miners had a mine on that mountain, where they dug out nearly pure gold. Sam Boyd prospected that mountain every chance he had, but he never found the Spanish mine; still he did come upon a place where low-grade ore out-cropped on the surface. With Benny Clark helping him, Boyd dug a long tunnel, where one of those dumps can be seen today. The other tunnel and mine dump was the work of Nobel Snyder, who dug at a place where a well-known geologist told him he would find gold. But Snyder's efforts, like those of Boyd and Clark, were to no avail. The question is, is it worthwhile to prospect that area any further?

Consider the New Park Mine, located only three miles straight-line to the northeast of those two old mine dumps. The New Park had its start as the Glen Allen Mine in 1870, and was worked steadily for nearly one-hundred years. During that time it produced well in excess of one-hundred million dollars! Initially, the New Park was a lead-silver property, but as its shafts approached one-thousand feet in depth, it became a leading gold producer, with ore in its pearl fissure assaying seven ounces of gold and forty-five ounces of silver. At that time it was the nation's sixth largest gold producer. The richest workings in the New Park Mine were deep under the volcanic cones which are a prominent part of the landscape at Midway. Streams of boiling mineral-laden water from those cones are now used to fill swimming pools at fancy resorts. The richest gold veins at the New park were found in volcanic fissures filled with boiling water. Temperatures in the depths of the mine were so high that miners were furnished barrels of ice water. What might those pioneer prospectors at Midway have found if they had sunk vertical shafts into those hot water fissures instead of running horizontal tunnels into the mountain side? What if someone today drilled a vertical hole into one of those fissures? What if . . . .

An examination of current road maps indicates that the shortest and most direct route from Heber Valley eastward to the Ute Reservation is via Daniels Canyon, the course of the present highway, but that was not the route followed by Indians or by Spanish miners. Before the first wagon roads were built, the old Indian trail from Heber Valley to the east followed Lake Creek to its summit, and then continued down the canyon of the West Fork of the Duchesne River. Settlers at Heber City noticed that Indians chose the easier grade of Lake Creek and the West Fork, even though the mountain pass between them was 9,500-feet in elevation, rather than climb the steep and twisting Daniels Canyon to its summit, at a much lower 8,000-feet. Old carvings of turtles, arrows and Catholic crosses cut into giant pines and ancient aspens along that old trail proves that Spanish miners favored that route as well.

Nearly all of the old-time cattlemen at Heber Valley agree that in the early days the main traveled horse trail, as well as the later wagon road, went east to the Ute Reservation by way of Lake Creek. John Crook, one of the first settlers, wrote that the distance to the reservation via Lake Creek was forty miles shorter than the route through Daniels Canyon. Journal entries made during the Blackhawk War of 1866 tell the same thing. Ute Indians made several raids on Heber Valley. William Foreman, the prospector who had a tunnel in Daniels Canyon, was posted as a lookout at the head of Lake Creek. On July 16, he observed a band of Utes who had stolen settlers livestock and gave the alarm. A pitched battle ensued, during which two settlers were wounded and one Indian was killed. Crook wrote that the wounded men were taken to Heber City by the shortest route, "down Lake Creek from the ridge between Heber and the reservation."

It wasn't until the 1870s that a rough wagon road was built up Daniels Canyon. A history of Uintah County states that when it was built, there was only one old and long abandoned log cabin between the Bethers and Boren Mine at the mouth of the canyon and the Murdock Trading Post at the Whiterocks Indian Village, more than one-hundred miles to the east. It would be interesting to know who built that "long abandoned" log cabin near the head of Daniels Canyon, and even more so to know why it was built there, so many years before the first Americans ventured into those mountains. That wagon road hadn't been improved much by 1877 when Pete Dillman reported that it took him twenty-one days to take a wagon from Heber Valley to

Gold was found here in 1831. More may still be hidden there!

Ashley, some thirty miles east of the Whiterocks Indian Agency. Dillman wrote that crude road crossed Daniels Creek thirty-six times, each crossing being worse than the last!

The old Spanish diggings discovered in Daniels Canyon by Aaron Daniels, William Foreman and others prove that Spanish and Mexican miners explored and dug for gold and silver there long before Mormon settlers built that first wagon road. Those early-day miners used pack animals, not wagons, and their trail of choice was through Lake Creek Canyon. Just above Curtis Crook's home along Lake Creek, there are the ruins of an ancient looking stone wall, built at the base of an out-cropping ledge of rock. Those stones are carefully fitted together, and the wall is strategically placed, as if it may have been a lookout point or a defense fortification. Perhaps it was a guard post along the Old Spanish Trail, for Crook was told by his grandfather that the wall was old and green with moss and lichen when he first saw it during the 1860s. If only those stones could talk, what tales they might tell!

Curtis Crook's home is located only a short way from the mouth of Pole Canyon, a side gulch leading off from Lake Creek. Some unusual finds have been made there during the past few years. While digging a hole near his sheep camp, "Whiskey Joe" Murdock uncovered a rusted metal box which had an old map in it. Those who have examined it say it is an old Mexican mine map. Dan Allison, a Heber Valley stockman, ran livestock in Pole Canyon for a long time. While following his stock one day, he picked up a fist-size chunk of white quartz, criss-crossed with stringers of yellow gold. He found it along the trail near the top of the dugway, about six miles up canyon. Allison spent some time looking for more of that quartz, but never found another piece. Is there an outcrop hidden somewhere in the oak brush nearby? That gold specimen had to come from somewhere.

Rob Dansie thinks he might know where Allison's quartz gold came from. A south-of-the-border sheepherder who was working for his family found a most unusual pick in Pole Canyon. It is very similar to a prospector's pick, except that it is much larger and was hammered out of wrought iron by someone who was skilled at blacksmithing. It has both a hammer-head and a chisel-point, but unlike manufactured picks, its head is secured to the rotted wood handle by pins driven through holes on each side of the metal head. It is a one-of-a-kind artifact, both unusual and valuable. If you want to prospect in Pole Canyon, start where "Whiskey Joe" Murdock dug up a buried map box, follow the trail to where Dan Allison picked up that piece of gold-laden quartz and pay special attention where that old Spanish mine pick was found. Three clues should be enough for anyone!

Curtis Crook's grandfather, Heber Crook, told him a story of some interest to today's prospectors. Back when Heber Valley was still being settled, an aged California prospector would come to the valley each spring, riding into town on the old Heber Creeper Railroad. He would rent a pack horse and slowly make his way up Lake Creek, often stopping at Crook's home to rest both himself and his pack animal. Crook couldn't help but notice the old fellow's advanced age, so he always offered him a meal or lodging for the night. Crook recalled how the old man would follow the long, open ridge on the north side of the canyon, a ridge which gave easy access to the headwaters of the West Fork of the Duchesne River and Currant Creek. Each year at the end of summer, he would come back down canyon, his pack horse loaded down with two saddle bags, each heavy with chunks of dark brown colored quartz rock, shot through and through with stringers of yellow gold.

After several years of making his yearly journey up the steep canyon, the old man told Crook that he would take him to his mine when he returned in the spring, for he knew his time was short and it would probably be his last trip. He said that he had no family, and that he wanted to show Crook his gold ledge in appreciation for the kindness and hospitality Crook had given him. Spring came again, but the old man didn't, nor did he ever return again. Crook later learned that he had died in California. The secret of the brown colored quartz died with him, but its source can't be very far from the head of Lake Creek, perhaps somewhere along the canyon of the West Fork, or maybe across the ridge at Currant Creek. Your search starts at the old rock ruins near Curtis Crook's house.

That high and windswept ridge which separates Lake Creek and the Provo River drainage from that of the Duchesne River runs roughly north to south for thirty miles and more. It gave Spanish explorers easy access to such rich mining districts as the Strawberry country, Currant Creek, Wolf Creek and the West Fork. Over the years, livestock herders working for Heber Valley ranchers like Fred Giles and Cal Giles have come across many Spanish artifacts in that sprawling area. A Spanish armor-helmet was found half buried in a giant cedar tree where bark and branches had grown around it. Apparently, it had been placed in a fork of that tree sometime in the long ago, where it was left and forgotten. Another herder found a strange looking metal boot or armored footgear along Trout Creek, while a badly rusted breast-plate was discovered near the head of Co-Op Creek. All of those relics and more were found across a wide section of rough mountain land. Surely they weren't lost by Ute Indians or Mormon pioneers. One can't help but believe they were lost by Spanish miners.

The mountain road which follows the Provo River from Heber Valley past the little hamlet of Woodland enroute to Wolf Creek passes by several old Spanish mines. Near the head of Riley's Canyon, only a few miles north of the Wolf Creek road, there is a shaft typical of those dug by Spanish miners. A row of shallow footholds cut into the side of that awesome pit leads down into unknown depths - - - unknown because no man dares descend that frightening spiral staircase. Other old diggings can be found just below the ridgeline which leads west towards Moon Springs. On one high ridge, rockhounds dig pieces of semi-precious horn coral, which they make into beautiful jewelry. While fishing along the Provo River near Woodland, Ed Grose picked up a chuck of white quartz containing visible gold. It is only one small piece of gold ore, no doubt washed by the river to where Grose found it, but one can't help wonder where it came from. An 1869 map of that area shows several old mines along the Provo River near Woodland, but little is known about them today.

Just east a few miles past Woodland, a narrow dirt track leads off towards the foothills and the edge of the cedars, to what is appropriately called Cedar Valley. An old logging road once led up the mountain side there, but hardly a trace of it remains now. Old mining claim maps from the 1860s show several sawmills along that road, as well as cabins and houses where mill workers lived, but there is no sign of them now. Several mining claims are also shown, with even less of them remaining. But just off that old timber trail, at the edge of a cliff overlooking the valley below, hidden behind a stand of cedar trees, Ronald Krause discovered an old mine, a vertical shaft he estimated to be about forty-feet deep. There isn't any waste dump to give its location away, for rock dug from it was thrown over the cliff, making a natural looking rock slide. When Krause first saw that mine, there were still pieces of an old-time hand operated "cigurna" or whim hoist standing guard over that dark pit, but over the years since it has rotted away and fallen into the depths. Krause knew he had found something unique, but he was hunting deer and didn't have the necessary equipment to explore its unknown depths.

Since that discovery, Krause has led the author and "Old Dan" Tucker back to that old shaft, where the author descended to its bottom. That shaft wasn't forty feet deep as Krause had thought, but over one-hundred feet, nearly straight down! It had been dug just off the vertical and in the manner of Spanish shafts, it is round without supporting timber. There are traces of silver and copper in the narrow ore vein at its bottom, but it was evident that those early miners had taken every bit of ore carrying any good values. Descending that shaft hadn't been easy, but trying to climb back out was worse. Its walls were slick and wet with moss, and there were few footholds. It was scary! There is an old miner's saying which pretty well describes it: "If you want to learn how to pray, go down into the mines!"

But that trip with Krause wasn't lost time, for less than a mile from that shaft we discovered another old digging. While crawling through oak brush we came upon a very old mine dump. It would have gone unnoticed had we not seen the squared end of a timber protruding above the ground. Examination revealed it had once been part of a whim hoist, a primitive hand-made affair turned by manpower. That recently found digging is without doubt located on the same ore vein as the Krause shaft. Although there is a waste dump, it can't be seen by anyone not actually crawling through that heavy brush. Its size indicates an extensive underground working, but there is no sign of either a shaft or tunnel portal. A depression in which century-old aspens now grow may be a shaft site, but probing with long metal rods has so far failed to break through rocks and timber which cover it. As this is being written, that depression is filled with four feet of melted snow water, but its secret will be revealed before long. One can't help but wonder why someone so long ago went to such effort to cover and conceal that old shaft if there isn't something of value still there.

Higher on the mountain, along a deer trail, Krause found still another mine portal, hidden at the base of a cliff. Its entry is only a narrow slit, about two feet wide, but it increases in width as it penetrates the mountain. Tool marks cut into the portal rock reveals that miners removed only the narrow ore vein, taking no more waste rock from the side walls than necessary, a method of mining found in all Spanish mines. There is a near-vertical shaft only a short way into the mountain, where steps or places to stand are cut into the rock at about four-foot intervals, just about the height one man could lift a sack filled with heavy ore to another on the step above him. When Krause found that narrow entry, there were several rotted pine poles covering the opening, so old that they were green with moss. It was obvious that they were placed to cover and conceal that portal. Krause replaced those poles just as he found them.

Although we had little difficulty locating the vertical shaft and were lucky to stumble upon the still uncovered diggings in the oak brush, our luck at finding the narrow slit which leads to the stone-steps mine was not as good. It would have been easier if Krause had not replaced the poles which cover the entry. Many years had passed since he last saw it, and the mountains had grown a lot steeper for him since then. They were just about as steep for us, and we are a lot younger than he is. There are lots of ledges and rocky cliffs in the area where Krause believes that old mine portal might be, and we've covered only a small part of it. We haven't given up yet, but if you find it first, it's yours!

Beyond Cedar Hollow, the Wolf Creek road winds its way up past Nobletts Creek and the old Carlile Mine, half-hidden in the aspens about a mile from the road. It continues past the Mill Hollow turnoff to Wolf Creek Pass, 9,500 feet in elevation. The road drops down the Duchesne River side of the mountain past Phelps Brook, where an old Mexican map shows a mine site. A half-dozen searches made during the past two years has failed to locate it, but that doesn't mean it's not there. Just before the road reaches the confluence of the West and North forks of the Duchesne River, a side trail turns to the north across a rocky side hill and into a narrow gorge, the mouth of Rhoads Canyon, one of the landmarks named for Thomas Rhoads. We know only that Rhoads often used that canyon to gain access to the mountain fastness beyond, but where he went in that

Bud Krouse led the way to this old Spanish mine at Cedar Hollow; a shaft 100 feet straight down!

high country isn't well known, for he talked little about his trips. Perhaps he had good reason not to talk about his wanderings.

Caleb Rhoads was also seen in that same country, and he was just as secretive as his father in matters pertaining to mining. But Caleb sometimes seemed to enjoy taunting acquaintances by dropping clues to where his mines were located - - - that is if his friends were astute enough to detect those clues, which were often hidden in normal conversation. On one occasion, he and John Powell were hunting deer along the Provo River in the mountains east of Kamas Valley, where they bagged a fine big buck. That deer was too heavy to carry, so they hung it in a tree until they could return with a pack horse. Quite some time later, Caleb asked Powell if he remembered the time they bagged the big buck. When Powell replied that he did, Caleb said: "One of my mines is not very far from where we hung that deer." He said no more about it, so Powell had no idea in what direction the mine might be from that tree, or how far "not very far" was.

Many years later, during the 1890s, Powell made a rough sketch drawing of the area where they had hunted and the place where they hug the deer, which place is now known as Pine Valley. (see Map E) That is where the North and Main forks of the Provo River converge, and only a few miles from its confluence with the South Fork. Powell recalled that they hung the deer at the foot of the mountains on the south side of the Main Fork. If you compare Powell's sketch with the 1826 Garcia Map (Map I), yet to be described, you will note that the Mine Of The Virgin Silver is located on the mountain front above the South Fork. That is "not very far" from where they hung their deer.

On other occasions, Caleb Rhoads was known to pull a packsack filled with gold ore from beneath his bunk, or perhaps from the edge of a haystack, just to show it to some friend, or sometimes give a piece of gold-laden quartz to a neighbor who was in need. All who saw or received that gold agreed that it was more gold than quartz. Sometimes he seemed to taunt friends or visitors by saying that his mine was somewhere in the mountains near where they had met, or perhaps in sight of some landmark each knew well, such as a trail fork or a lake. He never volunteered any further clue in which direction the searcher should look, or exactly how far away it was. Dozens of similar clues which he dropped over the years could be cited.

Thomas and Caleb Rhoads weren't the only early settlers to discover Spanish mines in the mountains near Kamas. In 1870 an old mine was found by a group of prospectors, but their description of it and the directions they gave weren't clear then, and they are no less confusing today. Even the oldest residents of Kamas Valley are puzzled by that report, which reads as follows:

> A remarkable discovery has been made at Kamas Prairie in the Weber Valley. In August (1870) a party of mine prospectors came upon a sort of shaft filled in with loose dirt and rocks. Their curiosity aroused, they excavated it and opened up a mine dug years before by Mexicans. The walls of the shaft had been cut with iron tools. A series of steps cut into the wall led down into the shaft. Debris was removed, and at the bottom a tunnel ran for a great distance along the ore vein. Among the rocks removed were some very good specimens of silver.

That report raises many questions, not the least of which is what was meant by "Kamas Prairie" and "in Weber Valley." Was that shaft on the valley floor or was it in the canyon of the Weber River? The tunnel at the shaft bottom "ran a great distance," but how far was that? Was it a few feet, one-hundred feet or more, and in what direction? Did the finders develop the mine, did they discover that it had been worked out, or did they simply abandon it? That same "remarkable discovery" was also reported by the Utah Gazetter, but with some differences in wording, but not enough to answer the questions raised. The Gazetter stated the tunnel "ran an indefinite distance" and that specimens of "tolerably rich silver" were found. Old-timers like 87 year old Gordon Taylor of Kamas or the late 95 year old Lavere Crandall of Peoa have never heard of that old report, nor do they have any idea where that shaft was located. One can't help but wonder if the same miners who dug that shaft weren't the same party which abandoned a large cannon which was found in the Kamas valley by soldiers from General Albert Sydney Johnston's army in 1858. That cannon, cast at Seville, Spain, in 1776, is now on display at the National Guard Armory at Salt Lake City.

An examination of mining claim notices at the county courthouse at nearby Coalville, where entries date back to 1870, raises even more questions. A man known as Marshall "Doc" Smith filed a claim to "The Old Spanish Mine," but all we can learn from his location notice is that it was located about two miles from Kamas in the West Hills. That there was such a mine and that he employed men to work there is proven by an item which was submitted to the Deseret News on March 23, 1885, by Bishop Sam Atwood of Kamas:

> About ten days ago, a young man named Moroni Pitt, who is employed at the Old Spanish Mine west of the valley met with a painful accident and narrowly missed being killed outright. He had two dynamite blasts to touch off. When he lighted the first charge, his candle went out, which he tried to relight. Before he could touch off the second charge, the first exploded, throwing him forcibly against the wall of the tunnel, bruising him severely and nearly blinding him. The wonder is that he was not blown to pieces.

It was common in those days for prospectors to file "floater" claim notices, so that anyone trying to learn exactly where their discoveries were located would be frustrated. One of the first prospectors on record to legally stake claim to an old Spanish mine near Kamas was Heber C. Bell, one of the earliest seekers of lost mines and treasures. Bell was a cagey fellow who discovered a number of Spanish mines and also found several Spanish caches, but his back trail is hard to follow. Many of his finds were made on the north slope of the Uintas, where few men then prospected. The location notices of Heber Bell and his brother, Joseph Bell, are still on file at the Coalville Court House, but they are all "floater" claim notices, which when filed gave only enough information to satisfy the recording clerk, but not enough to divulge the exact location of the claims.

One of the Bell brothers' old Spanish mine claims was located somewhere in Red Pine Canyon, but there are no less than a half-dozen Red Pine canyons in the Kamas area. One of them is located near Moon Springs, a place where several Spanish mines have been found. Another is located just over the ridge in Riley's Canyon, a steep gulch named for a prospector who had claims there. Two of those mines are located northeast of Moon Springs, above present day Samak. One of them was worked extensively by the Woodland Mining Company; by Douglas Simpson, Frank Smithies, Milton Stevens, Charley Bostard and others. Another mine located between Moon Springs and Red Pine Canyon was worked at a considerable profit. Owners of the Stallings Mine shipped high-grade galena ore which outcropped on the surface; however, few of that company's records remain today.

High on a steep ridge overlooking Beaver Creek, there is still another old mine, its waste dump shiny with tons of "fools gold," iron pyrites. It has been abandoned for so many years that a large tree grows up through what was an old engine at the mouth of the tunnel. Mace Foreman and a man named Starr worked there a long time ago. It is an extensive digging, and deep into the mountain the tunnel makes a sharp turn to the east, towards Pine Valley. Pieces of high-grade copper ore can still be picked up on the waste dump, some of them assaying a few dollars in gold. A little higher on the mountain, southwest of the pyrite dump, there is yet another mine dump with a few rotting logs of an old cabin close by. Even less is known about it. It might be well worth while for someone to take a closer look at those old mine dumps. A prospect hole which some dirt-poor prospector abandoned back when gold was worth only sixteen-dollars an ounce, could, with modern milling methods, prove to be a very profitable venture, especially with gold now selling at nearly four-hundred dollars an ounce.

If you investigate some of those old mines, be sure to check for platinum. A lot of old-timers threw platinum on the waste dump, unaware of what it was. Platinum is worth more than gold now. The older the diggings, the more likely it is that you might find platinum, for during the Spanish period, miners were penalized for having "platino" in their ore. It is on record that some Spanish coins were counterfeited with gold-plated platinum metal, and there were actual cases where platinum bars were dipped in gold to pass as gold bars. As early as 1510, the Spanish king decreed that counterfeiters and smugglers would be fined four times the value of a fake coin or bar, and be subjected to whatever other punishment the crown ordered. One officer in the king's service was sentenced to ten years slave labor for passing fake gold bars. Platinum is highly prized now, but during Spanish times it was considered to be worthless.

Exhaustive research into those old Spanish mines above Kamas Valley, including discoveries of mines and caches uncovered by the Bell brothers, has revealed facts which may come as a surprise to those who have not considered looking where they did for lost treasure. More about the Bell brothers will be disclosed in following pages. Charley Bostard, an old codger who was always known as "Beaver Creek Charley," located several Spanish mines near Kamas. According to his location notices, one of them was along Beaver Creek, another along Slate Creek and still another in Moon Canyon. Gordon Taylor recalled that "Beaver Creek Charley" uncovered a pretty rich pocket of ore at his diggings in Moon Canyon. He also discovered some "specimen" gold at a ridge-top outcrop just across the canyon. Jesse Nelson, who once lived at Kamas, recalled that when he was only a boy he worked for "Beaver Creek Charley," at a digging which was located near the head of Wide Hollow, just east of Kamas. The old man told Nelson that he was sure it was a Spanish mine because of the way it had been dug, arched and without supporting timbers. Nelson recalled that he always thought of Bostard as an aloof, lordly character, who he always called "Lord Beaver-Brook!"

There's still gold in this old tunnel along Slate Creek, but are you brave enough to dig for it?

John Urses, now a well-known metal detector dealer at Salt Lake City, had a summer cabin along Beaver Creek for many years, just below Wide Hollow, and over the years he explored just about every canyon and ridge in that area. Urses made an interesting discovery in Wide Hollow, on the trail to one of "Beaver Creek Charley's" Spanish mines. He came upon a place where a centuries-old trail could still be seen, at places cut back into the rocky side of the canyon wall. Where that trail crossed a narrow gulch, hand-fitted stones had been used to build a bridge of sorts. Urses believed it was part of a pack animal trail which led to mines above Wide Hollow and to the high ridges beyond, along the crests of Hoyt Peak.

Some of "Beaver Creek Charley's" most intriguing finds were made along Slate Creek. For more than fifty years, Gordon Taylor and his father, Moses, before him have bank-rolled prospectors. During that time he has noticed something unusual repeat itself. Every few years someone will find gold float along Slate Creek, but no one has ever traced that float to its source. Taylor has observed that float is always found in the spring of the year, and it is his belief that the high snow-water run-off is washing that gold down the creek from somewhere higher in the mountains. Of interest is a find made by Bill Winters, a geologist who was examining a mining claim on Hoyt Peak for the late Gale Rhoades, which is only a short distance north of Slate Creek. He picked up a piece of red-colored rock in which native gold could be seen without the use of a magnifying glass.

Just a few years ago, when the famous Silver King Mine was still operating at Park City, a sheep-herder brought a chunk of unusual quartz rock to mine assayer Tom Grose to have it analyzed. The herder told Grose that he found it while following his sheep along Slate Creek, and agreed to stop by the assay office to pick up his report when he brought his sheep back down canyon in a few weeks. When Grose prepared to assay that rock, he was startled to see that when it was roasted, it turned a red color, and tiny beads of yellow gold percolated to its surface. Several people have since examined that specimen, and all agree that it is fascinating, all speckled with tiny beads of gold. Time passed, but the sheep-herder never returned to pick up his assay report. Grose never knew who he was, not even his name. Maybe "Beaver Creek Charley" was on to something good at Slate Creek!

Wayne Handy discovered an old mine tunnel in Slate Creek Canyon, possible one of "Beaver Creek Charley's" mines. Its portal had long since been covered with shale rock sliding down from the mountain. The author and Handy dug into that covered portal, and later Cecil Dalton explored its depths for 250-feet and more to where it was caved. It was an arched tunnel without timber, as Spanish mines often are, nevertheless, it was badly caved and very dangerous. Near the end of the tunnel, a narrow vein of white quartz and iron pyrites continued into the caved area. Later assays indicate that quartz carries a half ounce of gold, which probably isn't enough to warrant reopening those long abandoned workings, considering the tunnel's isolated location and its dangerous cave-ins. Those old Spanish miners never had to worry about the high cost of timber, and their Indian slave labor was free. They didn't have to worry about OSHA or safety regulations either!

A while back, a man named George Bozinoff located and recorded a group of mining claims not far from Slate Creek, on Pitt 'n Page Hill. That is close to Rock Creek, a place where gold has been found, not very far from where it empties into the Provo River. According to the descriptions given on his claim notices, still on file at the county courthouse, Bozinoff's claims were located in reference to varying distances from a Spanish mine shaft. No one today recalls exactly where that shaft was located, although some say it was filled in to keep someone from falling into it. A careful search of the places described on those claim notices has failed to locate it. There seems to be little doubt that Bozinoff found what he was looking for, because "Old Dan" Tucker recalls that Bozinoff presented a

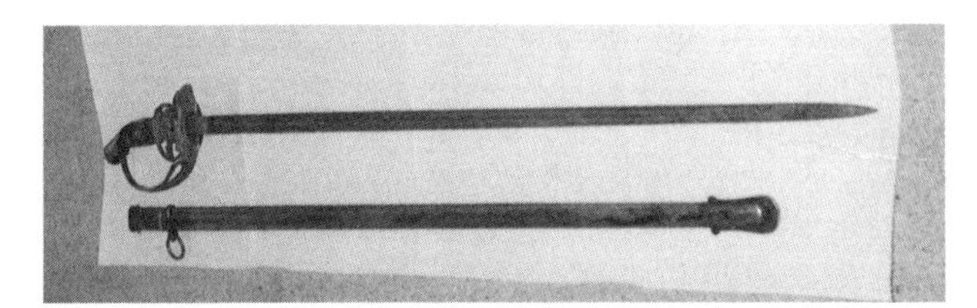

A Spanish sword was found along Rock Creek near Pit 'n Page Hill, high above the Provo River.

newly married couple with a poke of gold as a wedding present. Did those nuggets come from one of his claims on Pitt 'n Page Hill? Or could they have come from Mine # 3? Several people have reported seeing a caved tunnel portal in those hills. Only the top of the portal posts and the cap timber can be seen, but cut deep into that pine timber is the inscription "Mine # 3." A forest service employee, a rock-hound and a deer hunter have all seen that portal, but strangely, none of them have been able to go back to it. If there is a Mine # 3, is it logical to think there is also a Mine # 1 and a Mine # 2?

Bozinoff was a strange character. Over a period of several years he took in different partners, but he told them little about himself or his mining claims. Whenever a partner grew tired of doing the heavy work, Bozinoff would get rid of him and find a new helper. Those who worked for him said that he would locate old signs or symbols on trees or cut into ledge rock, from following an old parchment map which he had, and after carefully making hand drawn sketches and recording their location in a notebook, he would destroy those signs so others couldn't find or use them. He burned one large tree along Rock Creek which had the initials "C B R" and the date 1856 cut into it.

Leo Coe no longer lives at Kamas Valley, but he still recalls working with Bozinoff. Coe remembers going to several of his mines. All had been carefully covered and concealed with brush or tree limbs, so that hardly a trace of them remained. If Bozinoff hadn't pointed them out, Coe said he never would have seen them. Coe also said that the Spanish mine shaft used by Bozinoff as a reference point for his claims wasn't far off the Mirror Lake highway. A second shaft was higher on the mountain, at the edge of a small meadow or flat area. At one place along Rock Creek, Bozinoff drove an iron rod through a tree. Coe says that rod points directly across a canyon to a tunnel where Bozinoff's best ore came from. Its portal is now covered with pine poles. Coe now requires oxygen to live at a low elevation and can't get back into those mountains anymore, but he thinks those old covered mine shafts and tunnels might be pretty hard to find.

Another of Bozinoff's partners told the author how they discovered an old campsite, close to the foot of a rock-slide, where there is an old mine tunnel. There is no waste dump, the rock from that tunnel being dumped down that rock slide so that no sign of the diggings can be seen. That campsite below the tunnel was obviously very old, so that only someone actually looking for it would be likely to see it, and then probably only by accident. It is not very far from the 1856 pine which Bozinoff burned. Hanging from a limb under a giant pine at that camp, where it was protected from the weather, they found the rusted remains of an old umbrella. Only the iron spokes and hardwood handle remained, the fabric long since rotted away. Carefully cut into that wooden handle was a name: Sidsie Rhoads!

There are several old maps which point the way to Spanish treasure along a stream which is shown as Rock Creek. The Pine Mine Map (See Map H) and the Oro or 1851 Reinaldo Map (See Map D) both show such a stream, with a trail leading from a lake to a pine tree which has the date 1856 cut into its bark; according to some that being the date when Thomas Rhoads discovered a mine which is marked with an "X" on his map. Many treasure hunters have assumed that the stream shown on those maps is the Rock Creek in Duchesne County, and that the lake is Moon Lake. They have tried to fit the features of those maps with the landmarks of the Moon Lake area, but in vain. "Old Dan" Tucker thinks he knows why they can't do it: the Rock Creek shown on those maps is actually the Rock Creek which drains the Pitt 'n Page country into the Provo River. There are good reasons to think he may be right. The initials "C B R" were found on the 1856 pine which Bozinoff burned, and near the mouth of Rock Creek down-country from that tree, Tucker and Ed Siddoway found two artifacts which leave little doubt that Spanish miners were there long before them. They discovered the rusted hilt of a Spanish sword, and close by it, a cannon-ball; pretty good evidence that Spanish miners passed that way long before Thomas Rhoads began following their trail.

Over the past few years, several strange deaths of prospectors have been investigated in the Provo River country above Kamas Valley; not all of them adequately explained. Two prospectors were shot by unknown persons. Investigator's reports indicate they were accidentally killed by deer hunters, but there is one problem with those conclusions: neither man was killed during deer season. One young fellow who was known to be searching for the Rhoads Mine failed to return to his home. A search and rescue party found his body, where he had been hanged by the neck from a large tree. On the coroner's report, the official cause of death was listed as suicide, but that report leaves one question unanswered: his hands were tied behind his back in a manner which required another person to tie the knots. If you're going out into those Killer Mountains, keep a close watch over your shoulder. There could be someone behind you!

CHAPTER 4

THE OLD-TIMER'S TALE
Soapstone Mountain to Lightning Ridge

North and east of the Provo River headlands, near the top of Wolf Creek, there is an old Spanish mine. It is easy to find once, but I've learned that it is a lot harder to find a second time. Several other people have also seen it, but then couldn't find their way back. It is in sight of the Silver Meadows, but hidden in a thick stand of pine timber, not far from a small mountain lake. It may be the mine described in Heber City's Wasatch Wave one-hundred years ago:

> Two men have found an old mine east of Heber City, near the shore of a mountain lake. The ruins of an ancient arrastra (a primitive Spanish mill) were found near the mine. Ore samples carry sensational values in gold!

That description matches the old mine I found; but unfortunately there does not appear to be any follow-up reports in later issues of the Wave.

Back in the 1950s I went to Silver Meadows to spend a few days with Roy Peterson and Harold Giles, two old friends who had herded sheep in that area for most of their lives. Roy remembered seeing that old mine many years before, and he also recalled two men who once made camp near there. Roy said they built a small log cabin at the edge of the meadow below the mine, and he added that they were "shady characters." He said they were probably the two outlaws who robbed a store at the old placer gold camp on Currant Creek. My curiosity aroused, I decided to look for that old mine. I didn't find it, at least not then, but I did find something else even more interesting.

Along a small creek in a rocky canyon below where the mine was supposed to be located, I came upon what appeared to be the site of an old smelter. Among a pile of rocks which was probably part of its foundation, I picked up several pieces of metal. They looked as though they had been spilled or splattered onto the ground while still in a molten state. Trapped in that slag were small bits of rock, caught when the liquid metal flowed around them. That metal was corroded into a dull gray color, and it was very heavy. It could easily be bent. When I returned from Silver Meadows, I had an assayer at a Park City mine where I then worked run a test on those pieces of slag. His tests revealed that the metal was a silver-lead bullion.

There used to be a spring not far down canyon from that old smelter, but after six years of drought, it is now dry. But when that spring ran with a strong flow early in the year, it would wash out a lot of black sand, heavy with both iron and silver. Sometimes the water would almost be black in color. Could that water have been colored by running through slag under that old smelter where I found the pieces of bullion? What happened to the miners who worked that old mine and smelter? Did they leave when Mormon pioneers came into the country, or were they driven from the mountains by Indians? Did they take all of their bullion with them, or could some of it still be cached near that old mine or smelter? And what happened to the two outlaws who robbed the placer miners? Could they have cached some of their loot near their old cabin? Those are interesting points to ponder.

Dave Betts is a miner who now lives at Woodland. For the past several years he has been reopening an old mine close to the Silver Meadows. That mine's location has been known for a long time now, but recent finds during development work as well as research into its history suggests that it is probably a Spanish mine. Until recently, most folks simply referred to it as the Pace Mine, assuming that it had been dug by A.C. Pace some-

The Silver Meadows Mine shortly before it was vandalized.

time after 1906, but new information gleaned from mining records at the Wasatch County Courthouse and interviews with old-timers living nearby has proven it was an old mine when first discovered by Pace.

While researching old mining files, Betts learned that the mine was discovered and claimed by Pace, a Park City mining man, and his partner, a geologist named Gillespie, during the closing months of 1906. Almost by chance, Betts located documents which revealed Pace had been killed only four months later, early in 1907, so he couldn't have dug that shaft, which is more than 200-feet deep through solid rock. Geologist Gillespie couldn't have dug it either, for he was killed at almost the same time in a separate incident. It had also been thought that Pace and Gillespie built an old log cabin which was close to the mine, but testimony of several old-timers has proven that cabin was old and in ruins when Pace and Gillespie first saw them. One early settler remembered seeing that cabin more than seventy years ago, at which time their lower logs were so rotted and settled that the peak of the roof rested on the ground. Based on those recollections, that cabin was probably at least one-hundred years old when Pace first saw it.

Mining records reveal that about 1915, one of Pace's sons tried to reopen the Silver Meadows mine, and during later years others also worked there, including Mont Walker and Doug Simpson. They attempted to reach the old shaft bottom by digging an incline shaft, aimed at connecting with the older shaft at depth. All sorts of problems plagued their efforts, including caving ground, bad air and the theft of their hoisting engine. Their bad luck finally brought work to a halt. Dave Betts planned on reopening the original shaft rather than continue work at that jinxed incline. He has a lot of experience from mining at Park City, and the square-set shaft timbers he and his crew put in place testifies to their expertise.

While cleaning out the caved rock in the shaft, they came upon a peculiar boulder oddly out of place, a huge rock from an entirely different formation, as if it had been brought from another area and rolled into the shaft. Choosing to remove that large rock rather than work around it, they blasted and discovered that it had been purposely placed to conceal an unknown side drift running horizontal to the shaft. Crawling back into that ancient digging, they came upon a Spanish "zurron" or leather ore sack, and some wrought iron tools, so rusted that only an outline marked where they had been left. Samples from the ore vein in that drift were taken to the Ontario Mine at Park City for assay, and after testing was complete, the assayer asked from which level of the Ontario those samples were taken. When told that they came from Silver Meadows, more than forty miles away, he stated they were nearly identical to ore found at the Ontario.

Some mining engineers believe that the ore vein at Silver Meadows is a continuation of the silver vein at the Ontario Mine, which vein has produced more than two-hundred million dollars to date. To confirm their opinion, the Utah Geological Society's guidebook of the Uinta Mountains states: "The Park City Formation, a light-colored calcerous limestone, emerges from beneath the tertiary formation just east of Heber City and can be seen where it crosses the South Fork of the Provo River. It outcrops south of Wolf Creek and may be followed eastward to the Duchesne River."

Betts and Joe Smith with several other miners put in months of hard work during the short summer season at the mine's 9,500-foot elevation, buying equipment they couldn't afford and purchasing supplies with money needed at home. They installed a first-rate hoisting plant, laid track and put in an ore car and loading ramp for trucks. Many days the miners worked without pay, hoping they would be paid when times were better. The old shaft was reopened to below the one-hundred foot level and hopes were high that they would soon uncover the ore vein. But it may be a long time until that vein is reached, for during the last week of September, 1993, vandals dashed their hopes. The new shaft house was pushed over and equipment was thrown down the shaft. Everything of value was stolen, even the heavy eight-by-eight timbers atop the shaft. Years of work was destroyed in minutes. It was too late in the year to start over again, but next spring Betts hopes to erect a new gallows-frame and hoist house. There is a good lead on who the vandals were, but if you have any information about the vandalism at the Silver Meadows Mine, contact Dave Betts. He can use your help.

During July, 1992, several prospectors from the Duchesne area made a discovery not far from Silver Meadows. They are close-mouthed about their find but will say that assays show good gold values, in what appears to be a telluride type of ore. It was in that same area that a sheepherder saw an old mine working while he was driving his flock out of a heavy thicket. Hidden against the mountain front behind that thick brush, he could see the posts and cap timber of a mine portal. He said the waste dump is so old it is covered with large aspen trees, which are nurtured by a small trickle of iron-stained water which seeps from that caved adit. After he came down from the mountains just ahead of winter's first snow, he tried to tell me where that tunnel is located, but his Basque descriptions weren't exactly clear; still I'll be heading up that way come green grass. It shouldn't be very far from those claims with the high gold assay.

LaVar Thompson is part Ute Indian whose livestock work takes him into just about every nook and cranny of the Duchesne River country, including the Lightning Ridge area, which is part of the Soapstone or Rhoads Plateau, east of the Silver Meadows. Thompson found a Spanish mine along that ten-thousand foot high ridge, not far from where it drops off into Trail Hollow. The names he uses to describe certain areas may not match place names on a modern map, but it might be worthwhile to try to follow his back-trail. He described "two small hills near the end of the ridge," which place he called the "Bear Hole." You'll know you're in the right area when you come to a grove of giant aspens, their bark ripped and torn with deep claw marks. It was between those two small hills that he discovered an old mine. When Thompson first saw it many years ago, there were the ruins of a corral near by, where miners kept their pack animals. Only by careful search can you find evidence of that old corral today. Still, if you can find the "Bear Hole," you should be able to find that old mine shaft, between "two small hills at the end of the ridge."

Thompson isn't the only one who found evidence of Spanish miners in that area. Several years ago, Vern Muse, now deceased but who for years was the town barber at Duchesne, found a sword half-buried in a tree, about three miles up country from where the West and North forks of the Duchesne River meet. That sword is now in possession of a family member who doesn't want it photographed, but there is no doubt of its existence, for just about everyone at Duchesne has seen it at one time or another.

A monument to Jesse Bigler. He was killed on Lightning Ridge, September, 1917.

During the summer of 1992, the author and several friends climbed down from Lightning Ridge into the Bear Hole with "Rabbit" Maxwell, a man who knows those mountains well. He pointed out the rotted remains of the corral mentioned by LaVar Thompson, now completely overgrown with large pines. He also took us to a giant dead aspen tree, which has a large heart cut into its bark. Most prospectors agree that a heart is a Spanish symbol for treasure. Sometime long ago, a square wooden stake was driven into that heart. That stake is so old that it is now black with age, but if you sight along it, it points to the old mine shaft discovered by LaVar Thompson. Hidden under the branches of a gnarled old pine close by that shaft, we found the rusted remains of an old pick and shovel. That shaft is at the foot of a high cliff where winter snowslides dump tons of ice and rock across the base of the mountain where it is located. After sizing up those heavy boulders, we decided that reopening that shaft would have to be a job for another day.

The mine found by LaVar Thompson and "Rabbit" Maxwell, and the Spanish sword discovered by Vern Muse are all part of the legend of Lightning Ridge, a two-mile high promontory named for Jesse Bigler, a sheepherder who was struck and killed there by lightning in September, 1917. If you keep close watch as you hike along that ridge, you will see a small marker dedicated to his memory, placed there by Forest Ranger Ed Adair. It is a dangerous place, in more ways than one, an isolated high ridge where sudden and unexpected summer storms can erupt violently from a single black cloud. In July, 1991, another sheepherder, Felipe Roman, was struck and killed by a lightning bolt, in sight of the place where Jesse Bigler died.

A few of the Old Ones among the Utes tell a tale that in the long ago of their grandfather's time, Spanish miners brought a band of captive Navajo slaves north to work mines at Lightning Ridge. Those mines were at a place where an unusual red-colored rock was found. They say the Spaniard's brought Navajos because Utes were too hard to capture and often rebelled against their captors. For a time the Navajos proved to be docile workers, but they too wearied of their forced labors, until one day there was a terrible uprising during which most of their Spanish oppressors were massacred. But many Navajos were also killed. From centuries of working as silversmiths, the Navajos knew the value of gold and silver better than the Utes, so that when the survivors of that battle returned south to their canyonlands home, they took a heavy load of the precious red-colored gold ore with them.

The massacre at Lightning Ridge would have remained only a legend had not an elderly Navajo came to Roosevelt City in 1881. Following a map drawn for him by his grandfather, he found his way to the ridge; but as he approached the place where the mine was hidden, he was suddenly shot at by two Ute Indians whose job it was to guard the red gold. At that time the Utes still claimed ownership of the ridge. The Navajo fled into thick timber to make his escape, and several days later returned to Roosevelt. There he confided in one of the town's leading businessmen, a man he had been told he could trust. At first the merchant thought that the story of the Navajo massacre and the lost mine of red gold was nothing more than Indian folklore, but when the old man showed him a piece of the red gold ore, he began to have second thoughts. He agreed to ask his Ute friends about the legend while the old Navajo returned to his reservation home.

After careful investigation, the businessman became convinced of the truth of the Navajo's story, and that there had in fact been a massacre involving Navajos and Spanish miners sometime in the distant past at Lightning Ridge. He was even told by Navajo tribal leaders that they had long wanted some sort of commemorative monument erected in honor of the Navajo slaves who died there. He contacted the elderly Navajo as he had promised to do, and even sent a check to pay his fare back to Roosevelt. When several weeks had passed and the Navajo still had not arrived, he made a telephone call, only to learn that the aged Indian had been shot and killed by an unknown assassin just as he was preparing to return north. The check sent by the merchant was never cashed, nor was the map the Navajo had ever found. Some say that the Killer Mountains claimed still another victim far away in Navajo land.

A tale so strange it could be fiction had its origin in 1931; but that story is not contrived, it is true. Unfortunately, the author was given some incorrect information which appeared in the original edition of *Faded Footprints*. In relating that erroneous material, a man named Bill Nunley was unfairly characterized. For the record, Bill Nunley was a gentleman in every way. The true story of Bill Nunley and Wilford Binggeli and their lost mine on Lightning Ridge is as follows; the facts of the story obtained from Binggeli himself.

In November, 1931, Wilford Binggeli and Bill Nunley were part of a crew of foresters doing just about the hardest kind of work a man can do, using hand axes to cut down huge beetle-infested pine trees, in deep snow at ten-thousand feet in elevation. Jobs were hard to find during those early days of the great depression, so they were glad to be part of that crew, living in tents at Cold Springs, in the Soapstone Basin country some twenty miles east of Heber City. It was snowing when

A turtle carved into an ancient aspen in Rhoads Canyon; a sure sign of hidden treasure! (Courtesy: "Rabbit" Maxwell)

Binggeli and Nunley left their cold tent camp with the rest of the crew, to go to a forested area above Rhoads Canyon on Lightning Ridge. They worked all day cutting beetle damaged pines with axes, slowly making their way along that high ridge as the storm worsened.

As the day wore on they drifted further down the steep sided ridge, unaware that they were becoming separated from the rest of the crew. By late afternoon the storm turned into a blizzard, snow driven by a wind so fierce that visibility was cut to ten feet or less. The two men decided to return to camp while they could still find their way. Binggeli recalled what happened. He was breaking trail for Nunley, who even then was too old for that kind of work. They were hiking through a stand of pines and underbrush, just coming out into a small clearing, when he stepped off a low ledge onto what appeared to be part of the forest floor. But a carpet of pine needles covered by snow concealed a large weathered pine log. Binggeli said that log was quite large in diameter, yet it was so old and rotted that it broke under his weight, dropping him into what proved to be a well concealed mine shaft. He instinctively grabbed at a small tree with one hand as he fell, and used the axe in his other hand to hook onto the low ledge of rock he had stepped off. He disappeared so quickly and completely in the snow-driven blizzard that Nunley continued walking past that open pit without seeing him.

It took Nunley only a few steps to miss his partner's tracks, and he turned back to learn what had happened to him. He came upon Binggeli hanging onto the trunk of a small tree and pulling himself out of the shaft. As soon as Binggeli was safe, both men examined the shaft more closely, for they had been in that area before yet had never seen it. They discovered that the shaft was about four feet square in size, and was nearly vertical, having been dug at a sixty degree angle. It was probably that incline which saved Binggeli's life, for he had been able to get a foothold in a step-like notch in the wall rock, which broke his fall and helped him get out of that death-pit.

The two men broke loose enough of the rotted log so that they could look down into the shaft, and were surprised to see that it had steps or foot holds cut into the bottom, but those steps were dangerously slick with moss and dripping snow-water. They dropped pieces of rock into the opening and could hear them bounce and fall down that inclined abyss to a depth they estimated to be at least one-hundred feet. But what surprised them even more were the glistening white walls of the shaft. The area they had been working in was comprised primarily of a dull red iron-stained rock, but the quartz in that shaft was whiter than snow. That quartz vein was as wide as the shaft, about four feet, all dazzling white except for a streak near its center, which was pink in color. That pink colored quartz was so hard that Binggeli broke his axe trying to break a piece loose, but the white quartz on both sides of that pink vein was so soft he could break it loose with his hands. In his own words, "That sugar quartz was just full of gold!"

Binggeli and Nunley broke loose enough of that white sugar quartz to fill the pockets of their jackets, and then they covered the shaft with a half-dozen small trees they cut close by, mostly to keep someone else from falling into it, but also to conceal its location until they could return in the spring. Both men were experienced prospectors, so they took a compass bearing and noted the few landmarks they could see through the swirling snow. Both were certain they knew exactly where that shaft was located. By then it was nearly dark, and the blizzard had intensified, so that they became disoriented and couldn't find the trail which led to the ridge-top. Wading through the deepening drifts, they came to a ridge where they met several men from their work crew who were also returning to camp. It was well after dark when they got back to Cold Springs, where they found the rest of the crew had already struck camp and were loading tents, tools and other equipment onto trucks; hurrying to get out of those ten-thousand foot high mountains before they became snow-bound. It took two days of the hardest kind of work to get men and trucks out of those mountains and down to Heber Valley.

When spring came, it was the worst of the great depression. There were no jobs to be had anywhere, and neither Binggeli or Nunley could raise a grubstake to buy gasoline, groceries and prospecting tools. Several years passed before they could scrape together enough money to purchase equipment and supplies to return to their old camp at Cold Springs. A new growth of underbrush as well as several years of time had covered or changed the trails they once knew, and although they were certain they could go straight to the old shaft they had covered, days and then weeks of searching failed to locate it. It was almost as if someone had moved the landmarks they remembered. After a while they even began to argue about which way to go, uncertain of exactly where they had wandered during the blizzard. Binggeli thought they should go south along the ridge, while Nunley thought they should go to the north. Finally, after weeks of frustration and with snow beginning to fall again, they were forced to give up. Nunley went to southern Utah where he had mining claims on which assessment work had to be done, while Binggeli started a gravel business which soon took all of his time. Nunley died not long after that, while Binggeli's prospecting trips became fewer and more infrequent.

As the years passed, Binggeli said that if he had been alone when he fell into that shaft, it might have been possible for someone to convince him that it had all been a dream, but Bill Nunley had been with him, and he too saw that shaft. And they both saw the gold they dug from those sugar quartz walls. Nunley was an old-hand at gold mining, and he could tell what minerals an ore sample contained just by looking at it. He knew gold when he saw it, and he saw plenty of gold in that old shaft on Lightning Ridge. Binggeli later said that the only time he ever saw Nunley get excited was when he saw that gold. Nunley insisted that they never tell anyone about their discovery, for he was sure their find was on the old Ute Indian Reservation, so that if the Indians learned what they had found, they might kill both of them. Nunley firmly believed that they had discovered the Lost Rhoads Mine! In view of what happened later, he may have been right. During the next several years Binggeli told several friends about the old mine shaft he had fallen into, so that it wasn't long until others were seeking the lost treasure on Lightning Ridge. Then one day a Ute Indian came to Binggeli's door and in no uncertain language warned him not to tell anyone else about the gold-quartz shaft, nor was he to return to it. Not long afterwards still another man appeared at Binggeli's doorway and showed him an ancient looking map, obviously of the Lightning Ridge area, and asked him to identify several landmarks on it. That map showed the Duchesne River, Trail Hollow and a lake marked "Hierro," now known as Iron Mine Lake. Four crosses indicated the location of Spanish mines. More about that map will follow.

The waybill to the Binggeli-Nunley lost gold mine on Lightning Ridge is an easy one to follow. The Forest Service camp where they stayed was at Cold Springs, which is shown on any map of the Soapstone Basin. They traveled east from there to their work area, to Rhoads Canyon and Lightning Ridge. Neither man was certain of exactly how far they walked, or precisely in which direction, for the blizzard was so fierce that most landmarks were blotted out. They were cutting beetle-infested pines, so don't waste time searching on aspen covered slopes. Look for an area of thick pines, a lot bigger now than they were in 1931. You should come to a small clearing where they cut some small trees to cover the shaft. Those stumps should still be there. That's where the gold is!

Few of today's prospectors are aware that even before pioneer days, there was a wagon or carreta road up Rhoads Canyon, connecting Wolf Creek and the Duchesne River with Lightning Ridge and Soapstone Basin. It has only been since the advent of the four-wheel-drive Jeep that a vehicle road was built up that canyon, even though early settlers knew of the old carreta trail. Several years ago, government engineers built the Stillwater Dam on Rock Creek and the tunnel systems to bring water from that impoundment to the Duchesne River. During research on water rights, they discovered a file of old maps in the archives of the Library of Congress at Washington. Those maps are the hand-drawn surveys made by the military, dating from 1869 to 1903. They show all of the Uinta Basin and mountains before the Ute Reservation was opened to settlement. Examination of those rare and fragile maps shows great detail, such as the individual cabins of Indians like Bridger Jim, Towanta and Ike Shotnick. They also show mineralized areas, prospect holes, old log or stone ruins and trails or wagon roads, such as the one at Rhoads Canyon.

From the end of that wagon track in Rhoads Canyon, a trail continues along Lightning Ridge to two old mine dumps. They are difficult to find now, being almost melted away by time and weather, but pieces of native copper can still be found there. Assays reveal that ore is ten percent pure copper, but it was thrown on the waste dump by the miners who worked there. They weren't interested in copper; they were mining something far more valuable. Old "Many Whiskers" knew what they were digging. A few of the old-time stockmen who followed their herds around the head of Rhoads Canyon and along Lightning Ridge can still remember "Many Whiskers." He was a Ute Indian, very large in build, and mean when he was drinking, and they say he was drinking all the time! A government trapper showed me where "Many Whiskers" used to dig for nuggets along the base of a rock-slide covered ridge. Prospect holes where he dug along Mill Fork can still be found, if you take time to look carefully, although some say the Indians didn't dig them, that they are really old Mexican diggings. After he threatened several prospectors and shot at a few more, "Many Whiskers" was put away for a time, but after a few years he returned to the Rhoads Canyon country. Until a few years ago, there were still stories of prospectors and hikers having dangerous encounters with the old man.

One young fellow who wants to remain anonymous spent several years prospecting the Lightning Ridge area, from Iron Mine Creek to Trail Hollow and Soapstone Mountain. Last fall he made a camp just below the tree line in a gulch which drains into the North Fork. Late in the afternoon he spotted an old man riding a horse up that gulch. He was quite surprised, for he recognized the old man as someone he had talked to only a few days before at the General Store at Hanna. Since he knew the rider was ninety years of age, he watched him closely, wondering why a man of that age would be riding in such rough country so late in the day. Only a short way below his camp, where a small creek has its headwaters at a spring, the old man dismounted, and using a short-handled shovel began digging in the spring. On hands and knees, he methodically separated small stones from the sandy soil and dumped chosen ones into a small canvass bag. After the old man covered his diggings and rode back down canyon, the young prospector went to that spring and dug in it with a stick. Before darkness set in, he recovered a handful of small stones having a trace of gold in them. The following day he dug out enough to keep him in beans and bacon for a long time. He left for warmer diggings in Arizona a few days later, but he told me that as soon as the snow is gone, he'll be back on Lightning Ridge!

I have found a lot of Spanish signs in the mountains, some on century-old trees and others cut into stone ledge rock; and I have been given hundreds of photographs of similar signs, but I have doubts or reservations about the authenticity of some of them. I think a lot of them are nothing more than sheepherder "doodlings." But "Rabbit" Maxwell showed me a Spanish sign cut into an ancient tree on Lightning Ridge which I have no doubt about. It was made by Spanish miners. That sign is a turtle, a standard Spanish treasure sign. It is so old that it would have been lost long ago if the tree it was on, dead for a century or more when I saw it, hadn't grown in a thick grove where surrounding live trees continued to support it. Although its lower trunk had long since rotted away, it will probably stand for many more years, help upright by younger trees. Since its bark was breaking loose, Maxwell carefully removed that ancient symbol, rather than let it be lost forever.

Maxwell took me to other Spanish sign trees in that same area. We discovered an old digging after following a line of tree signs, a digging which still has an old-fashioned shovel partly buried in it. Down a steep slope leading off from Lightning Ridge, there is an odd depression in the ground, where sometime in the long ago someone covered a hole with pine poles. Some say it is a sinkhole, but I'm not so sure; perhaps a little digging there might change some minds. But be careful, for the pine poles covering it are badly rotted now, making a death trap if someone should walk out onto them. On a rocky point about a half-mile vertically above DeFa's Ranch on the North Fork of the Duchesne River, there is an unusual cavern. Some think it is a natural cave, only an oddity of nature, but that doesn't explain the timbers which support its roof or the old picks and shovels I found in it. There are a lot of strange things on Lightning Ridge.

Late one evening I received a telephone call from an acquaintance who told me about a friend of his who was noticeably excited and obviously scared. He said he feared for his life. He told of going to an old Spanish mine somewhere northwest of Hanna in the Lightning Ridge area, with a partner who is a Ute Indian. He said that shaft was deep and dangerous; nevertheless, he descended into it for about two-hundred feet, to where there is a side drift in which a hoard of gold bars were stashed. Using some mountain climbing expertise, with the use of strong nylon ropes, the two men were able to lift one of those bars from the shaft. Here their story differs from almost every other gold bar story I have heard. Most gold bars I know of or have seen are quite small, no larger than a common building brick, and are usually rectangular in shape. Most are what is commonly called "finger bars," about an inch square and six to eight inches long. But the bar the caller described is tapered in shape, being about two inches wide across the top but three inches in width at the base. and it was eighteen inches in length. He said the bar weighs sixty-two pounds, which suggests it is probably bullion rather than pure gold. Since it is an unusual reddish-gold color, it probably contains some copper and silver. There are several marks on that bar which he doesn't know the meaning of, but the date stamped on it is perfectly clear: 1662!

I filed the caller's story away and probably would have forgotten all about it if I hadn't received another call several days later, from another person who had seen the same bar, and he identified the man who found it. He too said the finder was genuinely in fear for his life. He said that when the finder and his Indian companion attempted to return to the shaft, they were shot at by an unseen rifleman, a person whose aim was so accurate that it was obvious that he could have hit them had he chose to do so. If you don't think such things still happen, remember the old Navajo Indian who was shot at on Lightning Ridge. The white man and his Indian partner ran down from the mountain, fleeing for their lives. It was only after they escaped that the Indian told his partner that another prospector had been shot and killed there not long before. Score one more for the Killer Mountains?

It isn't often that one is given substantial information about those Indians whose job it is to guard Spanish mines and caches, but I received just such a lead from a man who had been raised on the old Ute Reservation. Not long after the described incident in the mountains above Hanna where the Spanish mine shaft is located occurred, my informant was hiking in that very same area, not far from that shaft. He was stopped by a Ute Indian who he immediately recognized as a friend he had attended school with. Both were later in the army together, where each won sharpshooter medals. So it came as a surprise when his old school mate told him he could not allow him to go into those mountains. When he realized that his friend was serious, he suggested they continue up the mountain together. The Indian told him he would like to walk with him, but that the Old One would stop both of them. He hadn't seen anyone else until then, but looking up the ridge he saw another Indian standing on a rocky outcrop, watching the country below. His army buddy would only say that white men weren't allowed in that area, which incidentally, is where that sixty-two pound gold bar was found. But if you still aren't convinced and want to take a hike into those forbidding mountains, remember, that Indian guard was an army sharpshooter!

There's something else you might want to know about Lightning Ridge. Ed Twitchell used to live on the Indian Reservation, and he had a story to tell. Twitchell said that many years ago he saw a gold bar which had been found in a shaft somewhere north of Hanna. Actually, he only saw half of that bar. It had been found by two Indian boys and they had cut it in two pieces. The half which Twitchell saw was about three inches wide and nine inches long. As near as he could recall, that bar wasn't gold colored, but was sort of a reddish-gray. He remembered that when an assay office melted it down, it contained quite a bit of silver. The size of that half-bar of gold which Ed Twitchell saw sounds a lot like the bar recently found would look if it was cut in half.

There have been other fascinating finds in that still wild country between Hanna and Lightning Ridge. A few summers back a Forest Service employee was checking a tract of timber which was to be posted for bid. He was riding his horse atop Lightning Ridge when a sudden thunderstorm struck, and not wanting to be on that ridge close to the same place where Jesse Bigler had been killed by lightning, he turned his mount and headed down country towards Rhoads Canyon. He had ridden only a short way when he saw a shiny object in the leaves on the ground. Being trained to watch for any potential fire hazard, he stopped to see what it was. He was astonished when he saw that the object was a small bar of gold!

That bar was unusual in that it was only about an inch square and three inches in length. One end was jagged and rough, as if it had been broken from a longer bar. He never talked much about finding that bar, still, he never denied that he had it, so that within a short time quite a few people either saw it or knew of it. Then one morning two wild-eyed roughnecks pushed their way into his office. Both were wearing holstered pistols and appeared to be half-crazed and dangerous. The two toughs demanded to know where he found the gold bar, and when they were ordered from his office, one made a move as if to reach for his pistol. Being a life-long outdoorsman, he threw them out of his office before they knew what was happening. But that didn't end his troubles with the gold bar. "Crazies" began hounding him about it, and he began receiving telephone calls at all hours of the night. Tired of being harassed and fearful for his family's safety, he got rid of the bar. He doesn't talk about it anymore, and sometimes denies ever having it, but those who saw it know better. And those who know about the red gold of Lightning Ridge say that is where that gold bar came from.

You still don't believe gold bars are being found? I do, for I saw one which was found just last summer! On a July weekend a deer hunter made his camp on Soapstone Mountain, from which he hiked to various areas while looking for a place to hunt during the coming archery season. At a place almost in sight of the spot where the Forest Service employee found the half-bar of gold, in an area of large rocks and aspen trees, he spotted what he at first thought was a brick protruding above the ground. Wondering why a brick would be in such an out of the way place, he stopped to pull it loose and was startled to see that it was metal. Later measurements revealed that "brick" was two and a half inches in width, one and a half inches thick and five inches in length. It weighed just over five pounds and was reddish-gold in color.

That deer hunter knew nothing about prospecting or treasure hunting, so a gold bar was the last thing in the world he would think of. But he knew he had found something unusual, perhaps some sort of souvenir copper ingot from Kennecott, and brought it to the author. I immediately told him it was not copper, for it had not corroded to a verdigris-green as copper does when exposed to the elements, nor had it turned black as silver will. There were no marks on it, except for a curious mottled surface texture, as if it had been cast in a sand-mold. It had a definite look of antiquity about it, with a reddish-gold patina which only years of time could produce. It was also unique in that it had been poured in a mold which had not been level, so that one end was thicker that the other. At the author's suggestion, the finder had that bar analyzed by professional assayers at the American Barrick Mine at Mercur. That assay revealed the bar's metal content to be forty-seven percent gold, forty-five percent silver and eight percent copper. It's not every day that you see a real gold bar!

If you are still a skeptic, take a close look at the accompanying photo. Since that fabulous discovery, the finder has also become a believer. What's more, he has purchased a new metal detector! While detecting near where he found that gold bar, he uncovered a small brass bell, about four inches wide and high, of the type Catholic padres used at small mountain missions. You can be sure that while he is hunting deer, he will also be keeping a close watch on the ground, and not just for deer tracks. He may even use his new detector more than he does his bow and arrows!

But an interesting question remains. Why have so

A gold bar found by a deer hunter on Soapstone Mountain, July, 1992.

many gold bars been found in such a small area? The two bars found northwest of Hanna, the broken bar found by the Forest Service man and the deer hunter's bar found only recently. Remember the legend of the red gold of Lightning Ridge. Perhaps the bars found in that area fell to the ground from a pack saddle on an ore train coming from the shaft in the mountains above Hanna. Maybe some disgruntled miner who thought he wasn't getting his share of the gold dropped them along the way, then never had a chance to come back for them. Or perhaps they are all part of a cache stashed away somewhere on Lightning Ridge. As the Mexicans say: Quien Sabe - - who knows?

The old Indian and Spanish trail followed the West Fork of the Duchesne River downstream to the mouth of Rhoads Canyon and the North Fork of the Duchesne. One of the first homesteaders along the North Fork was Frank Savage. Near where he used to ranch, several small creeks tumble down steep, rocky gorges to meet the North Fork. The narrow canyon bottom is lined with near-vertical cliffs and ledges, but it is possible to cross the river below them with a Jeep, that is if it is not during the high water runoff of early spring. A boulder-strewn track twists its way into the mouth of Trail Hollow, while a side trail turns northward into Iron Mine Canyon. Hidden between boulder basins and rock slide ridges there are several old mine workings. They are well concealed and nearly impossible to find in the dense underbrush and down-fallen timber on the mountain front. At one there is a waste dump where there are tiny bits of silver ore mixed with the waste rock, indicating that someone dug for that white metal sometime in the long ago.

High above the south side of Trail Hollow, where a park of giant aspens and towering pines mark the boundary between that gulch and Lightning Ridge, a man who is one of the hardy breed of prospectors who frequent those rugged slopes made an exciting find in August, 1993. He found a small silver container about two inches in width and four inches long, a sort of tube having a screw cap on one end, not unlike an old-fashioned salt cellar. There were no markings on that container, but inside he discovered two old documents, folded several times and so fragile that they began to break to bits when he attempted to remove them. He took the container to town where with help those brittle papers were carefully removed and pieced together. With pains-taking effort they were traced onto carbon paper before they fell apart.

One of those papers had what were determined to be drawings or tracings of three Spanish coins on it. At present there is nothing to explain their meaning. The other faded document proved to be a map. It clearly shows a stream which is doubtless the North Fork of the Duchesne, and two smaller creeks in canyons which appear to be Trail Hollow and Iron Mine Canyon; the latter named "Hierro," or Iron. There is also a high promontory which is probably Lightning Ridge. There are several symbols shown on that map, including a bird and a turtle, both of which signs have been found carved on ancient aspen trees nor far from where that map was discovered. Four crosses, each within a circle, are thought to be mines, while near one of those crosses there is a square box figure, which may be a smelter. That map is an almost exact likeness of the Spanish map shown to Binggeli at his home almost sixty years earlier!

But even stranger than the map is a large silver coin found with it. That coin has lines scratched or engraved into its surface, which exactly match the drawing or the river and its side canyons. There are four small holes drilled through that coin, and those holes match precisely the four crosses shown on the map! That coin is actually a map which could be carried by someone for reference until more detailed directions were needed; then the paper map could be consulted. That silver metal case, the documents within and the map-coin are truly amazing finds. They could lead you to treasure! (See Map F)

That there were Spanish miners sometime in the long-ago where that old map and coin were found is further proven by still another recent and equally amazing find in the same area. A weekend prospector was hiking along an old trail through thick pine timber when he came upon what he at first thought was a dagger, driven deep into an ancient pine. Although the tree had grown around the dagger, he was able to cut and pry it loose. Examination indicates that "dagger" is actually a

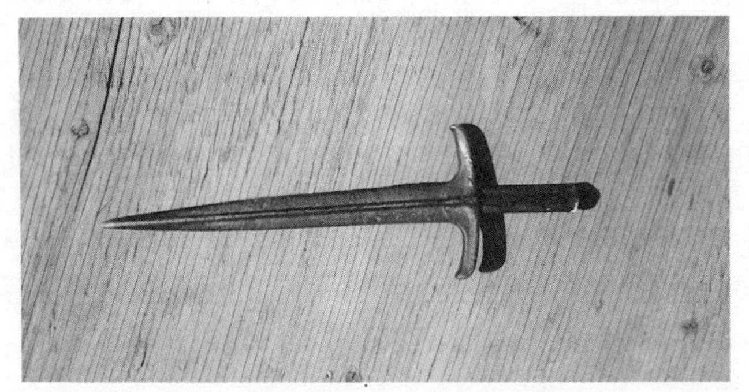

An ancient Spanish "barrena" or miner's drill, found driven into a giant pine close to the lost Stone Staircase Mine.

Spanish "barrena" or miner's drill, almost exactly like another found at an old mine along the Yellowstone River, some twenty miles further east. It is crudely made of wrought iron and has four cutting edges which could quickly cut through the hardest rock. It is well preserved, having been sheltered from perhaps centuries of weather by the giant pine. The finder, who wishes to remain anonymous, believes it was driven into the pine as a reference marker.

Somewhere in that wild country of cliffs, rock slides and down-fallen timber, between the towering ledges of "The Bluffs" above Iron Mine Creek, the dark depths of Trail Hollow and the sheer drop-offs of Lightning Ridge, there is one of the most fabled of all lost Spanish mines; The Lost Stone Staircase. During the hard times of the 1930s, a Civilian Conservation Corp (CCC) worker who was building forest trails and terraces to prevent slope erosion fell into an old mine shaft. He was pushing his way through thick brush when suddenly he stepped into an unseen pit. Fortunately, that hole wasn't deep and was inclined so that its slope broke his fall. There were wide steps cut into the floor of that incline, which he was able to use to climb back onto solid ground. He was an easterner who had frequently been ridiculed by his western-born companions, so he never mentioned the incident to them, not wanting to be taunted further. It wasn't until some weeks later when camp talk turned to tales of lost mines that he told of his harrowing experience. He described how the shaft was in thick brush, close to an especially large pine tree which stood at the base of a steep slope. By then the CCC Camp had been moved across the mountains to Wolf Creek, too far for any of the workers to walk back to search for the mine shaft. It is unlikely that any of those young men ever heard of the lost Stone Staircase Mine or had any interest in it. But plenty of prospectors have heard of it since then, and quite a few of them have spent years looking for it, but no one has stumbled upon it since the days of the CCC.

The late Gale Rhoades spent as much time as anyone looking for that famous mine. Dusty was a dedicated and conscientious prospector, and he made careful notes and sketches of the places he searched. He never found the Stone Staircase, but he did do a lot of research and made some interesting discoveries. He gave me many of his notes and sketch maps before he died, which describe the terrain where the mine is supposed to be located. Those sketches show the shaft almost exactly as the CCC worker described it, having a line of stone steps cut into its inclined course. They also show several sink holes, an old pack train trail in places cut into the rocky side of a canyon wall, and a gold-bearing ledge. That ledge could very well be the source of "float" gold which several prospectors have found in Trail Hollow. Not long ago a well known Utah television personality picked up some small gold flakes there, as did a Peruvian sheepherder who tailed his flock up that canyon during the summer of 1990. Still another hiker showed the author several water-worn gold nuggets he found in Trail Hollow.

Is it gold? The author checks an ore sample at a caved tunnel portal in Trail Hollow.

Those nuggets found in Trail Hollow are placer gold, recovered from the dry bed of an ancient river channel. Geologists know that during the last ice age, some twenty-five thousand years ago, a large river drained the Uinta Mountains pretty much along the course of the present North Fork of the Duchesne River. Similar ancient river channels have been panned for gold in the California Sierras with great success. That those same kind of ancient channels exist along the foothills of the Uintas has been well known to savvy prospectors for a long time now. One old prospector whose name would be well known to many wrote the following in his field note book: "Some of my companions thought that the river bottoms carried the gold, but that was a mistake, and much time and labor was wasted finding it out. Streams which ran thousands of years ago, their dry beds now high on the mountain sides, caught the gold in placer deposits, where it is now found." It may come as a surprise to some that those same ancient river channels are the very places where Thomas and Caleb Rhoads found a lot of their gold.

Near the head of Trail Hollow, where it levels off onto Soapstone Basin, atop an outcrop of ledge rock where it is difficult to find, there is an old vertical shaft, and on the mountain front below it, a caved tunnel portal. It is almost certain that sometime in the distant past that tunnel connected with the bottom of that shaft. There is a long, narrow waste dump of rock spilling down the mountain below the shaft, but you wouldn't recognize it from below, for it appears to be just another rock slide. Until last year, a weathered log still spanned that shaft, nearly worn through where a rope was used to hoist an ancient looking leather ore bucket. In a stunted growth of pines clinging precariously to the mountain above that shaft are the scant remains of a log cabin, so worn with age and the winds which scream across those rocky slopes that its gray colored logs now stand only a few feet above the ground. Only someone with a lot of perseverance, or someone who was digging out a lot of gold ore, could have stayed long in such a cheerless place. Across the canyon from that old cabin and shaft there is a jagged and towering line of broken cliffs, where only last year a professional surveyor from Ogden collected some colorful rock specimens. When a friend suggested that he have them assayed, he learned they carry about two ounces of gold to the ton.

The old-time Cornish miners who emigrated to the gold mines of the West from their home at Cornwall, England, were considered to be the best miners in the world. Those Cornish miners had a saying about where to find gold: "Gold rides an iron horse." What they were saying was that Iron was the mother of gold; that the place to look for gold was where iron was found. High in the head of Trail Hollow, E.P. Ferry, owner of Park City's Flagstaff Mine and Marsac Mill, located the Ferry Iron Mine. Extensive diggings can still be seen there, and if you know just where to look, the foundations of several old stone cabins. Iron ore from Ferry's mine was hauled by wagon and team to his mill at Park City, where it was used as a flux in smelting. Ferry's Cornish miners knew something about gold riding an iron horse, for it is on record that some of them found gold in Trail Hollow. Less than one mile from the Ferry Mine there is an outcrop of white quartz fully four feet wide, in a country where all the rock seems to be only gray colored limestone. There is a tunnel driven into the mountain below that quartz vein, and what appears to be a crude smelter stands at the mouth of that tunnel. Yes, Trail Hollow is gold country, and there is enough iron and quartz there to interest any Cornish miner.

At its upper reaches, Trail Hollow opens up onto the mesa-like flat top of Soapstone Mountain. If you go prospecting there, be careful that you don't step into a sink hole, or it could be your last step! All across that mountain top, from Cold Springs to Lightning Ridge, there are dozens, perhaps hundreds of dangerous sink holes. Some are only two or three feet in width, while others are so wide a man can't jump across them, and they are all death traps. Many are located atop barren limestone ridges where they are easily seen, but others are hidden in thick brush, just waiting for the unwary. All of them are deep, from one-hundred feet or more to depths which are said to be bottomless. Some of the easily found holes have been explored by spelunkers, but there are many others known only to a few prospectors or wandering sheepherders.

Al Slator, a one-burro prospector, for that is how he travels, with a burro to pack his gear, has explored many of those mysterious sink holes. I have been with him when he descended for one-hundred feet and more, and then followed some winding side drift for several hundred feet further. He says that some of those sink holes are beautiful, having stalactites and walls of colored flowstone. Some have underground streams flowing in them, while the walls of others are ice-covered until mid-summer. There are tales of Spanish gold as well as swords and armor being thrown into those holes, so who knows what might be found in them? Some say that a cache of gold bars was removed from the "Quigley" hole a few years ago! It could be, for that is

A forgotten mine on Soapstone Mountain. Note the massive quartz vein at its portal.

The Owl Hole, sacred to the Ute Indians. Who knows what secrets are hidden in its depths?

Gold bars have been found on Lightning Ridge. Are more still hidden there?

within a mile of where the Forest Service man and the deer hunter found their gold bars.

What some call the "Owl Hole" is unique in that its surface opening is separated into two halves by a natural stone bridge. It drops into a large underground room and then descends even further to a second chamber, one-hundred and forty feet below the surface. That lower chamber is approximately sixty-feet high, with a white quartz crystal ceiling. A million and more crystal facets reflect a lantern light in a dazzling prism of fire-light. Indians believe it is a sacred place, so if you go there you may find stick figures, colored strings or shells and bones placed around the entry hole. Please leave them as you find them.

Ray Miller, an old-timer who lives at Heber City, recalled when he herded sheep on Soapstone Mountain sixty years ago. There were two herds, and Miller took the lower one where it was harder to herd because of thick brush and trees, leaving the easier and more open high country around Cold Springs for his partner, seventy year old Bob McDonald. Miller talked about riding up the mountain to visit McDonald's camp, which was located on a low pass between Cold Springs and Trail Hollow, where breezes kept the mosquitoes away. Often he would find the old man working with pieces of heavy black rock, so soft it could be cut with a pocket knife. McDonald told Miller that rock came from a nearby sink hole. Many years later, long after McDonald had gone to that gold mine in the sky and Miller was working at Park City's Ontario Mine, he learned that black rock which could be carved with a knife was horn-silver!

Boyd Clyde was another herder who worked with Miller, and he too had a tale to tell about those Soapstone Mountain sink holes. Clyde said that he used several saddle ropes tied together to climb down into one of those frightening pits for one-hundred feet or more, to where the gray limestone walls of the cavern were cut by a metallic vein of soft black rock. Clyde said that when he broke chunks of that rock open, the inside looked like molten silver! Miller never learned which sink hole Clyde found that heavy black native silver rock in, and unfortunately, Clyde is no longer around to tell us. There are a lot of rock ledges, outcrops and sink holes on Soapstone Mountain. Most of them are in limestone, but there are also a lot of pieces of "float" rock there, some of it a heavy black rock. All you have to do is find the right piece and then trace it to its source. It's called prospecting.

CHAPTER 5

LORE OF LONG AGO
Waybills to Weber River Gold

During the 1870s, B.M. Froiseth published an excellent mining map of the Utah Territory. It was called Froiseth's New Section & Mineral Map, and it was updated several times during the 1880s. Froiseth's map was conspicuous among other early maps because of its accuracy and completeness. Every mining district of that period was shown, with mines at prominent districts like those at Park City and Alta carefully noted. On a ridge-like 9,000-foot peak high above the Weber River and overlooking Kamas Valley, another mine was shown. The inscription on Froiseth's map states simply, Old Spanish Gold Mine! (See Map G)

That mine wasn't placed on Froiseth's Map because legend or folklore claimed it was there, nor was it added as something of interest to help sell his maps. It was shown for only one reason: because that is where a Spanish mine was located. What does that mine's location say to those armchair historians who claim there are no such thing as Spanish mines in the Utah mountains, or that there were no Spanish miners in those mountains a century or more before the first Americans came into the territory? Quite bluntly, it says bunk! Of course there were Spanish mines in the territory before arrival of the first American explorers or Mormon pioneers. That is why Mr. Froiseth put them on his map - - because they were there. And one of them was located on a ridge high above the Weber River.

The history of the Weber River Canyon above Kamas Valley is a difficult subject to research, or on which to find accurate information. Aside from a few pioneer journals, little has been written about that area. The first homesteaders to penetrate that canyon in search of timber to build cabins discovered old mine workings, ancient in appearance even when they first saw them. The tales they told and the journal and diary entries they made about their finds may be considered folklore by some, but those who have taken time to search out those old records and investigate their contents have profited in more ways that one. Among my interview records, I have the transcript of an interview with Mr. William Peterson who lived at the little community of Peoa. That record was made by Wayne Handy many years ago, and Peterson was telling of a time long before that interview. He described several old mines found in the Smith & Morehouse and Gardners Fork canyons of the Weber River a century and more ago. But it is his description of a Spanish mine found by a neighbor of his, an old prospector named Louie Smith, which is of particular interest:

> A fellow by the name of Louie Smith found several old mines. He was a good miner, knew pretty well what he was doing. One mine he found was up on Black Mountain, above Smith & Morehouse. He took some good ore out of there, but he found something really valuable up in the South Fork Canyon. He worked there quite a while, drove a beat-up old Dodge Brothers truck, one of the first ever seen around here. Everyone though he was just wasting his time, but then one day he struck it rich, got himself a new truck and bought a ranch over by Oakley. He never worked much after that, never had to, hired all his hard work done. I guess that he found all the gold he needed!

If all we knew about Louie Smith's find was what Peterson related, there would still be some who would say is was just a story, more legend, but an entirely different source, and a most reliable one at that, verifies Peterson's account. Gordon Taylor, a businessman who has lived at Kamas for more than eighty years, told the

Louie Smith found treasure in South Fork Canyon, but did he find all of it?

author the following, unaware that I already knew about the Louie Smith story.

> I knew an old prospector named Louie Smith. He was quite poor, but then one day he bought a ranch, paid cash for it, what is now the Fitzgerald Ranch. It was generally known that he had found a lot of gold, but no one seemed to know where it came from, but he prospected a lot up in the South Fork. Some folks said that before he died, he cached a lot of that gold somewhere on his ranch. I asked the Fitzgerald boys about it a long time ago, and they told me they had dug a lot of holes looking for it. They said they hadn't found anything, but they still thought it was there, somewhere.

That Louie Smith found a lost mine or perhaps a cache of gold seems pretty certain. His story doesn't sound much at all like folklore.

The South and East forks of the Weber River described by W.W. Phelps in his narrative of exploration with Thomas Rhoads were actually the confluence of the Weber River with Beaver Creek. The major forks of the Weber River are much further upstream. Except for the South Fork where Louie Smith found treasure, the five main forks of the Weber River are found closer to its headwaters. On a map, the Weber River looks something like an out-stretched left arm, with its fingers and thumb corresponding to its forks. From the north, or left as one looks upstream, the little finger would be the Dry Fork, next the Main Fork, Middle Fork and Gardners Fork, with the thumb representing the Smith & Morehouse Fork; the latter located several miles downstream from the other forks. Spanish mines have been found in every fork of the Weber River.

Fish Lake is situated at the head of the Dry Fork, with an old mine located only a few yards from its shoreline. Thomas Rhoads called several of the mines shown on his maps, the Pine Mines. He drew a pine tree on one of those maps to identify them. (See Map H) Upon returning from one of his frequent prospecting trips into that area at the headwaters of the Weber River, Caleb Rhoads was heard to boast that he had located one of those Pine Mines, something his father had been unable to do. The late Gale Rhoades believed that mine on the shore of Fish lake was one of the Pine Mines, and recorded his claim to it at the county seat at Coalville, as Pine Mine #3, on July 25, 1978. Since his death, new locators have called that mine the Santa Ana.

The Pine Mines which Thomas Rhoads searched for, one or more of which he may have discovered at the headwaters of the Weber and Provo rivers, should not be confused with another group of Pine Mines located by Caleb Rhoads in the Rock Creek and Brown Duck Creek country of Duchesne County, which mines will later be described at length. The Pine Mine #3 located on the shore of Fish Lake on the Dry Fork of the Weber River, called the Santa Ana by some, requires some clarification. The person calling it the Santa Ana did so thinking it was the mine shown on the 1851 Reinaldo Map; but in fact the real Santa Ana Mine is located near Rhoads Lake on the Gardners Fork of the Weber, several miles to the south. Wayne Handy discovered the real Santa Ana Mine on Gardners Fork nearly forty years ago.

The upper Weber River is a glaciated wilderness of rock slides and rugged rocky mountains. Some prospectors have become discouraged and question whether there could be any valuable ore deposits in that glacial moraine and limestone structure. The Utah Geological Society's yearbook for 1969 answers that question with the following: "The Park City Formation, a strong ore-bearing structure, is exposed from Fransom Canyon, at the northeast corner of Rhoads Valley, and along the upper course of the Weber River to Moffat Pass, where it disappears under glacial deposits." Almost anywhere the Park City Formation breaks the surface, ore-bearing veins or fissures can be found. Not all are worth mining, but even low-grade veins could lead to valuable ore bodies at depth.

The author first investigated that old mine at Fish Lake during the early 1950s, before its portal had caved in. That tunnel was driven into red shale, and judging from the size of the waste dump, it was probably less than one-hundred feet in length. The waste dump extends into the lake, so much of it has been eroded away over the years. Some who have only seen that mine during recent years claim that several old log cabins close by were built by Caleb Rhoads, but they weren't; they were erected by pioneer stockmen who constructed a small dike at the lake's outlet. A few years

Slide rock has buried Wayne Handy's mine along the Weber River.

back a person who should best remain nameless carved the initials "C B R" on several large pines near those cabins, but to anyone familiar with old tree carvings, it is obvious that those initials have been recently cut. Just how old that mine by the lake might be is difficult to judge, but it is in an area rich with old Spanish mines. Standing silent guard over Fish Lake and the Dry Fork Canyon is Gold Hill, a towering peak honey-combed with ancient mine workings. Those Gold Hill mines will be dealt with in detail in a following chapter.

High on the Main Fork of the Weber, close to Abe's Lake, there are several old mines, their history long since lost and forgotten. Early day stockmen first saw them, but they are now caved beyond any practical or economical means of reopening them. On the Middle Fork there is a mine which appears to be of great antiquity. It is difficult to find, for it lies in a place where slide rock constantly moves to cover its portal. It was first discovered and claimed by Wayne Handy many years ago, but he was never able to penetrate its dangerous depths for more than forty feet, although it is obviously much deeper than that. Every time he got the portal dug out and support timber in place, more slide rock would move down the mountain front to cover it again. Able to work on only a part time basis and without heavy equipment, the task proved to be a formidable one. He hasn't given up yet, hoping that in time he will find the financing to continue work, but money doesn't come easy. Handy says that if you want to learn something about bankers, just try to borrow money to search for lost treasure!

The map drawn by Antonio de Reinaldo in 1851, one of two maps found on the body of a Mexican miner by Thomas Rhoads, shows the location of the Santa Ana Mine as being on Gardners Fork of the Weber River. Careful measurement and triangulation between that mine and known natural landmarks proves that conclusively. That map shows several rivers, including the Santa Ana (Weber) and a range of mountains designated as Sierra de los Timpanogos; the Wasatch Range where it butts up against the Uinta Mountains. Several lakes and trails are also shown, as well as three mines. Thomas and Caleb Rhoads may have been the first Americans to located one or more of those mines. Later prospectors like William Peterson and Louie Smith may have stumbled onto them without the aid of a map. It is quite likely that Smith might have discovered a cache left by Reinaldo, since he became rich overnight; a cache some still search for.

Another map which has surfaced in recent years is one drawn by Jose Joaquin Garcia, Captain, Mexico Municple, during the period 1826-1831. (See Map I) The origin of that map is uncertain, several people claiming credit for its discovery; but it is almost certain that neither Thomas or Caleb Rhoads ever saw it. Had they been aware of it, they almost certainly would have located some of the mines shown on it, for it is that detailed. Note that the Garcia Map shows some of the same mines which are on the Reinaldo Map, but in greater detail. It also refers to those mines as being among the lakes of the high plateau, just as the Reinaldo Map does. The Mina de St. Ana shown on the Garcia Map was known to both William Peterson and Wayne Handy. Both staked claim to it, separated many years in time, but neither had the means to adequately explore or develop it. More about that Garcia Map shortly.

Many mines have been found along the Smith & Morehouse Fork of the Weber River, or in its many steep and rugged side canyons. The author has a handwritten claim notice to the Apex and Last Chance claims, which were discovered in September, 1930. It is an interesting fact that the Old Spanish Trail which followed the North Fork of the Provo River crosses a high divide between that stream and the Weber River watershed. That trail descends the Weber side of the divide down Box Canyon, an impossibly rough and treacherous gulch leading down into Smith & Morehouse, and passes right by the Apex and Last Chance claims. That trail continues across Holiday Park to the Dry Fork and Gold Hill beyond. Ancient Spanish symbols on rocks and trees still mark that trail followed by Reinaldo in 1851, by Garcia in 1826 and by other unknown adventurers long before them. Many of the places where they dug for gold and silver are now lost to both memory and record. There is no doubt that Caleb Rhoads also followed those dim trails, for they are shown on a sketch map he made on the back of a Western Union message form. That rough drawing shows Rhoads Valley, the Weber and Timpanogos (Provo) rivers and an unnamed lake. Two trails, one from Rhoads Valley and the other from the North Fork of the Provo River, connect with the Bridger Road. Several words and numerals on the map are nearly illegible; their significance unknown. Perhaps you can decipher its hidden meaning. (See Map W)

The area shown in greatest detail on the Reinaldo and Garcia maps is the South Fork of the Weber. It is the only fork separated by some distance from the others, and it is nearest to the Old Spanish Gold Mine shown on the Froiseth Map. There are many accounts of gold and silver being found in the twists and turns of the South Fork Canyon. Tom Costas was a well known and respected geologist for the New Park Mining Company at Park City, and he later managed several mining companies of his own, including the Gold Queen, Lucky Mac and Sun Uranium. During the 1950s Costas and John Peezley were trying to locate a promising ore outcrop they had heard of along the South Fork when they unexpectedly came upon an old mine shaft, of such ancient appearance that they stopped to investigate.

That shaft was located beneath a huge pine tree, where lower branches and shady-place growth nearly concealed its vertical opening. On that giant pine standing directly over the shaft was a stout limb, extending out over the opening, its end cut squarely off. Marks could plainly be seen on that limb, worn deeply into the wood, where sometime in the distant past a rope had been pulled over it, no doubt to lift an ore bucket or some other heavy object from that yawning pit. Depth of the shaft could not be determined, since a splash could be heard after dropping a stone into it, proving that it was partially filled with water. I heard the story of that discovery when I worked for Costas, but he said that he and Peezley spent little time examining that old digging, for it was just another old mine, and not the ore outcrop they were looking for. Later Peezley told me that he doubted that anyone would ever find it again, except by chance, the way he and Costas found it.

A Jeep trail climbs up the South Fork Canyon past Pullem Creek. It is rocky and badly eroded by high creek water. Although it is a dedicated road, some property owners would like to close it, but if you can get up it, you will come to a high alpine meadow at the mouth of a pine filled canyon, in the shade of towering cliffs. Just above Pullem Creek, you can see where the mineral rich Park City Formation overlays the even richer Weber Quartzite, at some places in cliffs towering 1,700-feet high. It was from that contact between the Park City Limestone and the Weber Quartzite that Park City miners dug out more than half a billion dollars in silver ore! In that high alpine basin near where the road ends, on a grassy knoll at the edge of the aspens, there is a rock lined grave. On an aspen close by an inscription is carved: "Here lies Josey Wales, a good old friend." It is only the resting place of a sheepherder's faithful old dog, but for a moment it could make you wonder just who is buried there!

A dim trail leads further up canyon, past the old Lambert Sawmill log cabin, and higher up past two small mountain lakes, to Duke Peak and Castle Peak just across the ridge line. Just beyond that ridge, on the Provo River side of the divide, there is a mine shaft sunk into an iron outcrop which old-timers used to call a "blowout." For a hundred yards and more below that shaft the ridge is stained red from rusty iron ore. Somewhere before that ridge is crossed, according to tales the old-timers still tell, there is a mine tunnel known around Kamas Valley as the Black Bull Mine. From those stories, that tunnel is hidden behind a thick stand of choke-cherry bushes. If what they say is true, a cache of gold bars is hidden in that mine. Some say that cache is guarded by a great black bull, a spirit condemned to stand eternal watch over the Spanish treasure concealed inside that tunnel.

A few years ago, four veteran deer hunters, all of them well along in years, were hunting in the canyon below Duke Peak, not together but close enough they could keep in contact. One of them accidentally came upon that tunnel as he was pushing his way through some thick brush; choke-cherries according to some, but kinikinnick by other accounts. When he met up with his companions a short time later he offered to show them the tunnel, and since it was not far off, they agreed. Weary from the high elevation and dragged down by the weight of years, they trudged back up the mountain side. But as they broke through the brush thicket, a huge black bull suddenly stepped from behind some ledge rock and stopped in front of the tunnel portal. It was an unexpected and unnerving sight, and one of the partners, a superstitious Indian, jumped back in terror, saying that the black bull was the spirit of murdered miners. He turned and fled down the mountain, and his fear must have been contagious, for he was closely followed by his companions. According to the tale, none of them ever went hunting in South Fork Canyon again.

Wayne Nelson, an old prospector who admits to being a little long in the tooth and unable to get out in the mountains like he used to, tells a little different version of the Black Bull tale. Nelson claims that many years ago a logger fell into a hole which was concealed by thick brush, and that in that hole he found a pile of metal bars. They were small, about an inch square and six inches long, what some call "finger bars." Since there were several old mines close by he thought they might be soldering metal bars; still he put two of them in his jacket pocket. It wasn't until some months later, after the logging camp had closed for the winter, that a scrap

The grave of Josie Wales, on the trail to the Black Bull Mine.

metal dealer paid him a good price for them. That logger later said that there is a place where three large pines form a triangle, and that the "money pit" is in the center of those trees. He said those three trees can be found by following tree signs from the sink hole on South Fork Creek. It may be only a coincidence, but just last summer a treasure hunter using one of those new-fangled "doodlebug" electronic detectors on the cliffs above the South Fork told me that his instrument pointed out some sort of metal concentration in the pines across the canyon, at just about the place the logger described. I don't have much faith in "doodlebugs," but maybe someone ought to check it out.

Wayne Handy tells of a man who runs a few head of livestock in the South Fork Canyon during the summer, and like most stockmen, in order to break even he employs as herders several permit-carrying south-of-the-border aliens who do not speak English. A couple of years ago as he was visiting one of his stock camps, he noticed a chunk of rusty white colored quartz rock, heavily encrusted with stringers of yellow gold. Trying to conceal his excitement, he asked where the herder had found it. In broken pidgin-English the herder said something about it being a pretty rock, and tried to tell him about an old mine tunnel in a ledge of rock by some choke-cherries; away up there he said, the wave of

A rotted pole ladder leads down into a black abyss. Is this the lost Black Bull Mine?

his hand taking in about five-thousand acres of rock slide covered mountain side. The rancher figured that as soon as his fall work was caught up, he and that herder would find that mine tunnel, but it wasn't to be. Snow came early that fall, and during the winter the herder's visa expired and he returned to his home in manana-land. Who knows? I've heard stranger tales. Remember Louie Smith? He struck it rich somewhere along the South Fork. Maybe he stumbled onto the Black Bull Mine. I've been up that way a few times, but I didn't see any black bull. Maybe I'll give it one more try, next summer.

There is a strange rock slide in South Fork Canyon. It doesn't look unusual to me, but it so intrigues college professors that University of Utah geology students were assigned to study it each year for a half-dozen years. Back during the 1950s they established a base camp below that rock slide, and it proved to be so interesting that one year a team came all the way from the University of Minnesota to add their expertise to the investigation. Local residents had no idea what was so fascinating about that rock slide, but it must have been pretty exciting to spend so much time and money on it. Maybe it wasn't the slide rock they were interested in, but something more valuable. If you know just where to look, there are several old mines close to that rock slide.

Near the mouth of Maxwell Draw there are three old diggings which used to be called the Yellow Jacket mines, and close by are the remains of two old log cabins. Further up canyon, in a side gulch leading off to the left, there are two cave-like openings called the Bear Holes. There are also several old diggings in what old-timers call Mine Hollow, but what is now shown on modern maps as Welch Canyon. A few of those old-timers, DeVon Wilde of Oakley among them, remember an old prospector known only as "The Dutchman," a loner who had a mine high on the South Fork; according to Wilde, somewhere in Mine Hollow. The "Dutchman" worked in a tunnel which was close to his cabin, but no one has ever found either the mine or cabin, so well are they hidden. Apparently he found enough paying ore to buy bacon and beans, for they say he worked there for quite a few years. Wilde recalled that a geologist associated with Phillips Refining examined some of the "Dutchman's" ore, and announced that it "was pretty good stuff." When he finally got too old to climb the canyon anymore, the "Dutchman" covered his tunnel portal with heavy logs and rock. No sign of his cabin remains. Some say he tore it apart and scattered the logs, while others say it was burned. Unless the logs covering that tunnel portal rot away and expose its entry, or some bear or varmint digs into it, the "Dutchman's" hidden mine will probably remain hidden.

Another South Fork mine, the Peerless, reported high assays in silver ore, but company records are hard to find to verify the value of shipments made from that property. That a good sized crew of men was employed there is evidenced by the fact that about 1898, the

Deep inside the old Peerless Mine. Specimens of gold-leaf on quartz have been found near-by.

Peerless company built a boarding house for its miners. One old fellow at Kamas recalls that his mother was employed as a cook there. Some say there was also a one-room school house, at the mouth of Pullem Creek, which children of mine employees attended. If you're careful, you can still wiggle your way back into the Peerless tunnel. Although it is caving and dangerous, you can still crawl back more than two-hundred feet to where the roof has collapsed, leaving a huge stope or chamber. Just be sure you have your last will and testament made out if you venture back into those old Peerless Mine workings!

Besides the danger of caved mine shafts, there are sink holes all across that country, natural openings only two or three feet wide, but often hundreds of feet in depth. Some call them mine shafts, but they aren't; they are only natural limestone sinks, very similar to those on Soapstone Mountain, but generally much smaller and not as deep. One promoter of a diggings on nearby Hoyt Peak even sold stock in one of those sink holes, claiming it was a lost Spanish mine, but then he could sell ice-cubes to an Eskimo! There is a large sink hole at the end of the Jeep road in South Fork Canyon, where the entire creek disappears into the depths of a swirling whirlpool. That water bursts forth from a large spring about a half-mile down canyon, but somewhere between that sink and the spring, there must be an underground water course, for there is a lot more water coming out of that spring than there is going into the whirlpool.

A high, rocky and jagged ridge nearly twenty miles in length separates the Weber River from Kamas Valley and the Provo River. Its timber covered flanks hides hundreds of steep, brush-choked side canyons, gulches and draws. That ridge is the headwaters for dozens of creeks and small streams on both sides of the divide, and its skyline is marked with a dozen or more 10,000-foot peaks. One of those lofty peaks has looked down on more than 300 years of man's efforts to claim the golden treasure it hides. That mountain has witnessed the passing of Jose Joaquin Garcia and Antonio de Reinaldo, the map-makers, as well as the two Rhoads, father and son. More recently it has seen the coming of prospectors and treasure hunters. And it has wept at the death of one of its favorite sons, and one of my best friends, Gale Rhoades. This is the story of Hoyt Peak.

Who first dug for precious metals in the shadow of Hoyt Peak will probably never be known, but the earliest record is a waybill left by Jose Joaquin Garcia, penned in 1814, several decades before he sketched his now famous map. That document refers to work which was performed at mines on Hoyt Peak as far back as 1782. It was discovered in the Spanish Archives at Mexico City through the efforts of Russell R. Rich, Professor of Early History, Brigham Young University,

Provo, Utah. It is quite likely that Professor Rich also discovered the Garcia Map of 1821-1826, described earlier. Both the map and the document appeared at about the same time. Another document recently uncovered in the Archives at Seville states that the Garcia Mine was registered in 1722, and was worked until 1749, when it was abandoned due to Indian troubles, and wasn't reopened until 1782. The mine was owned by the Garcia family for nearly a century. The waybill discovered by Professor Rich has been authenticated by document experts who have examined it in great detail; their opinions stating that the language of that document is consistent with the period in which it was written, and that several now archaic words and phrases in it would be unknown to anyone not living at that time.

Documents written by the well educated class, such as the clergy, were actually written in a script known as "procesal," in which it isn't unusual to find words of Arabic, Greek and Latin origin. The procesal style of writing is a continuous flowing script of rounded Arabic-like letters and abbreviations, having little or no punctuation to indicate where sentences or phrases end. Bundles of documents called "legajos" may each contain thousands of pages, which for the most part do not relate to each other. A researcher must examine each legajo one page at a time, each one only part of an estimated fifty-million documents in the Spanish Archives at Seville. A document similar to the Garcia waybill reveals the many different forms which letters in procesal might take. For example: In that ancient script, the letter "A" is shown in fourteen different styles, "E" in twenty-one styles, "P" in sixteen and even the little used letter "Z" in thirteen styles! It would be virtually impossible for someone today to forge such a document.

The Garcia waybill when translated reads essentially as follows:

> Continuing Document, Period 1782–1814: This claim pertains to the Mine Of The Yutahs, also called the Josephine de Martinique, for the Empress. The mine can be found 12 leagues west of the Rio Timpanogos headlands and 2 leagues from the mouth of the Rio de Santa Ana, to the southeast. Travel one league south through the Valley Of Grass to a canyon which enters the valley from the east. Follow that canyon to a peak, rounded and barren of growth, and from that peak measure 1,600 varas to the northeast. In the direction of the mine there is a peak, small and timbered. The portal of the mine is covered with rocks and bark, and is at the base of a small dark ledge. The mine has two tunnels and one shaft. One tunnel runs 400 varas to the west and one tunnel runs 350 varas to the southeast. The tunnels and shaft are all one mine. The shaft runs 73 varas vertical and has four rooms, which served as

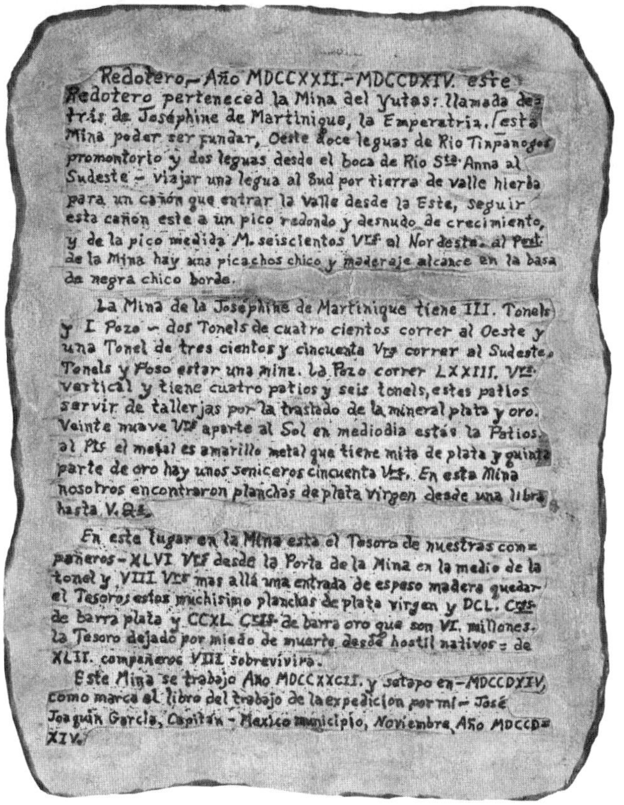

Redotero– Año MDCCXXII.–MDCCDXIV. este Redotero perteneced la Mina del yutas: llamada detrás de Joséphine de Martinique, la Emperatriz. /esta Mina poder ser fundar, Oeste doce leguas de Rio Tinpanogos promontorio y dos leguas desde el boca de Rio Sta Anna al Sudeste – viajar una legua al Sud por tierra de valle hierba para un cañón que entrar la valle desde la Este, seguir esta cañon este a un pico redondo y desnudo de crecimiento, y de la pico medida M. seiscientos Vrs al Nordeste. al Pest de la Mina hay una picachos chico y maderaje alcance en la basa de negra chico borde.

La Mina de la Joséphine de Martinique tiene III. Tonels y I Pozo – dos Tonels de cuatro cientos correr al Oeste y una Tonel de tres cientos y cincuenta Vrs correr al Sudeste. Tonels y Poso estar una mina. la Pozo correr LXXIII. Vrs vertical y tiene cuatro patios y seis tonels, estes patios servir de tallerjas por la trastado de la mineral plata y oro. Veinte nuave Vrs aparte al Sol en mediodia estás la Patios. al Pts el metal es amarillo metal que tiene mita de plata y quinta parte de oro hay unos seniceros cincuenta Vrs. En esta Mina nosotros encontraron planchas de plata virgen desde una libra hasta V. Qs.

En este lugar en la Mina esta el Tesoro de nuestras companeros – XLVI Vrs desde la Porta de la Mina en la medio de la tonel y VIII Vrs mas allá una entrada de espeso madera quedar el Tesoros estos muchisimo planchas de plata virgen y DCL. Cgs de barra plata y CCXL Cgs de barra oro que son VI. millones. la Tesoro dejado por miedo de muerte desde hostil nativos = de XLII. compañeros VIII sobrevivira.

Este Mina se trabajo Año MDCCXXCII. y satapo en – MDCCDXIV, como marca el libro del trabajo de la expedicion por mi – José Joaquin Garcia, Capitan – Mexico municipio, Noviembre Año MDCCD= XIV.

The cryptic Garcia waybill; your guide to 650 cargas of silver and 240 cargas of gold!

> workrooms for the transfer of metals silver and gold. The rooms are 29 varas apart to the sun at mid-day. By percentage, the metal contains half of silver and a fifth part of gold. In this mine we find slabs of virgin silver, from one pound to as much as five pounds. In this place is the treasure of our comrades. 46 varas from the entrance of the mine and in the center of the tunnel, 8 varas beyond a door of heavy timber, remains the treasure. There are many slabs of virgin silver, and 650 cargas of bar silver and 240 cargas of bar gold, which are six millions. The treasure was abandoned for fear of death by hostile natives. Of our 42 companions, only 8 survived. This mine was worked from the year 1782 and was covered in the year 1814, as is written in the journal workbook of the expedition, by Jose Joaquin Garcia, Captain, Mexico City, November 1814.

Even before the Garcia waybill was found, it was known that there were old mines somewhere in the deep canyons which furrow the brow of Hoyt Peak. Historian Frank Silvey interviewed W.P. Mecham, an early settler who lived along the Weber River. Mecham related that his father told him that for several years after arrival of the first Mormons, Mexican miners would bring "conductas" or bullion-laden pack trains down from Hoyt Peak in the fall of the year. During the summers while the miners were at their mine in the mountains, dozens of pack animals would be left to graze on the valley floor. In the fall miners would take those animals to their mine, and as they came down from the mountains, so heavy were their loads that the animals would stumble under the weight. Often the miners would camp overnight near Mecham's cabin, and would pay him for the feed and other supplies they used, with pieces of gold ore. He remembered one party which rested sixteen pack animals at his pasture, all of them loaded with bars of bullion. Mecham was only a poor farmer who had neither the time or interest to search for mines, so when the Mexican miners no longer came north from Santa Fe, their mines on Hoyt Peak became the lost mines old-timers in the valley talked about.

That Spanish and Mexican miners packed gold and silver bullion south to Santa Fe and the gulf coast seaports beyond was proven not long ago by a discovery made in New Mexico. The Hatch, New Mexico, Courier, issue of March 11, 1993, told of the discovery of a large stone having a \gun-sight notch cut into its top. When the finder sighted over that gun-sight towards the hills beyond, a peculiar rock formation could be seen. Investigation at that formation disclosed a small cave, the entrance to which had at one time been walled over with stones set in some sort of sand-mortar. A circle and cross had been found cut into the gun-sight rock, and above that cave entrance there was an identical circle and cross chiseled into a stone ledge. Several stones had been removed from that walled entry long before its recent discovery. The rancher on whose land the cave is located informed the finder that his father told him that many years ago a stranger camped near that walled cavern. The stranger had some ancient looking documents written in archaic Spanish, which told of a pack train carrying bars of bullion from mines in what are now the mountains of northern Utah. Those bars were cached there when the party was attacked by Indians.

Only a very few of those Spanish miners escaped that massacre, and none of them ever returned to their cache. Using a rough sketch map drawn by one of those survivors, the stranger spent most of the summer aligning its markings with the features of the land. Then one day the rancher noticed that the stranger had left his camp, but it wasn't until some time later that he chanced upon the place where he had dug into that walled over cave. What the stranger removed from that desert cavern is unknown, but the sighting rock with the circle and cross cut into it and the same circle and cross chiseled into stone above the cavern suggests that someone went to a lot of effort to make sure that he could return to that place. And the pieces of leather pack sacks the rancher found in the cave indicates that something of value was hidden there. That rancher thinks he knows what it was.

Probably the first Americans to stake their claims on Hoyt Peak was a party of prospectors led by Bill Bird, whose claim notices are on file at the county courthouse. In September 1898, they located a group of claims near the top of the peak on its east side, which they called the Josephine claims. Their claims were not named for the Josephine Mine named in the Garcia waybill, for that document wouldn't be discovered or known for another half-century. A word of explanation is probably in order concerning those so-called Josephine mines. Spanish miners named their claims just as American miners would later do. Just as an American might name his claim in honor of some popular political figure of the day, calling it for example, the Teddy Roosevelt or the Silver Queen, so Spanish miners did the same. Among Spanish mines in Utah there are probably dozens named the Josephine, in honor of the Empress. Only a year before Bird and his party staked their Josephine claims on Hoyt Peak, at least two different Josephine mines were reportedly found in the mountains east of Heber City, so their choice of a name was popular, if not original.

Gordon Taylor of Kamas remembers the first time he saw that Josephine Mine on Hoyt Peak. That was in 1937, the year he was married. He recalled that even then the old mine had a heavy wooden door over its portal, and that when the door was forced open, what looked like several stone steps led down into a caved tunnel which angled back under the ridge. The mine was not in a difficult place to find, and Taylor said that no doubt many deer hunters and other outdoorsmen had seen it long before he did. Jesse Nelson, who had worked with "Beaver Creek Charley" also recalls seeing that old mine, sometime back during the early 1930s. In September, 1939, a stock herder named John Young stumbled onto the old diggings and staked his claim to it. He knew that he had found a very old mine, but at least he was original in naming it. He called it the Mystery Mine. Over the next twenty-years or so, Young worked his Mystery Mine in an on-again, off-again haphazard fashion, without ever accomplishing much. An interesting description of Young's diggings was related to Wayne Handy by Jared Weller of Oakley:

> John Young has found the Lost Rhoads Mine on the east side of Hoyt Peak. It is still secured by a heavy wooden door, bound with iron straps, which no doubt was put in place by Spanish miners. Stone steps lead down into the mine, which consists of two entries with a main shaft between them. That center shaft is where the ore was dug. I have personally descended into that shaft for

fifty feet, but it is believed to be three-hundred feet deep. Mr. Young invited a team of geologists from Brigham Young University to examine the shaft, and they set off a dynamite charge which completely caved it.

That blast also destroyed a curious Spanish treasure sign, which those who saw remember well. On a ledge of stone above the shaft were several inscribed figures, showing what appeared to be pack animals similar to South American llamas. Their packs appeared to be empty as they approached the mine, but were heavily loaded when leaving it!

The effort of John Young to reopen the old Spanish diggings which he called the Mystery Mine is a real hard-luck story. More than fifty years of age when he first saw the mine, and often plagued with poor health and even poorer partners, none of whom ever had money enough to purchase needed supplies or equipment, Young tried for forty years to unravel the secret of his find. Of course he never had the benefit of the Garcia waybill to tell him where a fortune in gold and silver was cached. Much of his time was spent caring for his livestock and nearly as much fighting the Forest Service who slowed his work and even burned a cabin he built. There were years when he got little or no work done. Several times he let his claims lapse, only to relocate them later when some new partner was found. Young died in 1980 at age 92, no richer for the Spanish gold which was always just beyond his reach.

There are many old diggings on the slopes of Hoyt Peak and in the canyons below its crest. No doubt some were dug by pioneer settlers, but a few are obviously much older. About mid-way up Hoyt Canyon, shown as Hoyle Canyon and creek on very old maps, there is a tunnel with the ruins of two old log cabins close by. No one living today can recall who dug it, but ore specimens on the waste dump indicate that a good grade of silver was mined there. The Hoyt Peak Fault, one of the longest fault lines found in the Wasatch and Uinta mountains, crosses the canyon right at that old mine, and then drops down into Rhoads Valley, where it dips under an andesite formation near the town of Oakley. That fault is a known ore producer, from Rhoads Valley to Moon Lake and the Lake Fork River in Duchesne County.

At the top of Hoyt Canyon there is an old shaft timbered with pine poles notched to fit at their corners, not unlike the notched walls of a log cabin. They are green with moss and nearly rotted through, but it is still evident that whoever put them in place was a real timberman, a miner who took pride in his work. A few logs from a fallen cabin are just above that shaft, and there are strange signs cut into large aspen trees nearby. John Young knew of that old digging, for he claimed it as

The old shaft John Young called Spring Mine #1. Note the unusual notched-pole timbering.

Spring Mine #1. That claim name can still be seen carved into an aspen tree, near where the roads from Hoyt Canyon and Wide Hollow meet. Although Young filed his claim to that shaft, there is no evidence that he ever attempted to pump any of the icy water from it or do any real development work. In July 1993, mineral claims on that shaft were filed by Harry Ayala and Lonnie Mayhew. After an especially hard winter, that shaft was nearly filled with snow water, but they made a camp there and began to pump it dry. They have recovered some good ore samples by probing its depths, but like Young, they too are being harassed by the Forest Service, which agency will not allow them to camp at their claim. To live in the valley and drive into those mountains every day is time consuming and not economically feasible. If they are ever allowed to pump that shaft dry, it will be interesting to see what will be found in its depths.

Many old and intriguing artifacts have been found in the shadow of Hoyt Peak, one only a few months ago. Harry Ayala, the grizzled-looking partner in John Young's old Spring Mine #1 already mentioned, is an old fellow of Aztec descent. Ayala isn't a prospector who has to rely on modern electronic metal detectors or gizmos. He owns a set of Spanish dip-needles which have been in his family's possession for more than four-hundred years! While using those ancient instruments, Ayala discovered an old brass church bell of the type used by Jesuit priests in the long ago. Pat Giles of Heber City is another prospector who has mining claims on Hoyt Peak. They are located where a peculiar depression or sunken place was found on the mountain side. When that depression was dug into, an old tunnel portal was uncovered, and by it an interesting artifact was recovered; an old stone hammer, crude but no doubt effective. A B-L-M employee discovered several rusted pieces of a Spanish helmet on the Giles claims, and

Harry Ayala and Lonnie Mayhew demonstrate the use of 400 year old Spanish dip-needles.

there are stories about a small bronze cannon being found in that same area.

Gale Rhoades had shares in one of those old diggings on Hoyt Peak, and to protect his interests he became associated with work being done there, although only as a minority partner. Unfortunately, he became involved with con-men of dubious character. They knew nothing about mining, but Gale was able to influence them to hire several competent miners. But those in charge, in a hurry to make a fast buck, wouldn't let those experienced men do their work properly. Recently, knowledgeable miners have been in those diggings and have concluded that it is one of the most dangerous underground workings they have seen. Timbers of every size and length are propped under huge slabs of loose rock, placed with no regard for the angle of the foot and hanging wall structure. Like dominoes, if one timber slips or is knocked over, all will fall. Apparently, no real mining was accomplished, suggesting that attracting investors or promotion was the object, not development.

Places like Hoyt Peak attract con-men like carrion attracts flies, so that people having only the vaguest perception of geology and mining are easily taken in by them. The usual bait is the story of hidden treasure, often bars of gold or silver, cached in some old mine or cavern, perhaps behind a locked door or bulkhead of heavy logs; just waiting for the lucky finder. Tons of treasure, never just a few bars, was cached by those miners of old, who of course were all killed by Indians. All of those con-men have a map, always crudely made, on yellowed parchment or a piece of buckskin; said to have been found in an old mission somewhere in Mexico. Who made that map or how it got to Mexico is never explained. I don't know how many of those genuine treasure maps I've seen!

The con-man always knows just where the treasure is located, but it is in a difficult place to reach, so that a large sum of money will be needed to recover it. But not to worry, you'll get your investment back ten times over. I knew a man who made a pretty good living selling dozens of half-interests in just such a scheme. Before I met him, I was so naive that I thought there were only two half interests in a whole! And then there is the only person who knows exactly where the cache is located, usually some stove-up old cowboy in a rest home somewhere. Unfortunately, he too needs money. Or he is in a Mexican jail and bail money is needed to set him free. It goes on and on, ad nauseam.

Be warned: not all con-men wear silk suits and ties. There are a lot of them out in the hills wearing blue jeans and hardtoe boots. Don't fall for their pie in-the-sky stories, as full of holes as a piece of Swiss cheese. If you press the con-man about the discrepancies in his tantalizing tale, he may admit that he hasn't actually seen the treasure or vein of gold, but he knows it is there, for after all, hasn't he got that old parchment map? But what the heck, if that treasure or lost mine is no longer there, he has another map just as good, to some other cache or mine just over the next mountain. The moral of the story is, beware of con-men bearing lost mine or treasure maps. If it sounds too good to be true, it probably is.

And then there are assay reports. Any con-man worth his salt has a briefcase full of assay reports. They all show high values in gold or silver, and none of them ever show any values in some mundane metal such as lead or zinc. Remember, a single sample of ore does not make a mine. No matter if that selected specimen assays $50,000 to the ton; if that's all you've got, all you have is a piece of ore. I've heard hundreds of prospectors relate glowing tales about their assays, but you can't pay mining expenses with assay reports. Too many would-be mining moguls hand pick a sample from the richest part of the vein, but are later sorely disappointed when their first shipment doesn't even pay the freight bill. A picked sample is of no value, unless you're trying to con someone. You need a representative sample, a combination of many samples taken all across the face of the working, including low-grade and even no-grade from along both the foot and hanging wall, the sides of the vein, for you must pay the cost of removing that worthless rock just as you have to pay to mine the vein itself.

A word about assayers might be worthwhile. Not everyone who hangs out his shingle saying he is an assayer, is an assayer. To some of them, assaying ore is like reading tea leaves. Beware of the assayer who asks you what your sample might contain; chances are the report you pay him for will reflect just about what you told him! Spectrographic analyses are always suspect; they will tell you what is in a sample, but not how much.

It's encouraging to learn there is gold in a sample, but pretty discouraging when a fire-assay reveals that gold amounts to something like one one-hundredth of an ounce. Recently I had six different assayers test samples for platinum, and I received six different reports, ranging from no values whatever to more than one-thousand dollars to the ton. I'd like to buy at the first figure and sell at the last! Pick an old-timer, one who has been in business a long time. He wouldn't have lasted that long if he didn't know what he was doing.

For several years the author has had the opportunity to work with a group of professionals whose occupation is evaluating the real facts of lost mine and treasure tales and then following up the search for bonafide targets. Needless to say, they do not work from "genuine" treasure maps purchased from slick-talking con-men. Months were spent photographing promising sites from aircraft, using the latest technology in infrared photography and ground penetrating radar, and afterwards enhancing the images obtained with specially designed computers. The personnel involved are long time professionals, having located sunken treasure ships off the coast of Florida, lost mines in Central America and hidden caches in the Philippines. More recent work has involved tracing underground mineral veins in Mexico. Special photographic filters reveal old trails, excavations and mineral veins, even though they may be invisible to the naked eye. Much of the Uinta Mountains was investigated, including the Hoyt Peak area, where several previously unknown mine workings were discovered, as well as a suspect cache site.

We will return to Hoyt Peak, but one of the places we won't disturb further is the place where Gale Rhoades died. He was well aware of the Garcia waybill with its description of a treasure hoard containing "650 cargas of bar silver and 240 cargas of bar gold," hidden deep within that mountain. Gale Rhoades died of a heart attack while working on Hoyt Peak on September 27, 1988. Within minutes of his death, two of the miners working there saw one of his con-men partners steal all of his personal notes, maps and sketches, representing years of hard work. Some might say that he too was a victim of the Killer Mountains.

The iron-door Josephine Mine at Hoyt Peak. Gale Rhoades died there probing its secrets.

CHAPTER 6

SHADOWS ON THE TRAIL
Treasures of the Currant Creek Country

There is a high, windswept ridge which divides the Provo and Weber river drainage's from the country to the east. That grass and sage-covered highland which looks down on the Strawberry and Duchesne river headlands was a trail for pack animals and perhaps wooden-wheeled carretas during Spanish days, just as it is a Jeep trail today. From Heber Mountain and Currant Creek Peak one can see for fifty miles and more: north past Wolf Creek to the Provo River headlands, south to Daniels Canyon and the Strawberry Valley, or east to Currant Creek, Red Creek and the one-hundred miles and more of wild, broken plateaus and mountains drained by the Duchesne, the Rio Damian of Spanish days. Looking down into Currant Creek Canyon, one cannot help but be impressed by how large a piece of real estate it is, with the flat-topped mesa outline of Tabby Mountain barely visible in the blue-haze of distance. But it is only after hiking that country on foot, carrying a forty-pound pack with a prospector's pick in hand that one really learns just how big it is.

If you leave the Strawberry Valley and follow its namesake river up stream past Bjorkman's Hollow and Willow Creek, or take the old road past Co-Op Creek and the Red Ledges, you will come to know how big that country is. For a long time now, there has been a story about two Spanish cannon hidden behind thick brush at the foot of a rocky ledge, just below the ridge which overlooks the Little Strawberry Canyon. Several experienced treasure hunters have looked for those cannon, among them Mark Mason, Mark Josephson and Stan Brennan. Although they haven't found them yet, they have discovered much of interest in that rough mountain highland.

Stan Brennan has followed a line of Spanish trail markers from Heber Valley to the head of Center Creek Canyon. One of those markers is a turtle, believed by many to be a sure sign of treasure. It is on an age-old aspen which has the date 1756 cut into its bark. Brennan was able to interpret each symbol as it led to the next, to a place over-looking Strawberry Valley. There he came upon an old mine digging which even the oldest mountain people weren't aware of. In a small side-gulch in sight of that mine, Mark Mason discovered a strange square plot of ground where nothing grows, even though there is lush green vegetation all around it. Mason believes it is an old smelter site where quicksilver used in the amalgamation process has poisoned the earth.

You may not believe in the occult, but there are those who do. A psychic told Mason and his partner, Mark Josephson, that something of great value was buried at or close to that plot of sterile ground, and that it might help to solve the mystery if one of them was to under-go hypnotism. Josephson was placed under hypnosis and was asked: "What is buried there?" He responded, "He won't tell me." When asked who wouldn't tell him, Josephson replied: "Santos." When he was asked when the object was buried, he answered, "1791." Of course it could all be coincidence, but it is known that several caches were secreted away in 1791, including one at Rock Creek Canyon where the date 1791 is cut deeply into a stone ledge. There is also the date 1756 found nearby by Brennan. All three are planning next summer's trip to the upper Strawberry. Keep in touch to learn what they find.

There are old Spanish mines all along Currant Creek Canyon; too many have been found there to doubt that. One of those found mines has had a long and sometimes bloody history. Fifteen miles up Currant Creek Canyon from Highway 40, there is a dirt road which climbs steeply to the northeast. If you stop about two miles along that road and look to the east, a high outcropping of red sandstone ledges can be seen through the aspens. A short hike across a small creek and past some swampy ground leads to two caved mine tunnels driven into the mountain just below those red ledges. The waste dump at their portals is becoming difficult to find now, being heavily overgrown with white-barked aspens and second growth pines. If you take your time and keep a close watch, you might spot a few rotted logs where a cabin once stood. Part of that cabin was still standing and one of those tunnel portals was still open when I first saw them more than forty years ago. At that time there was a vertical shaft just north of those tunnels, close under the ledges and northeast of the cabin site. High on those ledges there is a large out-of-place square boulder which some say covers yet another shaft. Although those red ledge tunnels and shaft have been on Forest Service land for one-hundred years or more, just recently they were claimed by the Ute Indian tribe. The Forest Service granted the Utes a special mineral exemption to that area, placing those tunnels off-limits to non-Utes. I wonder why?

Stockmen from Heber Valley first discovered those old mines in 1908, as was reported in the Wasatch Wave at that time. That article described Spanish tools and other artifacts which were found in those tunnels. At first an honest effort was made to reopen those long abandoned diggings, but as the tunnels grew longer and costs escalated, financiers became anxious for some return on their investment. Many years ago I interviewed George Olson, the man who was in charge of work at that mine. He and his miners had succeeded in opening one of the tunnels for 125 feet when one of the bankers who was footing the bill came up with a scheme to sell stock. He salted the diggings by firing a shotgun loaded with fine particles of gold, and then brought a group of wealthy businessmen to the mine so they could see for themselves the color of raw gold glistening in the rock being removed. But several of those men were friends of Olson, so he exposed the fraud to them. With no funds to pay his miners, work soon stopped. Olson never had a chance to uncover the ore vein at the tunnel's end, and it wasn't long before those ancient diggings began to cave again. It's been a long time now since anyone tried to reopen those old tunnels under the red ledges. The ore vein which was nearly uncovered back in 1908 is still there, but those tunnels are on Indian land now.

Caleb Rhoads knew where those Currant Creek mines were located, for they were shown on one of the maps he had. It is pretty certain that he brought gold from those mines, and probably placer gold from Currant Creek as well. More about that placer gold shortly. In 1884, Rhoads made a commitment to pay for a new church house then being built at Price City,

The Josephine Mine in Currant Creek Canyon. Some believe ghosts of murdered miners still haunt its caved diggings.

where he lived. But unexpectedly, he broke a leg and was unable to go into the mountains to get enough gold to pay for its construction. Because of his promise to pay for that church, in the month of September, he told his half-brother Enock where that old mine on Currant Creek was located. It was Enock's first venture into the mountains in quest of gold, but he was confident he could find the way, for he had traced the trail to the mine from Caleb's map.

Enock's route from Price was over the crest of the Wasatch Divide to Indian Canyon, and down that long gulch to the forks of the Strawberry and Duchesne rivers. He went upstream along the Strawberry to its fork with Red Creek, which he followed to its junction with Currant Creek. His goal was the Red Narrows, at which place Caleb told him to leave the river and climb the mountain to the red ledges and the two old mines at their base. But where Caleb enjoyed some degree of rapport with the Indians because of the oath he had given to Chief Arropene, Enock had no such protection; neither was he the Mountain Man that Caleb was. Enock was not recognized by the Indians who followed him from the time he left the Strawberry River. At the Red Narrows he was attacked and severely wounded by Ute arrows. He tried to escape and was able to ride as far as Red Creek before he fell from the saddle and died. His body was discovered a few days later by two of the Murdock brothers from Heber valley who were running livestock along Red Creek. They buried him near where he fell, and the Killer Mountains claimed still another victim. Long after Enock's death, a map of those Currant Creek mines surfaced, possibly a copy of the tracing Enock had made from Caleb's map. It is usually called the Gleaner Gold Map. (See Map J) That map seems to pinpoint still other mines in Currant

The Red Narrows on Currant Creek where Enock Rhoads was killed by Ute Indians.

Creek Canyon, somewhere below Low Pass. If you check a modern map you will find Low Pass, but don't waste much time there, for it is not the Low Pass of pioneer days. More about Low Pass will be mentioned later.

Some say that Enock Rhoads' ghost still haunts the Currant Creek foothills. Some may doubt that, but Roland Rogers has actually been in the presence of ghosts there. He believes they are the spirits of Indian women, mourning the loss of their warrior-husbands, lost in battle with Spanish miners. During the summer of 1989, Rogers and his wife were camped in a tent while they were working an old mine not far from the red ledges. Late at night Rogers was awakened by owls hooting all around his tent, and as he listened more closely he heard the eerie sound of women and children wailing in the gulch just below where his tent was pitched. At first he doubted his senses, but a cold chill ran down his spine as the sound of wailing women and crying children came closer. It wasn't until then that he became aware that his wife had also been awakened. Both listened in terror as the shrieking became louder. They felt a fog-like damp as the sounds passed right through their tent. Neither dared move until the terrible sounds grew fainter and finally could no longer be heard as the spirits moved further up the gulch.

You may not believe Rogers' story, but you can believe that both Rogers and his wife believe it! So does Stan Sharkey and several of his friends. Bear in mind that Rogers and Sharkey have never met, nor until they read this have they ever heard of each other. One is from the Salt Lake City area, the other from near Ogden. Sharkey with three friends was hunting deer less that a mile from where Rogers had made his camp, although it was then later in the year. Each of the hunters went his own way, but Sharkey, who is somewhat of a rock-hound, followed a small creek in the gulch below where Rogers' tent had been. When he came to a small pool of water, he knelt down to drink. At the edge of that pool he saw a small gold nugget, glistening where sunlight reflected from the gravel. As he reached down to pick up that nugget, he felt the presence of someone behind him, but when he turned to look, there was no one there. Still, as he pocketed that nugget, an unseen presence seemed to whisper to him to put that gold back where he found it.

Only a few minutes later one of his companions came down the mountain to that pool of water, and to Sharkey's surprise, his hunting partner told him they were being watched. Even though they could see no else on the mountain, both had the feeling that someone, or something, was watching them. They started down the mountain and had gone only a short way when they were joined by their two companions. Even before they could exchange greetings, the other two hunters excitedly told them they wanted to get away from there, for someone had been following them and watching every move they made. Sharkey and his friends quickly agreed to end their hunt and get away from that haunted gulch. You may not believe there are ghosts or spirits at the old Spanish mines along Currant Creek, but I know about a half-dozen people who do!

A few days before his death, as Enock Rhoads made his way up the Strawberry River, he unknowingly passed by several mines which Caleb Rhoads probably never knew of. Just down stream from Lake Canyon and on the steep hillside above the river, at the edge of a rock slide, there is an old Spanish mine. Along the river, against an outcropping rock ledge, there are several large pines. If you look closely, you can see where marks have been cut into their bark, so long ago that they are now difficult to see. And if you follow up Lake Canyon, there are other old mines, one which even today keeps an Indian recluse in grub and horse feed.

Go up Lake Canyon to Witch Canyon, and follow up that gulch to where a small draw is crossed. Just up canyon and beyond a locked gate there is an old mine, in sight of the road. It is on land owned by an Indian, but not a Ute. Although he prefers to live in a crude cabin in a primitive style, without either running water or elec-

A weather-worn Catholic cross, a faded name and the date 1831; along the treasure trail.

tricity, those who know him say he can afford anything he wants, even the shiny new 4 x 4 pickup parked near his cabin. He lives off the land and wears hand-me-down clothes, but he is a good neighbor and will help anyone who needs help. But keep away from that old mine shaft, for he doesn't share that, and he carries a forty-four revolver to make sure you don't get too close to it!

Orson Mott herded sheep in the Currant Creek country for most of his life, and he was 97 years old when I last talked to him. He knows a lot about the old time Indians, and he also knows a lot about the Uinta Mountains which few white men know. He told me a different angle on the mine Enock Rhoads was looking for, as well as the place where he was attacked. Mott said Indians told him that Enock followed Currant Creek up stream to Little Red Creek, where he turned northwest to where there are two knolls, the ones which Caleb Rhoads often described as looking like a woman's breasts, one bare and the other brush covered, one of the mine clues he sometimes mentioned. From those two knolls Enock went around the head of Grass Hollow and down what stockmen later called the Raspberry Trail, to where a little spring of water seeps out from under a ledge at the base of a rock slide. Caleb Rhoads also mentioned camping at just such a place. There is an exposed vein of coal close by, and Rhoads once said there was a coal vein near one of his mines. Right by that spring there is an outcrop of green-colored rock which has a lot of iron pyrites in it, and that too is another clue Rhoads sometimes mentioned. According to what the old Indians told Mott, that is where Enock Rhoads was attacked. After his death, Indians covered that outcrop of quartz and yellow pyrites.

That there is a mine along Little Red Creek very similar to the outcrop described by Mott has long been known to a few old-time prospectors. If you follow up Little Red Creek you will come to a "marker" tree, a giant pine standing atop a low ridge. The top of that pine was cut off a long time ago, Mott says by Caleb Rhoads, to be a mark to watch for. If you keep a close watch you can still see that marker pine from the trail. There are several Spanish signs cut into that tree, and just a short way further along that same ridge there is a mine shaft. It has been caved in for a long time now, but it is easy to find because of a large depression in the ground, about thirty feet across and six or eight feet deep. Flecks of free gold can be seen in its wall rock.

Several years ago, a man doing some survey work in that area discovered a tunnel portal on the point of a ridge just a short way up country from the marker pine. He brought in a front-end loader to dig away the rock and debris which covered its entry. At about fourteen feet from the surface, the machine dug into a thick layer of wood ashes and chunks of slag, at what appeared to be a smelter site. A few feet deeper into the mountain side, it broke through into a tunnel which was completely filled with icy water. A tremendous torrent of water under pressure burst through that opening, pushing the front-end loader aside and eroding a deep wash below the ridge. Water spewed forth from that tunnel for several days, carrying pieces of rotted mine timber and rock, indicating the underground workings beyond that tunnel portal were extensive. As soon as the flood waters subsided, the mountain began caving into the tunnel, leaving only a muddy pit filled with rock and debris. It would take a well-financed company to excavate those old diggings. At about that same time, Bill Coleman, a rancher from nearby Fruitland, found some pieces of Spanish armor under a large cedar tree close to that tunnel. Orson Mott thinks that mine may have been the place Enock Rhoads was trying to find, not the two tunnels below the red ledges on Currant Creek.

The late Jerry Eagan was a backwoods prospector who lived for more than forty years in a log cabin he built near the head of Red Creek, near the foot of Tabby Mountain. He knew there was gold in Currant Creek Canyon. During an interview when he was quite old, Eagan told me about a young fellow who worked for the Bureau of Mines at Denver, who made his summer camp along Currant Creek for four years. Eagan recalled that he often stayed at the old cabin below the red ledges, but that he panned for gold somewhere along the river.

> He was going to college while working for the Bureau of Mines, panning gold to pay his way. He even sluiced gravel from beaver dams. He made enough panning gold to pay his tuition. I don't know how much gold he found; I guess it didn't cost much to go to school then, back in the 1940s when gold was still $35 an ounce. After he got his degree, he never came back, got a good job with the government. But I do know that he paid for his schooling with gold from Currant Creek.

I wonder if anyone else has tried panning gravel from a beaver dam? A dam along a small mountain creek would form a natural gold trap. A likely place to try might be near the head of Currant Creek. Eagan said that many years ago during the deer season, he and several friends were taken by auto to the top of the canyon. They spread out and began the long hike back to their camp, which was located far down country at the mouth of Layout Canyon. They hoped to drive some deer down canyon to their hunting partners, but all they found during their long walk were two old mine shafts. Both were in the bottom of the canyon near where the creek forked. Both shafts were pretty well filled with water. They never had time to investigate those old diggings, and Eagan never got back to them again. They might be worth looking into, for surely no one would have sunk those shafts in such an out-of-the-way place so long ago unless there was something of value there.

Just above the Red Narrows along Currant Creek, where Enock Rhoads was said to have been ambushed, there used to be an old placer mining camp. Not many people now know much about it, or exactly where it was located. Since the new Currant Creek Dam was built, heavy construction and modern campgrounds have just about erased every trace of it. Few people pan for gold there now, and probably even fewer think there is gold there, but a few old-timers know better. Sid Smith, an old Park City miner, used to pan for gold along Currant Creek back in the early days, and he recalled seeing the remains of what he called "ancient sluice boxes" along the Red Narrows. Sid told me they were made from sections of hollowed-out pine logs. He remembered that just after the turn of the century, two prospectors stopped at Al Murdock's trading post at Duchesne, where they paid for their purchases with placer gold they said came from Currant Creek. A few days later both men were found murdered in the mountains north of Vernal, their pokes of gold dust missing. Chalk up two more for the Killer Mountains?

Evan Smith, no relation to Sid Smith, was a shift boss at Park City's Judge Mine, and he told me that as a boy he lived with his father at a placer camp which was located on a sage-covered flat, just above the Red Narrows on Currant Creek. He said there was quite a good size camp there then, consisting of a dozen or more log cabins. Besides panning the creek, some of the miners dug coyote holes back into the hillside behind those cabins. I remember seeing several of those old cabins, but since the new dam construction, not a trace of them remains. Lee Fitzgerald, who used to run cattle along Currant Creek, told me that many years ago he employed an old herder who had been born at that placer camp. That herder told Fitzgerald that camp was a lively place when he was a boy, boasting a row of cabins and tents and a small one-room general store. That store was very likely the same one which Roy Peterson said was robbed by the two toughs who built their cabin near the old Spanish mine on the Silver Meadows.

There are too many tales of old mines along Currant Creek for all of them to be just folklore. Orson Mott, the near-century old sheepherder from Duchesne, told me that as a young man he herded sheep in Layout Canyon, which is just downstream from the Red Narrows. There is a big spring of water in that canyon, and carved into the bark of a large tree which stood just above that spring was the name E.H. Payne, with the date 1863. Who Payne was is now unknown or forgotten, but that was long before there were any settlers in that area, and it was just about the same time that Caleb Rhoads was bring gold from those mountains. It would be interesting to know who Payne was, and what he was doing in Layout Canyon in 1863.

Mott said that close to that tree with Payne's name on it, there are several graves outlined with stones, nearly hidden under a thick growth of choke-cherry bushes, which had grown long after those graves were dug. Indians don't outline their graves with stones. Mott recalled that coyotes or badgers had dug into those graves long before he saw them, and that scattered on the ground were bits of rusted iron and some old brass buttons. Were they the graves of Spanish miners? Lee Fitzgerald talked about an old inclined mine shaft along the creek near the mouth of Layout Canyon, but even when he first saw it, it was filled with water and the creek bank had caved over its portal. Could it have been the mine where whoever is buried in those graves worked?

Emery Smith ran sheep along Currant Creek eighty years ago. He told Orson Mott that during the 1930s a Mexican came up the canyon, riding one horse and leading another which carried his camp outfit. He camped in the canyon all that summer, moving his camp every week or so. Smith noticed that he always kept one horse saddled and ready to ride, like an old-time outlaw ready for a quick get away. Sometimes that Mexican would ride down to McNeill's Station at Fruitland to buy supplies. He was pretty close-mouthed and never had much to say, but Smith could speak his lingo so they got along pretty well. One day the Mexican showed Smith a map drawn on a piece of goat-skin, but even though he could read Spanish, Smith said he couldn't make much out of it. Apparently the Mexican never found what he was looking for, for he left when the first snow fell.

The following year Mott herded sheep for Smith again, and he recalled that one day one of his horses came up missing. He made quite a search for it, but never found a track. Then one day as he was walking along the mountain side following his sheep, he smelled something foul and noticed that his dogs were walking around in a circle a little further up the mountain. He climbed up to where his dogs were, expecting to find a dead sheep, but instead he found his horse, where it had fallen into an old mine shaft. He could see where the horse had tried to get out, where his front hooves had dug at the side of that pit. The shaft had been covered over with pine logs a long time before, but time had so rotted them that when his horse stepped on them, they broke and collapsed under its weight. A few days later he found a place a little higher on the mountain where those poles had been cut, so long before that the stumps were little more than piles of sawdust.

Mott told me that he discovered that shaft had been an air vent connecting with a tunnel which had been driven into the mountain from lower on the slope, a tunnel he never suspected was there, even though he had walked past it many times. There was no waste dump to reveal its location, only a small trickle of water seeping from under some rocks which had been placed to cover its portal. Later Mott found where the waste rock from that tunnel had been carried to a draw a little further around the mountain, purposely dumped there so there would be no sign to disclose where the tunnel was located.

Jerry Eagan, the old prospector from Red Creek, came close to finding the treasure of Currant Creek. He and a partner were in the same area where Mott found the concealed tunnel and shaft. Eagan was using a willow-stick dowsing rod while his partner employed a deep-seeker metal detector. Eagan recalled what happened:

> The stick led us to a place in Big Hollow, and the detector confirmed there was metal there. We dug down for eight feet but never found a sign of color, still the detector signaled that we were close to something metallic. There was no sign of any earlier digging; that place looked as if no one had ever been there. We stopped to rest and I picked up a piece of rock which looked just like all the other rock we had dug, but it was heavy, so I broke it open. The inside of that rock was straight silver! We never found another piece of it, although both the stick and the detector said there was more there. Now you know that nature doesn't make just one piece of silver, there has to be more there somewhere, a vein or an outcrop where it came from. I had that rock assayed and was told that if I could find more of it, it would be worth 600 ounces of silver and 4 ounces of gold to the ton!

East from Currant Creek, at the head of Red Creek, there is a large flat topped mesa called Tabby Mountain. It marks the eastern end of the Wasatch Plateau; beyond are the rolling prairie's and white-clay badlands of the Ute Reservation. That mountain was one of the last strongholds of Chief Tabby's Utes, and was named for that Indian leader. Not long before the reservation was opened to white settlement, a plan was devised to pay the Utes a few cents an acre for the lands the white man wanted. Indignant at being offered only a pittance for his ancestral home, Chief Tabby scorned Washington's offer:

> You may give presents to other Indians, but the Utes do not want your presents. You want our land because there is plenty of gold there, but all you are willing to give us for it is copper. Do you think we are fools?

Was Chief Tabby saying there is gold on Tabby Mountain? Some believe that is exactly what he meant. But though he pleaded to keep his namesake land from being homesteaded, Tabby Mountain was lost to the Utes; that is except for one small area atop that high mesa, only recently withdrawn and returned to the Utes as a mineral reserve. That withdrawal is very similar to the one at the red ledges along Currant Creek and still another on nearby Blacktail Mountain, where Rock Creek empties into the Duchesne River. Could there be some sort of pattern to those land withdrawals?

Tabby was the white man's name for the Ute Chief, but his given name was Tabby-To-Kwana, which interprets as "Child Of The Sun." He was also known as Tabian, Tabba, Tabinaw, Tobiah and Tabiona. He was a son of old Chief Uintah, and was a half-brother to Wakara and Arropene. Tabby was born about 1800 and died at the Whiterocks Indian Village on November 22, 1903, at which time he was well over one-hundred years old, some said one-hundred and five years of age. He was buried along Rock Creek at a secret and sacred place known only to the Utes. He had many children, the best known being Nauhnan, who was also known as Tabby. Nauhnan also lived to be very old, believed to be one-hundred and six.

Chief Tabby of the Utes. White men wanted to trade worthless copper for his gold.

Gene Keene of Kaysville has spent a lot of time prospecting on Tabby Mountain, and he knows just about every ridge and canyon there. He has discovered many old signs and symbols. At one place he had to cut one tree down in order to see a symbol cut into another, so closely had those ancient aspens grown together since those signs were carved. Not far from Volcanic Springs, Keene discovered an old shaft. He engaged Mary Horne, a dowser of some renown from Georgia, to dowse a map of Tabby Mountain. Horne pointed out a place where there is a mine and perhaps a cache, at almost exactly the same place where Keene found that shaft. Jerry Eagan spent half of his life in the shadow of Tabby Mountain, and he also found that same area of interest:

> There's a place on Tabby Mountain, over around Volcanic Spring; there's a little canyon back up there above the spring, and there's gold there. I've panned it, but it's real fine and hard to keep in the pan, but it's there if you can figure out how to save it. Orson Mott, the old sheepherder, also knows that country: Do you know where Dutchman's Flat is? It comes down from Tabby Mountain, right in there by Mud Spring. There's an old inclined shaft which goes down real deep.

Caved rock and rotted timbers are all that remain of this old mine on Tabby Mountain.

> There was a hand-made wooden windlass hoist there when I first found that shaft, seventy or eighty years ago now. There's an ore vein which comes down from the hogback just above the flat. I've found a lot of good gold float there.

In 1926, Reed Lewis and Guy Bronson were herding sheep along Tabby Mountain near the head of Red Creek. A mannerly and obviously well educated Spanish gentleman rode up to their camp one day, leading two pack animals loaded down with a camp outfit and prospecting gear. They shared their meager dinner with their unexpected guest, who camped near them for several days. The Spaniard made no secret that he was looking for a mine where his grandfather had worked many years before, and he showed Lewis and Bronson a map, beautifully drawn on a roll of parchment-like time-yellowed paper. That map showed six Spanish mines along the northwest flank of Tabby Mountain, as well as a trail which was marked by symbols cut into trees and ledge rock. As time permitted, Lewis and Bronson helped the Spaniard search for the treasure signs shown on his map, but they failed to locate any of them. After a few weeks they had to move their camp, for the sheep had grazed higher along the ridge leading towards Red Creek Peak. They never saw their Spanish visitor again, but over the years they kept a close watch for the tree signs he had shown them, and the six mine shafts on Tabby Mountain. They never found them, but that doesn't mean they are not there, it only means they never found them. As Lewis observed, if they are not there, why did that Spaniard come so far and work so hard to find them?

Jerry Eagan told me about a mine near a hidden campsite in a cave, somewhere along Red Creek and not far from his cabin:

> That mine tunnel was dug right at the side of a

> small side stream, and all of the waste rock and dirt dug from it was dumped into that creek, so there was no trace of a waste dump. Whoever worked there, and that had to be a long time ago, filled that tunnel full of cobble rock carried from the creek. It's hard to understand why anyone would work that hard to conceal their diggings if there wasn't something of value there, something they wanted to conceal. Back around 1938, a couple of fellows tried to clean that tunnel out, but they only got back in there about fifty or sixty feet. They worked all summer but they never got to the end of that cobble rock plug. It was tough times then, so they never came back the next summer. I guess that I'm the only one left who knows where it's at, but I'm too old to go digging there.

Eagan also told about a cave his brother discovered at about that same time, probably in 1939; a cave which could have some connection with that old cobble rock filled mine.

> My brother climbed up out of Red Creek in a narrow canyon right across from that mine; a sort of box canyon. He said it was just about impossible to get up over the rim-rock at its head. Just under that rim-rock there is a cave, one of those worn by the wind, and in that cave he found a miner's camp. From the smoke-blackened roof, it looked like someone had lived there for a long time. There was a place where a fire-ring had been used, and by it there was an old-fashioned iron pot, one of those old-timers with three legs, kind of like a dutch oven, but a lot bigger. It was too heavy for him to carry, so he left it there. As far as I know it's still there in that cave.

Stan Sharkey told about a time when he and Lee Dennis were cutting cedar trees for firewood up on Tabby Mountain. They were using a chain saw to cut a gnarled old tree when the saw suddenly encountered something which broke the chain. Deep inside that giant cedar, the blade had hit the barrel of an old-fashioned Spanish match-lock musket. They figured that it had been left leaning against that tree a long time before, and the tree had grown around it. Parts of that ancient firearm were salvaged and examined, and on the barrel the name Seville, Spain, could be made out. Judging from the size and age of that cedar, Sharkey estimated that musket had to be at least two-hundred years old. Near that place, the Forest Service had been tearing trees from the ground by chaining, pulling a heavy log chain between two tractors. At one place one of those tractors broke into an Indian grave, and scattered lots of small copper beads across the hillside. Was that old Spanish musket left there when that grave was dug? Could it have been the reason that grave was there?

Tabby Mountain helped the Killer Mountains add to their score. A tale told around the Fruitland area claims that back around 1925, two prospectors came from the mountain with enough gold dust to slake their thirst at McNeill's Station, a local roadhouse. One beer led to another, and when one partner started talking about finding an old mine and cache, the other tried to quiet him, and a fight broke out. A knife was pulled from a boot sheath and in less time than it takes to tell, the talkative partner was on the floor, dead. His drunken partner was hauled off to the local calaboose to face a manslaughter charge, but when dawn came he too was dead, stabbed to death. It was never determined who killed him, but among the drunks sleeping it off that night were several Indians who were often seen on Tabby Mountain. Who knows?

The death of James Hartzell was no mystery. He was related to the Rhoads family through marriage, since his father, Edward Hartzell, had married Caleb Rhoads' widow after his death. James Hartzell spent a large part of his life looking for the Lost Rhoads Mine. During August 1961, when he was seventy-two years old, Hartzell persuaded his nephew, Charley Williams, to take him to the Red Creek Peak area, a high ridge which separates the West Fork of the Duchesne River from Currant Creek and Red Creek. Hartzell had a map with him, one which his father had obtained from Sidsie Rhoads. He believed that map would lead him to a mine tunnel close to the Bobby Duke Trail, a pack animal route connecting Red Creek with the West Fork. His father had told him of seeing a chunk of gold ore which Caleb Rhoads had given to his wife, which gold he said came from that old mine near Red Creek Peak. It was said to be in a place where the morning sun would shine into the mouth of that tunnel. Hartzell took enough food and supplies to last him for two weeks, at the end of which time Williams was to return for him. The last time Williams saw his uncle, he was hiking up Little Red Creek, heading towards the peak, accompanied by his little pet dog.

The allotted time passed and Williams returned to where he was to meet his uncle, but the old man wasn't there. He notified the Wasatch County sheriff's office and a search party was organized. George Sorenson, one of the searchers, discovered Hartzell's body when his little pet dog growled as Sorenson passed by a thicket of brush. The decomposed state of the body indicated that Hartzell had died about two days after his nephew had left him. His body was found along the Bobby Duke Trail, about one mile from the West Fork, not far down canyon from an old log cabin hidden in a thick stand of pines. Hartzell may have been trying to reach that cabin, which was one of the landmarks on his map. That old cabin can still be found if you know just where to look for it, what old-time stockmen call the

Dude Young cabin. You can still see a name carved deeply into one of its walls: Caleb!

Sheriff Floyd Witt responded to the site, where a preliminary investigation was conducted. It was determined that the old man probably had died of natural causes. Joseph Olpin, a Heber City mortician, made a brief examination of the body, during which he found Hartzell's map under his shirt. He opened the map and told the others that it appeared to be some sort of treasure map. He replaced it under Hartzell's shirt, but what happened to it after that time has never been determined. It was probably released to some family member, but no one today seems to know who received it. The body was loaded onto Cal Giles' horse and was taken down canyon to the West Fork. Death came to James Hartzell a lot easier than it did to a lot of other prospectors who died in the Killer Mountains.

Down low on the West Fork of the Duchesne River, on the morning side of 10,500-foot Red Creek Peak, near where Sand Creek has its headwaters in a series of small beaver dams, there are two red-colored knobs which break the forest green of the mountains. They look out of place, and they are. They are outcroppings of iron ore on a contact between gray limestone and blood-red sandstone. Geologists tell me those knobs shouldn't be where they are, but then what do geologists know? Have you ever heard of a mine being found by a geologist? Prospectors find mines while geologists and engineers develop them. Orson Mott, the old-timer who herded sheep along Sand Creek before any of us were born, found a vein of ore down by those red knobs which assayed 65% manganese. That same geologist told me that manganese is a pretty good indication that there might be some gold there. This time I think he might be right!

About the time the Ute Reservation was opened for homesteading, back in 1905, some of the Fred Giles family from Heber City were grazing cows along the West Fork and Sand Creek. One of them told me that on several occasions he saw an old Indian woman digging in the rocks by one of those red knobs. The next time he saw her there, he took time to watch more closely. At first he thought she might be digging camas roots, but then he noticed that whatever she had tied up in her apron was pretty heavy, so he rode down to talk to her. Bundled up in her faded linsey-woolsey apron were several pieces of red iron-stained rocks. When he asked what she was going to do with them, the old woman replied that she would take the rocks down to the creek where she would break them up and then pan the fine sands in a clay pot. It was slow work, but in a few days she would recover enough gold to pay for her goods at the trading post.

At about that same time, some of the Jeremy's began ranching just below where Little Sand Creek washes into the West Fork. An old Indian woman, maybe the same one Fred Giles talked to, lived in a dugout on a hillside across from the Jeremy cabin, just north of Wolf Creek. She was nearly blind and obviously very old, some said more than one-hundred years old, but her memory was still good. She could remember when Caleb Rhoads used to camp just down stream, where Wolf Creek and the West and North forks come together. She recalled that Rhoads used to stop at her dugout, sometimes to give her a piece of venison or some flour, and sometimes to give her a few small pieces of gold; enough for her to pay for supplies at Al Murdock's trading post. She said that Caleb Rhoads was a good man.

That old Indian woman knew a lot about that country. She remembered that Keith Bird, a grandson of old Henry Harris, the Ute interpreter during pioneer times, once told her that in the old days his grandfather would go to the "Bird's Nest," where he would get gold which was used in a Ute ceremony. After that ceremony, the gold was always returned to Mother Earth where Bird got it. Whatever Henry Harris told Bird was probably true, for Harris was a close friend and confidant of Caleb Rhoads, and it was said that his word was as good as gold. Harris was actually part Pi-Ute, and came to Fort Duchesne around 1888. His father was Jimmie Reed, a white man who traveled west from Kentucky about 1827. If you should come across one of the pioneer maps of the West Fork country and examine it closely, you will find the "Bird's Nest" almost exactly where that old Indian woman dug for gold by the red knobs. Some may tell you that formation doesn't carry gold, but I think I'll side with that old woman; like her, I think that gold is where you find it!

Back in 1909, a sheepherder named Leonard Atwood observed two Indians paying for trade goods at the Murdock store at Duchesne with pieces of raw gold, gold in a dark brown quartz rock. Knowing that it would be useless to ask them where they got that gold, he decided to keep an eye on them. A few weeks later Atwood was at the Jeremy Ranch, known to most people now as the Joe Bush Murdock Ranch, where the West Fork of the Duchesne joins Wolf Creek. Atwood was working with some livestock when he saw those same two Indians ride past the ranch and continue up the West Fork Canyon. He was only one of several working cowboys, so he was satisfied that they hadn't recognized him.

Atwood asked Jeremy if those two Indians often passed by his ranch, and learned that for some time Jeremy had been aware that they were getting gold from somewhere higher up in the canyon. He told Atwood that after they passed his place, they wouldn't be gone long before they returned, usually only a day or two.

They made no secret they were getting gold somewhere along the West Fork, but Jeremy had little time to wonder about them, except once when they bought a saddle horse from him, paying for the animal with pieces of bright yellow gold in quartz. He remembered that it was small wires and pieces in a dark brown colored quartz.

Atwood waited until the two Indians were out of sight and then followed them. It was just about sundown, so he made camp when night came, since he didn't want to ride upon them unawares. He later told how he sat in the dark all night, holding the reins of his horse until daylight, afraid to make a fire which might give his position away. At first light he started on the Indian's trail again, and discovered that they had followed the West Fork for only a few miles before turning up a steep canyon to the south. He lost their trail in an area of lush grass and willows, where beaver had built their dams across a small creek. He carefully followed that stream up canyon, watching for an ambush, but he never cut their trail again. He said that even though he couldn't find their tracks, he still had a strange feeling that those Indians were watching him.

Atwood went back down to the West Fork where he made a concealed camp and waited for the Indians to return, but they never came back that way. He concluded that they must have seen him and crossed the ridge leading down into Red Creek. He had a job to return to and couldn't spend any more time on their trail. He returned to the West Fork Canyon several times during the following years, trying to find the Indian's mine, but to no avail. He used to tell his grandson that he was sure they got their gold very close to where he lost their trail, for Jeremy said they came back past his place within a day or two. Also, the small beaver dam canyon they went up was steep, but it was not very long, only about two or three miles to the ridge line. He was sure that their mine wasn't on the opposite side of the ridge, for had it been, they would have gone to it from Red Creek or Currant Creek, a shorter and easier ride. Unknown to Atwood, only a few years later another sheepherder would find the Indian's mine, only to lose it.

The old Indian and Spanish trail which follows the West Fork of the Duchesne to Wolf Creek and Rhoads Canyon passes within sight of a Spanish mine unknown to most treasure hunters. Without doubt, it is the same mine Leonard Atwood searched for. In 1920, a sheepherder named James Woolsey rode away from his camp just as the morning sun touched his camp on a ridge just east of Low Pass, a notch in the high ridge-line which separates the West Fork from Currant Creek. He had ridden his horse only a short way from camp, roughly following what was then called the Bobby Duke

Ed Grose at the landmark clue-cabin near Low Pass. Can the lost Woolsey Mine be far from it?

Trail. He was riding down a steep, brushy hillside which slopes off into the depths of the West Fork when he noticed the ends of a row of pine poles sticking out of the hillside. They were in a place where he had passed before, but on that day the early morning sun was just at the tight angle to penetrate the brush so that he could see the ends of those poles.

Woolsey stopped to investigate his find and discovered the outline of a tunnel portal in a shallow depression close to some slide rock, with the shape of a small waste dump below it. That dump was so old and overgrown with brush and aspens that he couldn't see it until he was actually standing on it. When he tried to pull one of those rotted poles loose, it broke, leaving a badger-size hole where it had been. His sheep being on the move, he had no time to investigate further, but later in the day he returned with a shovel to dig away some of the rock which covered the entrance. It seemed to him that sometime in the past, dirt and rock had been purposely slid down from the hillside above to cover that portal. Crawling into the small hole he had opened, he carefully felt his way to the end of that tunnel, which wasn't more than twenty or twenty-five feet from the portal. Using a candle for light, he saw a vein of shiny, brown-colored quartz rock crossing the lighter colored limestone country rock. In that quartz were stingers and bits of yellow metal he suspected might be pyrites.

Woolsey hadn't always been a sheepherder, for during the winters when sheep were turned loose on the West Desert, he sometimes worked at mining. He also knew a well-to-do doctor who paid him a small grubstake every now and then to keep an eye open for ore and old mines. He and the doctor had an agreement that Woolsey would prospect for minerals while herding sheep, and that if he found anything of value, the doctor

There is a name carved into the log wall of this old cabin below Red Creek Peak: Caleb!

would bankroll him. Woolsey collected a small bag of that quartz and pyrite rock, for he knew that pyrites sometimes carried gold values. He then covered the hole he had dug with rocks and pieces of rotted pine slabs so that no one else would find the tunnel before his samples were assayed. Several months passed before Woolsey brought his sheep down from the mountains just ahead of the season's first snowfall. Several days later he took his samples to his doctor-partner, but he was out of town, so he left the sack of ore and a note with his office nurse, explaining where those samples came from.

Unexpectedly, Woolsey got a job offer to herd sheep during the winter in New Mexico. He took that job and stayed with his new employer for several years. When he returned to Utah, one of his first stops was at his partner's office, but to his surprise he found another doctor in charge. No one could tell him where his partner had gone. Over the years Woolsey took other sheepherding jobs, and in time forgot all about the old mine tunnel on the mountain side near Low Pass. He was not a young man, and as the years passed he became unable to work in the high country anymore. Then one day he met his old partner's nurse, and learned what happened to the doctor.

After Woolsey left his ore samples at his partner's office and then moved to New Mexico, the doctor had the samples assayed. His nurse told Woolsey that the doctor became terrible excited when he learned those samples assayed $57,000 to the ton in gold! Unable to learn where Woolsey had gone, the doctor abandoned his practice and spent all of his time trying to find Woolsey's campsite near Low Pass. The only clue Woolsey had given him in the note he left with the samples was that he had carved his name on a giant aspen on the ridge where he had camped, and that the mine was only a short distance down country from that tree, in a place where it could be seen from the old Bobby Duke Trail. As soon as snow left the high country, the doctor began searching every ridge and gulch. Those who met him in the mountains said that he searched from dawn to dark, never stopping to rest, and often babbling to himself like a madman. Always exhausted and often hungry, and with the quest for gold driving him mad, he became so crazed that he had to be committed to a mental hospital.

By the time Woolsey learned of his partner's fate and that the ore samples he had taken from the old tunnel were not pyrites, but gold, he was too old and feeble to go into those mountains, even though he never doubted he could go straight to that gold. Not long before he died, Woolsey's nephew took him part way up Currant Creek canyon, to where they could look up and see Low Pass and the ten-thousand foot high ridge where he had camped nearly fifty years earlier. The road was still blocked with snow, but even if it had been open, the high elevation prohibited Woolsey from going any further. But he pointed the way to the gold:

> Go east of Low Pass, to where you will find my name carved on a large aspen tree. The mine is only a short way down the mountain from where I camped, in some thick brush and trees. You can see the mine from the Bobby Duke Trail.

Woolsey's waybill may sound simple, but I can tell you there are a lot of aspen trees east of Low Pass where he might have carved his name. "Old Dan" Tucker and I found one of Woolsey's name trees east of Low Pass, but there is no way of knowing if it marks his 1920 campsite. Although sections of the Bobby Duke Trail are still shown on some old maps, no one now is sure just where it crossed Low Pass Ridge. But if you can match the name tree to the trail, the puzzle pieces might fit together. And if they don't, there are still miles of tree covered mountain side to search. But for gold ore valued at $57,000 a ton, it's worth it.

Since the first edition of Faded Footprints, the author has researched the Woolsey tale from every angle, and has hiked every inch of the area where he carved his name and made camp. Several things have been learned which may help those seeking his lost gold. Interviews with many old-time stockmen has revealed that the Bobby Duke Trail climbed up out of Little Red Creek, passed through Coleman Basin and crossed the southwest side of Red Creek Peak. If you are using a map made before 1900, and you should be, Red Creek Peak will be shown as Tabby Mountain. The trail then angled down country past a corral and an old cabin heading towards Vat Creek, which is the beaver-dam creek where Leonard Atwood followed the two Indians. Cal Giles, a Heber City stockman, recalls part of that old trail, but there were several corrals and cab-

Jim Woolsey pulled a pole from the portal of an old Spanish mine and found gold, somewhere east of Low Pass.

ins between the ridge and Vat Creek. There are some rotted poles from one corral and one cabin still standing, the one James Hartzell was heading for when he died, but there is no mine near either of them.

Another point to ponder is the location of Low Pass. Locating that pass would seem to be an easy thing to do, since any good forest map clearly shows the pass as being between Currant Creek and the West Fork; but that is not the Low Pass of pioneer days. All of the old-time stockmen who herded stock along that high dividing ridge agree that when they were young, Low Pass was the divide between Little Red Creek and Vat Creek. That is where the Bobby Duke Trail crossed the ridge, and that is very close to where "Old Dan" Tucker found Woolsey's name on an aspen tree. A 1907 map, probably the time period map Woolsey would have used, shows Low Pass right where those stockmen remember it, not where it is shown on modern maps.

As mentioned earlier, the author spent some time recently with a California based treasure search crew photographing the Uinta Mountains with infra-red. One of the places investigated was the Red Creek Peak area. When films of that area are viewed using the proper filters, the Bobby Duke Trail can easily be seen, but its course is not where one would think a trail would be. Several cabin sites, old corrals and other places where the ground has been disturbed can also be seen. During searches on the ground, no sign of those cabins or corrals can be seen. There is a large corral high atop the dividing ridge between Currant Creek and the West Fork. It is shown on old maps as the King David Corral, yet not a man alive knows why it was given that strange name.

On a steep ridge, in a dense growth of huge pines, the author found the inscription "14S11" cut deeply into one of the largest trees. At first glance it might appear to be a section marker or a township reference, but a quick check of a topo map indicates that is not the case. Does that inscription have some special significance? Hopefully, these new clues combined with the use of modern technology, a lot of hard field work, and maybe a little bit of luck will be all that is needed to lead some treasure hunter to the Lost Woolsey Mine!

CHAPTER 7

CAMPFIRES GROWN COLD
Along the Duchesne to the North Slope

The Antonio de Reinaldo maps used by both Thomas and Caleb Rhoads show the lakes of the upper mesa and four rivers whose headwaters have their source at those high mountain lakes. Although each river flows from the mountains in a different direction, all begin as a tiny trickle from melting snow within an area of less than one square mile. From the Mirror Lake Basin, the Weber flows to the northwest, the Provo southwest and the Bear due north. The Duchesne has its headwaters at Mirror Lake and flows to the southeast. On the Reinaldo maps, those rivers are the Santa Ana, Antonio de Padua, Cumanchi and Damian.

With the possible exception of some now forgotten fur trapper, Thomas Rhoads may have been the first American to stand at the place where all four of those rivers have their origin. Since he followed both the Provo and Weber rivers upstream to their source, he could not have missed seeing the Bear and the Duchesne. Long before the turn of the century, when the first stockmen and timber cruisers made their way into the Mirror Lake Basin, they came upon an old log cabin which proved someone had been there long before them. That old cabin had the initials "K R" cut into one of its logs. Caleb Rhoads was often called "Kale" by his family and friends. Some consider the location of that old cabin to be critical in finding one of Rhoads' richest mines, as will be described later in the search for the Lost Cabin Mine. Several old mine workings have been found in those high lake basins, surrounded by towering peaks like Mt. Watson, Bald Mountain, Hayden Peak and Gold Hill. Most have proven to be little more than prospect holes, for at elevations near 13,000-feet, where winter snow often remains deep until mid-summer, the brief season was too short for Spanish miners to develop extensive diggings. Nevertheless, there are many old Spanish treasure signs at the base of Haystack Mountain or on timber-line pines below Murdock Mountain. On the back (east) side of Mt. Watson, narrow stringers of high-grade copper can be found, some of them assaying a few dollars in gold.

Spanish miners may have worked a gold placer high on the Duchesne River, in the shadow of 12,433-foot Mt. Agassiz. The late Cliff Roberts of Tabiona was one of the first men to herd livestock in that rough, boulder-strewn wilderness, back when a lean-to of pine boughs was the herder's only shelter and his supplies those he packed in on horseback. Close to one of those lone mountain camps, Roberts discovered some rotted sections of long abandoned wooden sluice boxes along the river, and faint but well-worn trails leading off into the rocky canyons which scar the broken sides of Mt. Agassiz. In an interview I had with Roberts when he was well over ninety years of age, but his mind still clear enough to describe landmarks by the names he knew seventy years before, the old man gave me the following directions:

> Go up the Duchesne River until you come to the forks. Follow up the East Fork to Fish Lake #3 and pass it on the north side. Go on to Fish Lake #2, but go past it on the south side. When you come to Fish Lake #1, go back on the north side. Go straight down that creek, don't ever cross it, to where it meets the next stream, then go right up there. That's where the gold is! I know gold when I see it, and that's where it's at. There's a big rock ledge which the stream washes against, and I could see gold shining in the water there. I picked up some pieces, and they were heavy and shiny. I saw it, but I guess I was a damn fool, for I never kept any except for a few pieces I put in my pocket. I was young then and thought I could go

> back any time I wanted to, but before long I had a large family and time just got away from me. But I know gold when I see it, and that's where it's at!

Just reading his words doesn't convey the emotion and conviction I heard in his voice when I talked to Cliff Roberts. There's no doubt that he knew that high Duchesne River country intimately and could describe in detail its every mountain, lake and stream; even though he might not have used the place-names we know those places by today. The river where he found those remnants of ancient wooden sluice boxes so long ago still looks the same as he described it. His memory was sharp, and I knew that what he said was true, for he also talked of other times and places which only someone who had lived in that country when he did would have known. He talked about the old Indian prospector white men called Pick Murdock, and of his nephew, Al Murdock, who was killed in a gunfight near Duchesne in 1911. Unknown to Roberts, the Murdocks were my great-uncles, so I knew that the things he said were true. I back-tracked Roberts' trail of seventy years ago, and it is just as he said it would be. But times have changed, and the gold placer he talked about is now in the Uinta Primitive Area, where countless government regulations and restrictions stifle today's gold miner, even though it is no secret that weekend prospectors do wash out a few pans of gravel when big-brother isn't watching. Still, it's just like Cliff Roberts remembered; well, almost like he remembered.

Frequently in treasure tales such as those found in adventure magazines, the author tells how after many years a prospector returns to locate his lost ledge or recover a cache, but the country has changed so that he no longer recognizes it. His lost ledge or cache becomes another lost treasure story. Some question how that can be, how the country could have changed in so short a time, but my interviews with Cliff Roberts illustrates exactly how that might happen. In describing a section of the Duchesne River as he remembered it, Roberts stated: "The river goes straight down there, and it's all covered over. It goes down a chute." Roberts accurately described the river as he knew it long ago, and his description remained accurate when the author used to fish that stream forty years ago. The river did flow through a chute, a place where it was squeezed in by the towering cliffs of a deep gorge, the bottom at places no more than ten-feet wide, below cliffs hundreds of feet high. And where it burst forth from that cataract gorge, it was covered, most of the stream flowing beneath a wide rock slide of huge boulders. That chute was impassable for pack animals, so Spanish miners built a twisting, treacherous trail across those unstable rock slides.

Long after Roberts left those mountains, a reclamation dam was built in Little Deer Creek Canyon, a side gulch whose waters enter the Duchesne just upstream from that gorge and chute. Not long after it was built, that dam broke, sending a raging flood of muddy water down the Duchesne River Canyon. That deep, narrow chute described by Roberts was filled with boulders, broken logs and debris; while the rock slide which had "covered" the river below was washed away by the torrent. The trout stream where I had fished was turned into a desolation of mud, rocks and broken trees. If Cliff Roberts could return to those mountains, he might say: "Nothing looks the same. The country has changed!" Floods, forest fires, snow slides and the works of man can alter an areas appearance. If you are trying to follow an old Spanish map or hand-written waybill, take into consideration that in over two centuries and more of time, the country has changed.

What message does this stone map along the Duchesne River have to tell?

Recently taken high-altitude infra-red photos reveal two mineralized areas in almost exactly the same areas Cliff Roberts described, but they are outside the Primitive Area. Field work on the ground was slow in that rugged terrain, but one of those ore zones was found last fall just as the first snow of winter whitened the ground, leaving its investigation for another day. But while searching that area, what may be an even more important find was made. On a ridge over-looking the Duchesne River Canyon, with Mt. Agassiz standing guard from within the Primitive Area, a strange and enigmatic treasure map was discovered, its features cut into a giant granite boulder! Lines and figures on that stone map clearly show the forks of the Duchesne River and the physical features of the mountains from Naturalist Basin to Murdock Mountain. Not as clearly understood are the meanings of several yet un-deciphered symbols, among them several letters, numbers and a square box figure. Could that strange stone map be connected in some way with the quartz ledge and placer discovered by Cliff Roberts?

Even more intriguing than the stone map is a crude sketch map drawn by Caleb Rhoads on a scrap of brown paper shortly before his death. That map shows what appears to be the Mirror Lake Basin and some "glacial peaks," with several lakes and the major streams which drain that high country. Several mines are marked with "Xs" and there is what appears to be a cabin site. That map bears the inscription, "This country is dangerous!" (See Map K) One of Rhoads' Indian prospecting partners, Happy Jack, was given that map by Sidsie Rhoads after her husband's death, and he in turn gave it to the outlaw, Matt Warner, for whom Rhoads had sketched it. Before doing so, Happy Jack copied that map in one of the small journal notebooks he always carried. The Rhoads-Warner connection will be explained later. Because the landmarks on that map do not fit the Mirror Lake Basin exactly, some have tried to orient it with the Rock Creek and Brown Duck Lakes country, but it doesn't match that area any better. Because it is hard to tell what area it depicts, some have called it the "Mystery Map." But if you study it closely, you'll see it really isn't a mystery at all.

Many geologists are in agreement that the most likely places where the Uinta Mountains might be mineralized are those few places where quartz intrusions cut through the limestone-sandstone core of that range. Not many of those ore-bearing intrusions have been found, but the 1969 Intermountain Geologists Association map of the Uinta Mountains pin-points one of them. The text accompanying that map states: "Near Mt. Agassiz there is a fine-grained quartzite, seventy-five to one-hundred feet thick, thinning as it approaches King's Peak." That's very close to the area shown on the stone map, very similar to Caleb Rhoads' "Mystery Map," and not far from where Cliff Roberts found gold!

Before turning north into the high country, the Duchesne River flows past the tiny hamlets of Stockmore, Hanna and Tabiona; the latter named for Chief Tabby of the Utes. Late in the summer of 1993, a crew of men were hired to build a fence around some ground an out-of-state businessman had purchased. While cutting a path through some cedar trees, the small tractor they were using broke through into a mine tunnel which no one had suspected was there. The land owner was summoned and after he crawled a short way into its depths, he ordered a heavy door set in concrete placed over that portal. It seems odd, for if he wanted to cover it to keep someone from going back into it, why wasn't a concrete slab poured over the opening. Installing a heavy locked door over the opening seems to suggest that he intends to go back into that mine tunnel again sometime. What other purpose would a door serve? If you watch closely you can see that door from the highway, at the edge of the foothills between Hanna and Stockmore. Like I said, it's strange!

Not long after the Ute Reservation was opened to settlement, Frank DeFa established a ranch along the North Fork of the Duchesne River. Several old men in that area still say that DeFa knew of a secret place where Indians who worked for him obtained gold; they say from an odd looking blue-colored rock. They say that DeFa would ride his old gray horse from his sawmill at the ranch and would go up into the mountains. A few hours later he would return with a grain sack about half-full of that odd blue rock. He would never talk about it or reveal the Indian's secret, so no one knew just where he got that unusual rock, but all of those old-timers say there was plenty of gold in it.

During the recent oil-boom in the Uinta Basin, it wasn't unusual for oil company geologists to hike into the back country. During the closing days of those prosperous times, two company men came upon a mine tunnel of ancient appearance, along the North Fork, and not far down country from where DeFa found gold. If you drive up the North Fork, you will see towering cliffs on both sides of the canyon, and it was high upon one of those ledges that they found a place where stones had been carefully fitted together to make a rock wall, a wall which was obviously man-made. Their investigation disclosed that wall concealed an old tunnel portal. Behind that wall they uncovered timbers which had once been a portal set, but for some reason beyond their understanding, the tunnel beyond that portal had been pains-takingly filled tight with stones and soil. Bear in mind that there is very little soil on those cliffs, so that it had to be carried there from far down the canyon. Wondering why anyone would have gone to such great labor to hide whatever lay beyond, they began the slow process of digging out those ancient diggings. With time to work only when their employment permitted, their excavation made slow progress. It came to an end when they were transferred to a more promising oil exploration area.

If you would like to take over their work at that old tunnel, just take a drive up the North Fork. Watch for a turnout where you can see a small mine dump low on the west side of the canyon, across the river. If you scan the opposite (east) side of the canyon with binoculars, you can spot their diggings high on a rocky point of land, just above some natural caves. Of course there might not be anything of value in that old mine; yet one can't help but wonder who someone filled it with rock and dirt and then built a rock wall to conceal it further if they weren't trying to hide something. Let me know what you find!

During a taped interview, Frank DeFa talked about a Spanish treasure. He said that one day many years ago while he was at Tabiona, he stopped to visit with Cliff Robert's father. They were sitting on the front porch of his homestead cabin when Louie Apprats, one of the old Utes from Rock Creek, joined them. It was generally well known that Apprats was one of the few "Old Ones" who knew the location of Spanish mines and caches. He lived in a small cabin along Rock Creek, just up stream from Utahn inside the reservation boundary; a place where he could keep a close watch on anyone going up Rock Creek. Apprats didn't join in their conversation, but he listened closely as they talked about lost Spanish mines and treasures which were said to be hidden somewhere in the mountains close to Tabiona.

While the white men talked, Apprats removed one of his moccasins, and then to DeFa's surprise, he pulled one of DeFa's boots off. He put DeFa's boot on his foot and then placed his moccasin on DeFa's foot. Then he said: "Frank, my moccasin fits your foot and your boot fits my foot, that makes us brothers." DeFa thought for a moment and then said: "well, Louie, if we're brothers, tell me where the Spanish gold is hidden." Apprats hesitated and then replied: "Sure, Frank, I know where it's at. See that big mountain? It's right up there." DeFa then said: "Since we're brothers, let's go up there and get some of that gold." Apprats suddenly turned pale, as if he suddenly realized that he had said more than he should have, and looked around to se if anyone else might have heard him. Then in a frightened voice he answered: "No, Frank, No! I wouldn't go up there for anything!" "Why not" DeFa asked? In a shaken voice Apprats said: "If we went up there and I showed you where that gold is, neither of us would ever come back!" When DeFa asked what he meant, the Indian replied: "The spirits would kill us both. We'd both be dead men!" Later DeFa said that he was certain Apprats knew where Spanish gold was cached, up on that mountain above Tabiona, but he also knew that his Indian friend would never take him to that place. "You see how superstitious those old Indians Are," he added, but was it only superstition?

Just upstream from Tabiona and Hanna, the Duchesne River joins its West Fork and then turns northward across what used to be the old Savage Ranch. Brothers Frank and Ranch Savage owned homesteads along that river a long time ago, but most of their land has long since reverted back to the Forest Service. Frank Savage operated a sort of dude ranch, guiding fishermen into the Granddaddy Lake Basin. He employed several Ute Indians as guides, and he was closely acquainted with many more. Those Indians trusted Savage and told him things they wouldn't have told anyone else. Savage learned a lot about old mines and cache sites in that wild country. During a taped interview made in 1970, Savage talked about some of the things he had been told and the places he had seen.

> I don't think any of the young Indians know where the Rhoads Mine is. Years ago some of the Old Ones told me that one of his mines was on the mountain side you can see across from their ranger station, back in those thick quakies (aspens), right where they can keep an eye on it.

Savage remembered a Spanish mine being found in Rock Creek Canyon:

> Over on Towanta Flat, pretty close to where old Towanta had his cabin, you can find rock which looks like pieces of animal horn, which carries good gold values. When they were building the road along through there, they uncovered an old tunnel. They went back into it, and found a great big room hollowed out. So far as I can recall, they didn't find anything in there, but it was an old mine, and no one would have known it if they hadn't cut into it while they were building that road.

Tales of Spanish cannon have been heard by many treasure hunters, and Savage knew something about them also:

> A few people who I knew told me they had seen a Spanish cannon. There was that old Indian, Julius Murray; he and one of his boys told me about a cannon. And Mel Thatcher, he said he saw one. Jed Hickman told me he was herding sheep up along Pigeon Water, up above Mountain Sheep Pass and back up in a little draw. He saw the barrel of a cannon sticking up out of the ground, in those cedar trees back up there.

Today a well-trod mountain trail leads up Hades Canyon, across from Iron Mine Creek on the North Fork of the Duchesne. Trail outfitters take fishermen around the base of Granddaddy Mountain on a pack train trail which leads to the shore of Granddaddy lake, one of the largest lakes in the Uinta Mountains. When the Geological Survey was exploring those mountains, they met George Beard, a prospector from Coalville, and asked about trails and lakes in that area. Beard replied, "Over against that mountain you will find a large lake, the Granddaddy of them all," from which statement the lake received its name. Within sight of Granddaddy Lake, a Ute guide sent by Chief Wakara took Caleb Rhoads to a place where Spanish miners dug gold before they were driven from the mountains during the Ute rebellion of 1844. According to Ute tradition, Rhoads was shown a place on the watershed divide between Hades Canyon and the South Fork of Rock Creek, where he could find gold in a deep crack or

Granddaddy lake as it looked in 1912. The Hathenbruck Trail to the Rhoads Mine passed this way.

crevice in the mountain top. It was a place where he had to lower himself down into a yawning chasm or fissure. Because of the danger involved should a rope break, or if he was caught alone in a position where he was unable to defend himself from some of the younger Utes who were beginning to question his right to be on their land and take their gold, Rhoads decided to take in a partner.

In 1894, Rhoads approached a man he knew and trusted, Frederick W. Hathenbruck, a merchant from Utah Valley. While Rhoads had been given only a meager education, Hathenbruck was a very learned man, having degrees or expertise in geology, surveying and medicine from the University of Heidelberg in Germany. He came to the Utah Territory with Johnston's Army as a physician during the Utah War of 1857. After he lost his right arm, he opened a mercantile at Provo City. During his army career, and later as a topographer, he learned the Ute language and became an interpreter and an acknowledged authority on Indian relations. But more important to Rhoads, Hathenbruck was a shrewd businessman, with many friends highly placed in government, including Reed Smoot, Mormon Apostle and United States Senator. Rhoads hoped that Hathenbruck would be able to cut through government red tape and take the shortcuts necessary to help him gain legal access and title to his mines on the Ute Reservation.

Both Thomas and Caleb Rhoads had made sacred vows to never reveal the location of Carre-Shin-Ob, what some called the Brigham Young or Church Mine; the place where gold was obtained to mint coins at the Mormon Mint. But Caleb had made no such promise regarding any of the Spanish mines or caches found by either him or his father, or shown to them by friendly Utes. Chief Wakara died in 1855, and Chief Arropene died only five years later, on December 4, 1860. Arropene's successor, Chief Tabby had made it clear that after Brigham Young's death in 1877 he would no longer be a party to the agreements made with Wakara and Arropene, nor would his braves be bound by oaths made by Young and Rhoads. Because of ever increasing hostility from the younger Utes, Rhoads went to great lengths to avoid being seen when he went into the mountains. When traveling alone he could follow the back trails, but it became more precarious when he began taking Hathenbruck with him. Later Hathenbruck would tell how they would go into the mountains by night, moving through deep canyons or going through thick pine forests to avoid being seen. There is a place on Rock Creek where the few white men then in those mountains sometimes saw Rhoads and Hathenbruck crossing the river, and to this day that place is still known to old-timers as the Hathenbruck Crossing.

Being a trained geologist, Hathenbruck was greatly impressed with the Rhoads mine near Granddaddy Lake. Ore samples which he dug himself were assayed at twenty thousand dollars to the ton, and that when gold was worth only sixteen dollars an ounce. He also dug up pieces of rotted leather ore sacks, rusted remnants of wrought iron tools and brass buttons from Mexican army uniforms; adequate proof that deep and dangerous crevice had been worked by Spanish-Mexican miners long before. Although that mine is often referred to as the Granddaddy lake Mine, recently discovered notes kept by the Indian Happy Jack and statements made by Pick Murdock make it clear that mine was not close to Granddaddy lake. Pick Murdock related how he and Happy Jack were taking gold from that mine to the Whiterocks Indian village. They had no packs to carry that gold, but they each had an extra pair of trousers on their pack horse. They tied those trouser legs at the cuff and filled the legs with gold money-rock. Unaware that there was anyone else in those mountains, they left those pants-leg packs of gold ore at their camp at Granddaddy Lake. On their return to camp they discovered that someone had been there, had found the gold ore, and had taken some of it. Within weeks a regular gold rush started, with prospectors hurrying to Granddaddy Lake, but the rush was short-lived, with no gold being found. Later Murdock and Happy Jack had a good laugh at the white man's gold rush!

A few years later when relations between the Utes and white men weren't so strained, Hathenbruck built a log cabin not far from that mine. It has never been a secret where that cabin located, but its location may come as a surprise to those who have spent long, hard days searching for it in the rough, rocky slopes along the shore of Granddaddy Lake. Cliff Roberts, who first took sheep into the Granddaddy Lake Basin, remembered how that cabin looked at the turn of the century.

Frederick Hathenbruck at his cabin near Lodgpole Lake. Would he have built it far from the Rhoads Mine? (Courtesy: Mr. Harold Hathenbruck)

He also made some interesting discoveries about that old mine worked by Rhoads and Hathenbruck.

> I've been to the old Hathenbruck cabin lots of times, at the southeast end of Lodgepole Lake. When I first saw it, it was real sturdy and a good shelter, but there's not much left of it now. There were two cabins, not far apart, a large one and a small one. That smaller cabin was built a long time after the Hathenbruck cabin was built, by a man named Behunin. A big pine now grows right where it stood. Rhoads and Hathenbruck mined quicksilver there. They had a big rolling crusher stone wheel, and some heavy iron bottles. They would turn that heavy wheel to crush their ore, and quicksilver would run through a hose into those iron bottles.

Roberts' description of the Hathenbruck cabin places it at the correct location, but his impression of the mining operation there is incorrect. He was a stockman who had little knowledge of mining, so the mistake he made is understandable. What he thought was a quicksilver mine was in fact a gold mine, although the ore did not come from the immediate proximity of either Granddaddy or Lodgepole lake, but was packed there because it was convenient to both water and fuel. The heavy stone which was used to crush the ore was a sort of arrastra, a crude but efficient Spanish mill in which a heavy stone is turned or dragged along a circular path lined with closely-fitted flat rocks. Ore is crushed to powder between the rolling stone-wheel and the rock floor. The drag stone is usually pulled by a horse, but Spanish miners weren't reluctant to use Indian slave labor to power their mills.

Although agonizingly slow in its operation, the arrastra is very effective at breaking gold loose from quartz rock. Quicksilver, which was carried in seventy six pound cast iron bottles was sprinkled or fed from that iron flask into the arrastra, where it readily adhered to the gold being separated from the crushed rock. It was not removed from that ore as Roberts thought. Each grain of gold would be absorbed by the quicksilver, which mixture is called amalgam. When the amalgam is heated, sometimes in a common frying-pan, the quicksilver evaporates or steams away in a vapor, leaving the pure gold behind. Quicksilver fumes are highly poisonous, so knowledgeable miners like Rhoads and Hathenbruck used a retort to condense the fumes and save the silvery liquid for re-use. Often when a retort wasn't available, the amalgam would be placed inside a hollowed out potato, which would then be roasted in a fire. The potato's meat would absorb the quicksilver while the gold would be retained as a button or nugget inside the hollowed out cavity. The quicksilver could then be recovered from the potato.

Several years ago, one of the old Utes who knew about such things, took a prospector into the Granddaddy Lake area where he showed him the location of an old mine. The Indian did this because he owed the white man a great debt; the white man having saved his son's life. The mine was a shaft driven at a forty five degree incline at the base of a small knoll, at a place where a growth of thick brush concealed its portal. The prospector later said that the entry was so well hidden that he didn't think anyone had been there in many years. He also said that the face of that old digging was criss-crossed with tiny wires of yellow gold! That prospector camped near Lodgepole Lake for several months each summer, digging enough wire gold to live comfortably for the rest of the year. But then a few years ago he noticed that he was becoming ill, especially after his yearly trips into the mountains. He died recently of what was called an unknown ailment, but an autopsy revealed that he had died of mercury poisoning. An investigation disclosed that the water he consumed while at his mountain camp came from the same small stream where Rhoads and Hathenbruck washed their ore after it was amalgamated with quicksilver. None of his family knew exactly where his mine was located, and his widow said she didn't want to know. Recently while reviewing and editing infra-red photos of the Granddaddy Lake Basin, a "hot spot" which looks suspiciously like the portal of a shaft was noticed not far from Lodgepole Lake, while a narrow line of color follows the course of that small stream. That water course might very well show as mineralization if it is impregnated with quicksilver.

What secrets are hidden in the old tunnel behind this five-ton stone door?

Ray Miller of Heber City is an old-timer who herded sheep in the Granddaddy Lake area a long time ago. He remembered just such a knoll as the prospector described. It has been sixty summers now since Miller made his camp near there, but he still recalls finding places where even before his time someone had cut long trenches across that knoll, cuts like those made with a "fresno" scraper. Miller also remembered that there were several places there where the ground was sterile and barren, as if it had been burned, as if by an acid or perhaps quicksilver. There was a small stream of water trickling from under the brush on that knoll; perhaps coming from the mine portal the prospector found. There is an outcrop of igneous rock which breaks the surface between Granddaddy Lake and Lodgepole Lake, an indication that area might be mineralized.

There is another thing about the Lodgepole Lake area which is interesting. Gene Keene of Kaysville visited the Bureau of Mines mineral display while he was at Denver. On display he saw what museum curators told him was one of the largest gold nuggets ever found in the Rocky Mountain area. That nugget is about 7" x 12" x 18" in size, and is worth a small fortune for its specimen value. It has been on display for so long that none of the present employees can recall exactly where the Bureau obtained it, but the curator told Keene that it was found somewhere in Utah, at a place local residents call the Granddaddy Lakes!

Dee Allred of Altamont has tried to have the old Hathenbruck cabin declared a historic site, but time seems to be working against him. Time has taken its toll, but vandalism has caused even more damage. Not long ago a "crazy" threw a stick of dynamite into that cabin, severely buckling its walls. Still, Allred believes it can be saved. One of his problems is the Forest Service. which takes a dim view of miner's cabins in the forest, even historical cabins. Allred has recovered several valuable mining relics from the Hathenbruck cabin area, which he hopes to put on display if the cabin is preserved and protected. Among them is a small mold used for casting bars of bullion. He also discovered an amalgamating riffle, similar to a sluice box but designed to use quicksilver to amalgamate with the gold. Ruins of a 300-foot flume which angled down slope to empty into a small pothole lake was still in place only a few years ago, but hardly a trace remains of it today.

There used to be a small smelter furnace where ore was roasted close by the cabin, but it too was destroyed by vandals. Allred has recovered pieces of slag from that smelter site, slag which assays one ounce of gold and seven ounces of silver, a high assay considering that slag was a waste product of the smelting process. Both Allred and the author request that if you visit the Hathenbruck cabin, or any other historical site in the Uinta Mountains, that you don't damage it further. That cabin is one of only a few physical proofs that Rhoads and Hathenbruck mined gold in those mountains. When you see that old cabin and see the site of their arrastra and smelter, you are very close to one of the lost Rhoads mines. Can it be very far away?

Gold Hill, mentioned earlier, dominates the headwaters of the Weber River, but it is more easily reached from the headlands of the Duchesne River. State Road 150 from Kamas Valley northeast to Wyoming passes through the Mirror Lake Basin and by Gold Hill. When American prospectors were first beginning to explore that high country at the headwaters of the Weber, two brothers now long forgotten were already mining on the north side of the divide, digging treasure from Spanish mines and caches at Gold Hill. They began their quest at the turn of the century, but while others sought Spanish treasure and the Rhoads Mine on the south slope of the Uintas, Heber C. Bell and Joseph B. Bell were finding their fortunes on the north slope. No doubt they uncovered Spanish gold long before they thought it necessary to protect their finds by legally filing and recording claims on their discoveries, since their earliest claims on record weren't recorded at the County Courthouse at Coalville until 1912. They continued prospecting, mining and cache hunting for at least another thirty years, the date of their last claim filed in 1942. Examination of those old records reveal that dozens of claims were filed on many of the old mines they discovered, at least five of them of Spanish origin.

Century-old pines grow among the ruins of what may have been a smelter, near where the Bell brothers dug for gold.

Like both Thomas and Caleb Rhoads, the Bell brothers were close-mouthed and secretive, so that little is now known of their finds. They spent entire summers in the mountains, but they filed claims only on discoveries they thought might be found by others, and even then they gave only the vaguest descriptions and directions to the recording clerk. Clues to the exact locations of their finds now taunt researchers and prospectors with such vague wording as "east and a little north of the Bare (sic) River," or "three miles from the old sulfur mine." Even at that early date it was thought that Spanish miners obtained sulfur from that old diggings to make gun-powder. About a mile from that sulfur mine, the Bells discovered several old stone cabins, so old that giant pines grew inside their walls. Those old structures can still be seen on a ridge leading down from Gold Hill, as ancient looking now as they were when the Bell brothers first saw them. Fortunately, some of their personal notes, sketches and records have been preserved by a son of Heber Bell, who while only a boy accompanied his father and uncle to some of the secret places where they dug for gold. The author was privileged to have access to those records.

It is unknown whether the Bell brothers had a map or waybill to help them locate the Spanish mines and caches they found, but apparently theirs was not a hit and miss sort of quest; their results were too good for that. Both were experienced prospectors and knew what type of formations were conducive to forming ore deposits. They studied and understood the complex

A curious smelting pot and retort, found by the Bell brothers. There is still a slab of silver in it! (Courtesy: Mr. Larry Bell)

geology of the north slope of the Uintas, and sought the igneous intrusions which makes ore bodies in that limestone country. Nearly one-hundred years later their skills and knowledge were confirmed at the sixteenth annual conference of the Intermountain Association of Geologists, whose report states in part: "A large variety of mineral resources is present along the north slope, including manganese, gold and copper." Their search took them along the north slope from Gold Hill and Hayden's Peak at the head of the Duchesne River eastward past 12,000-foot Kletting Peak to Deadman's Mountain, the Black's Fork country and to the 13,000-foot peak Indians call Tokewanna. It's still a wild and rugged place today, but it must have been a whole lot tougher place to prospect a century ago.

In the wind-blown and rock slide country between Hayden's Peak and Kletting Peak, over-looking McPheter's Lake Basin, the Bell brothers discovered a Spanish mine from which they recovered an old and valuable relic. The came upon a place where a tunnel had been driven into a vein of high-grade ore, and from the ancient artifacts they uncovered in that tunnel . . . wrought iron and bronze tools and leather ore sacks . . . it was evident that those earlier miners intended to return, but for whatever reason, now unknown, they never did. The most fascinating artifact found was an odd-shaped iron melting pot and retort, used by Spanish or Mexican miners to melt their gold and silver so that it could be cast into bars or ingots; what the Spanish called a "crisol." The author has a similar melting pot, which some call a "midden," which he found in the Pigeon Water country near Rock Creek long ago, but the find made by the Bell brothers may be unique. It is obviously very old and no doubt of European manufacture. It is approximately eight-inches in both height and width, round in shape and very heavy. What is yet an unexplained oddity is that the melting pot has two spigots or pouring spouts, something like a tea-kettle with two spouts. Several persons expert at identifying such antiques have surmised that one of those spouts may have been attached to a cooling coil to recover quicksilver, while molten metal was poured from the other.

That ancient melting pot has a tight fitting lid through which finely ground metal ore could be poured; and to verify its use, in the bottom of that pot there still remains a round slab of silver bullion, about one inch thick and five inches in diameter. There is also a trace of white residue inside, probably borax or some similar flux used in refining silver. It is only one of two such antique melting posts found in the Uinta Mountains, and as such it is a priceless relic of Spanish mining days. It is fortunate that when the Bell brothers carried heavy sacks of gold and silver ore down from those rugged peaks, they had the interest to save and preserve that old "crisol" or midden.

On a high ridge leading down from Kletting Peak, the Bells discovered a vertical shaft, ancient in appearance and too dangerous to descend into. Remember, that shaft was found at the turn of the century, so it could not have been dug by pioneer settlers, few of whom had ever entered that remote region. From lower on the mountain, they drilled and blasted a tunnel for 300-feet through hard rock to make a connection with that shaft. When they broke through into it, they discovered mine workings so old that what once had been tunnels large enough to walk in and work in had been so compressed with the weight of years that timbers were crushed and only crawl spaces big enough to squeeze through remained. They found plenty of rich ore there and began to clean out that old shaft, but unexpectedly both were called into the army during the summer of 1917. When they returned to the mountains after the armistice, they found that something, perhaps an earthquake, had so shaken the land that their hard-won tunnel had collapsed. Without funds to start anew, they moved to more promising prospects nearby and never tried to reopen that old shaft again. But it's still there on Kletting Peak if you want to give it a try!

Heber Bell's son recalls that his father and uncle found two old mines close to their cabin above the Stillwater Fork of the Bear River. At one of them they uncovered a squared post set in the ground, as if it had been a claim boundary marker. Cut into one side of that post were the initials "T R." Could that marker have been placed there by Thomas Rhoads? One of those mines located close by the Bell's cabin must have contained very rich ore, for Heber Bell's son still remembers that as a boy he saw four of the old-fashioned large size blue-colored Bromo Seltzer bottles and two fruit jars filled with nuggets and pieces of yellow gold! As nearly as he can recall, it was his impression that gold came from a cache found in one of those old mines on the mountain above Stillwater. After his father filled one of those bottles with gold, it was so heavy that young Bell couldn't lift it!

The north slope was a lonely place back in those turn of the century days, few making the long pack train trip from Kamas Valley to prospect there. But the Bell brothers did have one neighbor, an old recluse now almost as little known or remembered as they are. Old Lee Christmas, for whom Christmas Meadows along the Stillwater Fork was named, was also a north slope prospector. Christmas had several mines on Gold Hill, so rich that he could dig enough gold during the short summer season to allow him to spend his winters in more temperate climes. Christmas was a loner who shunned even his nearest neighbors, but still the Bells learned a few things about his hidden mines. He dug an open pit shaft on the Christmas Meadows, but they quickly learned that it was only a decoy to fool those who tried to find the source of his gold. The Bells discovered that his real mines were on Gold Hill.

Heber Bell said that Christmas never built a cabin or had a permanent camp, preferring to sleep in a different place every night. All of his one night camps were strategically placed so that he could keep a close watch on his back trail. Christmas kept an especially close watch on the Bell brothers to make sure they didn't follow him, but they did observe one habit he had which may someday lead someone to one of his mines. Christmas owned two burros, and whenever he hobbled them for the night, he would find a place back in the shadows of the forest where he could sit and watch them. While he kept an eye on his back trail and catnapped, he would bend small aspen or willow saplings and tie them in knots, so that over the years those saplings would grow into odd-shaped trees. Fishermen and hikers unaware of who Lee Christmas was, sometimes mention seeing those curiously knotted trees in the mountains. If someone were to follow his trail of a century ago and find where he made his lonely one night camps, how close might they be to one of his mines?

Heber and Joseph Bell were highly successful treasure hunters, locating old Spanish mines and enough hidden caches to keep them in the mountains for more than forty years, but they missed one of the richest finds ever made in those mountains, at a place they knew well but never recognized the value of. Between Gold Hill and the old ghost town railroad camp at Beartown, there are a number of old kilns where logs were burned to make charcoal; however, unknown to most people, one of those kilns was there long before the railroad was built. Today the ruins of at least thirty-two kilns can be found between Evanston and the upper Bear River. The Bells were aware that one of those kilns appeared to be much older than the others, but they never realized that it was a Spanish smelter and not a kiln. That fact was discovered some years later when another prospector dug up pieces of gold and silver ore from the grass covered mounds around its crumbling shell. He dug into those mounds and found they covered piles of slag, some of it worth thousands of dollars to the ton. He recovered enough of that slag to fill two railroad cars, which he shipped to a Salt Lake City refinery. Heber Bell said he saw the check paid for that slag; it was in six figures!

Pat Werner, a highly respected attorney at Rock Springs, Wyoming, is as savvy a prospector as you are likely to run across. He has spent a lot of time in the Bell brothers territory, much of it along Black's Fork. He discovered some very old Spanish diggings high on the West Fork of Black's Fork. He also learned that ranchers from the little Wyoming towns along the base of the Uinta's north slope occasionally found pieces of high-grade ore along the cattle trails there. Werner

Summers are short at the Bell brothers old mine on Kletting Peak.

An old stone ruins near the Bell brothers mine on Gold Hill. Was it one of their smelters?

informed the author that during the western land surveys back before the turn of the century, there was a meteorological station located high atop a peak above Black's Fork. A Forest Service employee later ran a single-wire telephone line from Robinson, Wyoming, to that station. That line ran from tree to tree, but where it crossed a barren ridge where there were no pines or aspens to hang that wire from, he had to dig several post holes. When the surface rock was broken, a vein of quartz was exposed, laced with wire gold; so heavy with gold that it was more gold than rock. Being a civil service employee, he knew that if he made his find known, that gold would be claimed by the government, so he erected the poles and piled pieces of heavy rock around them for support, and went on with his work.

That lineman carried several chunks of that wire gold ore in his saddle bags for years, during which time quite a few trusted friends saw and handled them. He waited for the time when he could retire and stake claim to his hidden find, but it wasn't to be, for before that time came he was stricken with emphysema, so that he could no longer go into those high mountains. He carefully described that telephone wire trail to his son, thinking that he would have little problem finding that old wire and follow it to where it crossed the ridge where he uncovered the gold. But that line hadn't been used in many years and had fallen from the trees it had been attached to, while new growth trees and brush covered the way so that the ridge with the poles on it was no longer bare. None of the country was as the old man remembered it.

A few old maps still show that meteorological station, but it is no longer standing. It was abandoned many years ago and was later burned in a brush fire. Livestock and wildlife have become entangled in that lone strand of wire and have pulled sections of it far from where its course once lay. But perhaps one day a prospector or a deer hunter might come across a strand of that old telephone line hanging from a large pine or aspen and follow its course to where a couple of rotted poles still stand atop a ridge, supported by piles of rock. Will he just continue on his way or will he pause to wonder why those poles were placed there? If he takes a few minutes to take a little closer look at those piles of rock, it could make a lot of difference to him!

For a long time now people have asked about what has come to be called the Lost Cabin Mine. Notes left by the Bell brothers should resolve those questions once and for all. Back in 1920, a man named Caleb Landreth organized a search for a mine supposed to be located somewhere near the head of the Bear River. Those accompanying him on that quest gathered together near Evanston, Wyoming, but their search was to be made on the north slope of the Uinta Mountains in Utah. Landreth's party consisted of a motley crew of nearly a dozen men, strangely, none of them were prospectors or miners. Only one man among them could recognize gold if he saw it. It was an undertaking destined for failure from the start, but it was one which would make the Gods of the Killer Mountains smile with glee.

Landreth alone knew the general location of the mine, the finding of which depended largely on locating an old log cabin. In his party were several gun-fighters from Castle Valley, including Matt Warner, one of Butch Cassidy's long riders. Harold Mossman was another of that breed, as were three would-be bad-men known as "Old Man" Warren and his two half-simple sons, named Thursday and Friday Warren. Edward Hartzell, the man who married Sidsie Rhoads after the death of Caleb Rhoads, was also taken along, supposedly because he could recognize the cabin and the area where it might be located. Ernest Roberts, a stove-up one-time horse rustler was the party's camp cook. The only "straight" man in the group was Amasa Davidson, a school teacher from the Evanston area who could recognize gold ore and assay it if necessary.

That curious crew followed the Bear River upstream to its confluence with the Hayden and Stillwater forks. Unknown to them, they made camp in sight of the Bell brothers cabin, which was high above the Stillwater on the north side of Kletting Peak. From that camp they followed an old trail which led to the ruins of the log cabin they were seeking. The location of that cabin, then and now, is the vital clue to locating the lost mine. Landreth told his followers that the mine would be found somewhere close to a trail which could be seen leading higher into the mountains beyond that cabin site. From that point versions of what happened vary, depending upon who is telling the tale. According to most accounts, Landreth led one group in search of the mine, the three Warrens went another way while Hartzell went off on his own. Matt Warner stayed in camp to keep an eye on Davidson, the assayer, and Roberts, the camp cook. It wasn't expected to take long, for the mine was supposed to be close to that old cabin.

While the searchers were away from camp, Warner had a change of heart about his cohorts and warned Davidson and Roberts that Landreth would probably kill both of them after the mine was found and the gold assayed, so there would be fewer men to divide the gold among. He took an old six-shooter from his saddle bag and gave it to Davidson, warning him and Roberts to be ready for anything. Late in the day the search parties returned, each denying they had found the mine, and each accusing the others of finding it. One of the Warren boys produced a chunk of rock which he said was nothing but pyrites, and tossed it to Davidson. Every man there could tell from the look on Davidson's face that the rock was gold, and angry accusations quickly erupted. A vicious argument exploded and in a twinkling six-guns were drawn and bullets were flying.

Both of the Warren boys were killed in the first exchange, with Roberts and one or two of the others hit and seriously wounded. Davidson fled into the forest and escaped the murderous fire. Nearly a week later he walked into Evanston, being the first one to bring news of the gun fight. A few days later Roberts staggered into town, all shot to pieces, where he later died of his wounds. It was later learned that Warner returned to Castle Valley where he became City Marshal at Price. No one ever learned if there had been any other survivors of the gun fight at the Lost Cabin. For years afterwards, Davidson kept that chunk of gold rock the Warren boys had found, but they alone knew where it came from, and both were among the first killed at Lost Cabin.

Over the years Davidson refused to talk about the shoot-out, but finally in 1951 he tried to return to the old cabin where the fight occurred. After weeks of search he discovered what he thought might be its ruins. But if the story as related is correct, then Davidson was mistaken, for what he identified as the shoot-out cabin was on the shore of Steiner Lake, east of the Mirror Lake Basin and at the head of the Weber River, not on the north slope on the Bear River drainage. To further confuse matters, in 1989 Lee Davidson, a grandson of Amasa Davidson, attempted to locate the old cabin from notes and a crude map left by his grandfather. His account appeared in the Deseret News. On his grandfather's map, an old sulfur mine was a key landmark to locating the lost cabin. Very likely, it is the same sulfur mine the Bell brothers located some

Deadman Mountain. Well named, for a dozen men have died there while searching for the Lost Cabin Mine.

of their mining claims in reference to. That sulfur mine is on Gold Hill, a long way from Steiner Lake and on the opposite side of the mountains. Lee Davidson concluded that the lost mine had to be somewhere north of Steiner Lake and south of Gold Hill. That's a pretty big piece of real estate!

During a recent interview with Howard Davidson, a son of Amasa Davidson, the author learned that Amasa Davidson would never talk much about the gun fight; according to his son because he believed he had killed one or more of the outlaws with the gun Matt Warner had given to him. After the battle at Lost Cabin and his escape, Davidson kept that weapon, not knowing the identity of the man who had given it to him. Then about 1937, he saw a photo of Warner in a newspaper and told his son, "That is the man who saved my life!"

Shortly afterwards, Davidson made the trip to Price City where he returned that revolver to Matt Warner. Warner died less than a year after their meeting. Davidson told his son that the reason he waited so long before trying to return to Lost Cabin was because he knew he was always being watched, even though many years had passed since that gun fight. He said that on one occasion while he was working in a field near Evanston, someone shot at him, so he thought he had good cause to fear for his life if he ever attempted to find the lost mine. They're all dead now, Landreth, Matt Warner, the Warrens and Davidson, so it should be safe to go into those mountains now.

Heber Bell told his son that he and his brother were prospecting in the mountains near the head of Bear River when the gunfight at Lost Cabin took place. They already knew Davidson and were surprised to see him with such a bunch of tough-nuts. They also recognized one of the outlaws, a man they called "Blackie," most likely Harold Mosslander. Several days after the battle, they ran into "Blackie" and he told them he had killed at least two men, and maybe more. He left the country and they never saw him again. The Bell brothers said that the gunfight was on Deadman's Mountain, at the head of the Bear River, not anywhere near Steiner Lake.

Verifying Bell's version of the shoot-out is an account given to the author many years ago by Jan Bodily of Woods Cross. Bodily said that a rancher named McKay employed a sheepherder who discovered an old mine rich with gold ore at the headwaters of the Bear River. Not long after his find, someone shot and killed the herder, reportedly for the gold he had in his possession. His killer was unaware that the herder often stayed at a line camp cabin owned by McKay, and failed to find a small sack of gold ore he kept under a bunk at that cabin. McKay later discovered that sack, which contained pieces of white quartz full of yellow gold. McKay became so excited at the prospect of finding the mine that he moved to the mountains, to live in the same cabin where the herder had stayed.

Although he was a well-to-do rancher, McKay became something of a recluse, spending all of his time searching for the old mine and its gold. Then, according to Bodily, McKay too was killed, strangled by someone who also wanted that gold. Angered at his father's murder, and perhaps to keep others from risking their lives in search of the Lost Cabin gold, McKay's son burned his father's cabin. The mountains where McKay and the herder were killed came to be known as Deadman's Mountain. It is the same mountain where Heber Bell said the gunfight took place. If you decide to search for the Lost Cabin Mine, your search begins at Deadman's Mountain. One other thing; keep a close watch over your shoulder. A dozen or more men have died there already!

CHAPTER 8

TWISTED TRAILS
The Mystique of Farm Creek

Two roads connect the tiny Duchesne River hamlet of Hanna with Rock Creek. One of them is rough and rocky, the other a little worse. That first road heads northward following Blind Stream, while the second leads eastward along Farm Creek. One of the clues Caleb Rhoads let slip, perhaps purposely, about one of the places where he found gold was that the trail to it passed between two knolls, one rock and brush covered, the other bare of growth. That trail crossed a small creek and turned up a canyon which led to the left just after those knolls were passed. No doubt similar knolls might be found elsewhere in the mountains, but there is just such a place near the head of Blind Stream.

Several hikers have come upon an ancient looking mine shaft in the area of those two knolls. That shaft is said to be located in Lake Basin overlooking the South Fork of Rock Creek. It is deep, but no one knows just how deep, for many of the foot-holds which had been cut into its vertical walls have sloughed off over the years, so that descending into it now could prove to be a perilous pastime. One brave, or perhaps foolish young prospector, recently told me that he had made his way down those crumbling stone steps while clinging to a rope he had tied to a tree growing back from the edge of that shaft. He descended nearly one-hundred feet before he ran out of foot-holds, from which level a dislodged stone disappeared into the black depths below, echoing as it bounced from side to side into that bottomless pit.

Wayne Murdock of Heber City is just about the last of the old-time Murdock family of pioneer ranchers. Some years ago he ran a herd of sheep in the Lake Basin country, near the head of Blind Stream. Mace Foreman, the same man who prospected the Kamas area with "Beaver Creek Charley," herded those sheep for Murdock. Murdock recalled that Foreman was a good herder, except that he spent more time prospecting than he did watching his flock. One summer Foreman discovered an old vertical shaft which had a rotted pole ladder sticking out of it. Murdock told me that shaft wasn't far from "the big hole," out "on the ridges." It might be the same shaft which Foster Rhoads of Hanna used to talk about. It too was near the head of Blind Stream in Lake Basin. Rhoads said that when he first saw that shaft, there were still some stone steps cut into its wall rock.

Foreman became so excited over his find that he quit his job with the sheep and spent the rest of that summer exploring that old digging and prospecting the area around it. He sometimes brought pieces of dark-colored rock to Murdock's camp, which they would break up with a camp axe. Foreman would pan the fines in an old cast iron Groswald frying pan. Louchious Miles, one of the real old-time prospectors in that country, once stopped by Murdock's sheep camp where he examined Foreman's ore samples, and it was obvious that he was quite impressed with them. When summer ended Foreman went back to herding sheep, since prospecting wasn't putting any bread and meat on the table for his family. After more than a half century, Wayne Murdock doesn't recall exactly where Foreman found that ore, other than that it was close to "the big hole," out "on the ridges." With the price of gold pushing four-hundred dollars an ounce, that old shaft might be worth looking at a little closer. You can find it at the edge of a dry lake bed, right on the 10,600 foot topo-map contour line. It's not hard to find. I did it, you can too!

There are a lot of old diggings along Blind Stream, some so old they are obviously of Spanish origin, but

Prospector Ed Grose at the old Foreman Mine at Lake Basin.

there are also some dug by pioneer prospectors. One old codger who has become infamous around Hanna because of his bluff and bluster has staked several hundred claims in that area, and he has been known to take pot-shots at anyone who trespasses on his claims. Even Forest Rangers, deputy sheriffs and Indian police have looked down the barrel of his six-shooter. Some say he once shot a man at Price and more recently shot another man in the leg. Others say he is only senile and shouldn't be taken seriously, but the man he shot still walks with a limp, just the same as if that old codger really meant to hurt him!

An abandoned logging road leads northward from the Blind Stream Road just before it crosses the summit ridge-line to drop down a series of steep switch-backs into Rock Creek Canyon. Trees with treasure symbols carved into their bark mark an old foot trail which continues after that road ends. Where that trail passes below a line of high cliffs, you can see the portal of a square-cut tunnel on one of those cliffs. It is difficult to climb up to, and one can't help but wonder who drove that tunnel into those cliffs so long ago. At the base of the cliffs there is a caved mine shaft, said to be more than one-hundred feet deep. A few years back an old mortar and pestle was found near that shaft. The man who found them isn't telling exactly where they came from, for there were still several pieces of rich gold in that mortar, and he wants to do a little more exploring, for he thinks that gold probably came from some place close by.

Most modern maps show the Farm Creek Road leading eastward from Hanna to cross Farm Creek Pass on a high divide between the Duchesne River and Rock Creek, but that isn't the Farm Creek Trail of pioneer days. The original pass shown on early maps or mentioned in pioneer journals or miner's waybills is actually located several miles to the north of the present day pass. Those few miles can make a lot of difference when you're trying to fit the pieces of a century-old puzzle together. A lot of people have wasted a lot of time searching for a mine and cache which some believe is along the present Farm Creek Road, when it is actually several miles further north along the old Farm Creek Trail. Another thing which many people don't know is that not all of that country is owned by the Red Rock Corporation, a group of half-blood Utes, for most of the pass area is the property of the mixed-blood Murdock family Utes.

Back in the days of Thomas Rhoads and Chief Wakara, the Farm Creek Trail went east from the present site of Hanna to the Red Ledges, and from there followed the Main Fork of Farm Creek, not the Right Fork as the present road does. It went northward through the Red Narrows, called the Red Gate in some miner's waybills, or the Red Spot on some of their maps. The trail continued up canyon to Round Grove, where

It's been a long time since anyone dug for gold at Mace Foreman's forgotten mine.

A prospector's mortar & pestle, found on the treasure trail along Blind Stream. There was still gold ore in it!

it turned to the right into Rough Canyon, known to old-timers as Sink Canyon. That trail crossed the ridge-line divide on the south flank of Farm Creek Peak, and passed through 10,000-foot high McAfee Basin. From there it turned down Dick Hollow to the Hathenbruck Crossing on Rock Creek, which crossing was located just down stream from Mount Albert. Old Dick Wanrodes, the Ute Chief for whom Dick Hollow was named, had a cabin at the mouth of that hollow, just below a high white rock outcrop which can be seen against the red sandstone cliffs from miles away. Caleb Rhoads was sometimes seen at the Hathenbruck Crossing below the Wanrodes cabin. If you hope to use old Spanish maps or follow the waybills left by those miners of yesteryear, you will have to think like they did, and use the landmarks they used, that is if you expect to follow the trails they followed.

It's been a long time now since anyone has used that old trail which winds down into Dick Hollow, and it's been even longer since strangers have been welcome there. Even livestock men who used to run cattle in those mountains weren't welcome in Dick Hollow. A few of them knew there was an old mine located high in that hollow, above the white ledges and just below Farm Creek Peak, but few of them ever went near it, at least not when they might be seen. Most of those old range riders called it the Bullet Mine, because back in the old days, Indians used to dig lead there to cast into bullets. It wasn't until white men began homesteading nearby that those Indians learned their bullets were made of silver!

"Old Dan" Tucker used to explore around that country when Wallace Jones ran sheep there, and he once came across that old mine. Even then it was covered with pine logs to keep its location a secret. Tucker said it was a shaft sunk on an incline, and that there was no waste dump to give its location away, the rock taken from it being scattered across the mountain side. LaVar Thompson also saw that old shaft many years ago, when there were still dried up rawhide ore sacks and pieces of strange looking leather boots strewn around. Thompson said that he recalled seeing a lot of trees with Spanish mining signs on them. Two of those trees had holes drilled into their sides, so that when sticks were placed in those holes, they pointed across the canyon and crossed right where that mine shaft was located. Many of those trees were burned about forty years ago, according to Thompson purposely burned to destroy those Spanish symbols.

One of the old range riders said that every time he got too close to that old mine in Dick Hollow, an Indian would ride out to meet him and turn him in another direction. The first few times that happened, he thought it only a coincidence, but then he noticed that he was always guided away from that mine. Mark and Bill Bleazard, two brothers who grew up along Rock Creek, recalled that every time they rode too far up Dick Hollow, old Walt Daniels, a Ute Indian who had a cabin on Daniels Flat just downstream from Dick Hollow and the Farm Creek Indian Ranger Station, would suddenly appear to accompany them. They noticed that he always managed to steer them in some other direction. Just a short time ago an incident occurred which goes a long way in explaining why strangers aren't welcome in Dick Hollow.

During October 1987, reports were heard that a young Indian boy had been shot and killed near the mouth of Dick Hollow, in some way as the result of a search for buried treasure. Inquiries failed to disclose any further information until December, when a

The Red Narrows or Red Spot; gateway to the lost treasures of Farm Creek Canyon.

A Spanish trail marker sign, on the forbidden trail in Dick Hollow.

prospector learned that the youth had been bragging at a local tavern that if anyone wanted to see some real Spanish gold, he could show them where to find it. According to that report, he had been stabbed, not shot. Several months later, a Ute Indian confided to one of his white friends that the youth who had been killed had learned the location of a cache from an older Indian. Local law officers who have no jurisdiction on Indian lands would only say that the investigation was an Indian matter.

Then late in 1988, another story was heard that Ute Police had apprehended two male Indian juveniles near Dick Hollow on Rock Creek with four bars of Spanish gold in their possession. Those bars allegedly came from a cache hidden in an old mine which the youths had heard older Indians talking about, a cache which was said to be sacred and which no Indian was allowed to go to. Although there is a similarity between the two accounts, it is almost certain they are separate incidents. Those stories became more than just rumor when several reliable persons were shown a photograph of two young Indian men standing at the rear of a pickup truck, under arrest by Indian Police and wearing handcuffs. On the open tailgate of the truck were four bars of gold! Those who saw the photographs were cautioned not to reveal the identity of the two youths, since both were juveniles and both were closely related to Ute tribal leaders.

One official who was contacted acknowledged that he was aware of the incident, but stated that since it was an Indian affair, it was being investigated without Duchesne County assistance. Soon after that, a tribal leader stopped two white prospectors who were in the Farm Creek area and warned them not to go into Dick Hollow. There is something else happening at Dick Hollow which even the oldest nearby residents say is unusual. Recently an old Ute from nearby Roosevelt has been seen digging at that old mine there, no doubt with the tribe's knowledge. It is unusual, for no one can recall when a Ute was allowed to dig around one of those old Spanish mines. There has always been strict religious reservations against doing so. The question many local people are asking is he digging for treasure, or is he moving treasure? Consider the following, and then you decide.

A peculiar coincidence recently occurred near Dick Hollow. Two men camping and prospecting in that area both saw the same thing, but from entirely different vantage points; yet neither man was aware of the other or even suspected there was anyone else near him. A California prospector had been hiking in the Rock Creek area for several days. To escape from the millions of mosquitoes along the river bottom, he climbed atop a lone hill on Daniels Flat to make his camp, where he hoped evening breezes would keep the insects away. About midnight he was awakened by a helicopter passing low overhead. That copter had no running lights, but he could see it clearly against the night sky, and he watched it drop down into Dick Hollow. Just before daylight he saw that copter again, as it left Dick Hollow and again flew low over his unseen camp. There wasn't any mistake about what he saw, for about a week later he saw that same copter again, flying into Dick Hollow late at night and leaving again just before dawn.

He had no idea there was anyone else in Rock Creek Canyon, but unknown to him, a man from Alaska who was vacationing in that area had ridden his horse into McAfee Basin, overlooking Dick Hollow, where he made camp. He said that late at night, about midnight, he heard a helicopter coming into Rock Creek Canyon from the east, and he heard it drop into Dick Hollow, where it landed. He thought it strange and probably unsafe for a copter to be flying at night with no running lights, but he gave it no further thought until just before daybreak when that copter flew out of Dick Hollow. About a week later he saw that same copter return to Dick Hollow. Once again it was late at night, but the moon was bright and he walked to the edge of the basin where he look down into the canyon below. He recalled thinking that the pilot must be skillful to land there at night. That copter left the canyon at the first light of day.

The intriguing thing about those helicopter flights is that both of the men related almost exactly the same story. Both versions were related shortly after the flights occurred, yet neither man was aware there was anyone

Two Indian boys uncovered four gold bars at the old Spanish diggings at Dick Hollow.

else in that area. Investigation of those flights revealed something else of interest. A family who lives at nearby Talmadge, just east of Rock Creek, also told of hearing those midnight flights over their home. Further, three men in a fixed-wing aircraft who had flown into that area to take infra-red photos reported seeing a black colored helicopter flying out of Rock Creek early in the morning. It is interesting that very few people in the Uinta Basin own helicopters, and one of them is a Ute Indian. Some have wondered where he gets enough money to live in the style he does, or how he can afford a helicopter. Not long ago he flew several people on a charter flight along Rock Creek Canyon. As they passed over Dick Hollow, he pointed down and said, "That's my bank! Right down there!"

There might be another explanation of where those two Indian boys got the four gold bars they had. Years ago there was a story told by some of the Old Ones, among them Mountain Sheep and Bridger Jim. They said that in the long ago of their grandfather's time, Spanish miners brought carettas to the old Catholic Mission on Rock Creek, where they were loaded with gold bars. Those bars were packed in boxes, some say in metal boxes. After leaving the mission, the Spaniards were ambushed and killed by waiting Utes. That attack occurred near a small mountain lake not far from Farm Creek Pass. To rid themselves of the cursed gold, those heavy boxes of gold bars were thrown into the lake. Swords and muskets carried by the miners and their soldier-guards were piled at the edge of a rock slide near the lake, after which rocks were slid down over the weapons. One of the Old Ones said that a musket barrel still protrudes from the rocks to mark that cache.

All of the personal property of the miners and soldiers; gold and silver, rings and coins were gathered into a large pile. A deer was killed and all of the treasure was sewn up in its hide. That buckskin wrapped treasure was buried in a shallow hole dug at the foot of a large pine, and was then covered with rocks, soil and pine bark. That tree was said to be close to an old pole corral not far from that lake and rock slide. An Indian who lives nearby once said that over the years erosion would expose that cache, so that Ike Shotnick, one of the Utes designated to watch over that area, would have to recover it. Several early settlers used to tell how one old Indian obtained Spanish coins and Catholic rosaries and crosses from a place where he dug down into loose bark at the foot of a large pine tree. Even today one Ute Indian well known in public affairs wears a large Catholic Cross, one not manufactured by modern means, but hammered out of raw gold in the manner Spanish artisans fashioned jewelry in the long ago. Those who have seen it say that it looks a lot like an old mission cross.

Not long before his death, the late Gale Rhoades with a partner from nearby Roosevelt spent some time building a raft and using a long rod to probe the bottom of a small lake near Farm Creek Pass. They discovered that the lake bottom had about eight feet of black mud and sediment on it. Those who know that area well say they probed the wrong lake; that the lake in which the gold bars were thrown is only a shallow mountain pond, one which has a long, narrow neck at one end. The gold was thrown into that narrow neck of water. If you go there to look for those bars, remember, most of those shallow water ponds are only snow water run off, and that after six years of drought that pond might be only a mud hole now, or maybe just a marshy meadow. One other thing, don't forget that it is being closely watched by Utes who are there to guard it.

There may be more than just one old mine in Dick Hollow. Wayne Nelson and Earl Garrison were among the first prospectors to search the Rock Creek area, and they recalled seeing a large Spanish cannon in what they called "Deep Canyon," a north-south lying gulch

Mysterious Treasure Hill at Rock Creek. Gold bars have been found there!

located near the head of Dick Hollow. Both agreed that it was larger than most cannons which have been reported or recovered in the Rock Creek area, having a bore of about three inches. If it ever had wheels or a carriage, no sign of them remained when Nelson and Garrison saw it. One time when they were approaching the upper part of Dick Hollow, they were stopped by an Indian Ranger who demanded to know what they were doing there. Garrison, who had married a Ute woman and therefore had trespass privileges, knew that ranger well and countered by asking what he was doing there. Since it appeared to be what Garrison called "a Mexican stand-off," that ranger, who knew Garrison wasn't the kind of man he wanted to have trouble with, walked away and left them to their exploring. They looked that cannon over carefully, but decided the only way it could be moved would be with a helicopter. Garrison later told the author that some think the old mine in Dick Hollow is the Brigham Young Mine, but it isn't. Still, Garrison said that one of his Indian relatives told him, "Plenty gold there!"

In Footprints In The Wilderness, the authors described a map which they called the Mine In Timber Map, but declined to publish a copy of it. That map was described as showing a stream, a natural stone arch, the "Red Gates," several sink holes or small lakes and a mine with the notation, "Mine in timber, two miles." That map has a most fascinating history. Louchious Miles, the already mentioned early settler and prospector in the Farm Creek area, was married to Nettie, a Ute Indian woman. For reasons now obscure, Nettie was acquainted with Caleb Rhoads, and when he died near Price City, she traveled there to be with his wife, Sidsie. From Sidsie she obtained a copy of one of Rhoads' maps, which she traced from the original. It was the Mine In Timber Map.

Confirmation that Rhoads had such a map and brought gold from the mine which it led to is found in a statement recorded by Glen Lambert, the Kamas Valley pioneer, in part as follows:

> Old Tom Rhoads had a son named Kale. He grew up to be a very big man; he had the biggest feet I've ever seen. My dad told me that one time he went over into the Rock Creek country to trap beaver, and that as he was returning to Kamas he came upon a place where Kale had made camp, where the North and West forks of the Duchesne River come together. Dad knew it was Kale's camp because of the huge boot tracks he found there. Before Kale left Kamas to settle in Castle Valley, he married Malinda Powell, a daughter of James Powell. Later Powell visited us and he told dad that Kale had a gold mine he went to every summer. When Rhoads died, Powell went to his cabin to help prepare the body for burial. There was an old metal trunk in that cabin which had some of Kale's things in it, and inside he found a buckskin map and about ten or twelve pieces of gold, each about the size of a hen's egg. They were pure gold, all of them!

Bill Bleazard. He and Hap Stevenson followed the old trail shown on the Tunnel In Timber map.

After the Mine In Timber Map was published in the first edition of Faded Footprints, the author was contacted by the daughter of Louchious and Nettie Miles, a woman now eighty-nine years of age. He was surprised to learn that she still had the map traced by her mother from the original Rhoads map. He was even more surprised when she gave him that original tracing, since she was too old to use it anymore. Map "L" in the Map Index of this revised edition of Faded Footprints is exactly as it was traced by Nettie Miles. It has not been seen by anyone for more than fifty years. It is as good as gold for the treasure hunter who examines it in close detail!

That buckskin map copied by Nettie Miles corresponds exactly with landmarks along the old Farm Creek Trail. It shows a natural stone arch, the Red Narrows, Round Grove and several small ponds near the head of Blind Stream. Instead of "Mine in Timber, that original is titled "Tunnel in Timber," with the notations "gold panning," "glenina (galena) mountain and "gold lead." Several who knew Caleb Rhoads well said he brought two kinds of gold ore from one of his mines,

A weekend treasure hunter uncovered this ancient Spanish smelter high in Farm Creek Canyon.

which might well be explained by the fact that on the map there is a tunnel where hard rock ore was dug as well as a placer gold diggings. Louchious Miles was said to have carried a tracing of that map for forty years, always trying to find that old mine. After his death, Nettie sent that tracing to Harold "Hap" Stevenson, a lifelong friend who lived at Mountain Home. When Stevenson received that map, he became so excited that he telephoned his partner, Bill Bleazard, and shouted, "I've finally got it! I've got the map to the Lost Rhoads Mine!"

Hap Stevenson was one of several brothers living in the Duchesne area, all of them interested in lost mines and Spanish treasure. Stevenson moved to Mountain Home with his parents in 1908, when that pioneer settlement was still called Lake City. Bill Bleazard also grew up on a homestead near Mountain Home, on Towanta Flat, and spent much of his life working livestock in the hills along Rock Creek. He knows that area as well as anyone, so after examining Stevenson's newly acquired map, it didn't take him long to relate it to the Farm Creek country. He and Stevenson prospected that area whenever they had the time to do so. It took them only a few trips to orient the places shown on the map with the actual landmark features of the canyon itself. They located the natural arch, the Red Narrows, Round Grove and the small ponds. They knew they were getting close to the tunnel in timber.

Hap Stevenson was superstitious, a trait he may have acquired from living among Ute Indians for most of his life. But for whatever reason, he was afraid of being caught in the mountains after dark, so he and Bleazard always made it a point to return to their camp or vehicle before nightfall. On their last trip into Farm Creek Canyon during October, 1962, they climbed high into Sink Canyon, following the old Indian and Spanish trail. Their climb took them longer than usual so that they were coming back down canyon late in the afternoon. In order to cover a larger search area, they were walking through the timber several hundred feet apart from each other. Knowing Stevenson's fear of the dark, Bleazard called down the mountain to him, saying they better hurry if they were going to get to their truck before nightfall.

Stevenson answered back, saying something so puzzling that Bleazard has wondered about it ever since. Stevenson called: "If we get too close to that gold, the spirits will get us!" Those were the last words he ever spoke. A few minutes later Bleazard called to him again, but there was no reply. When other calls went unanswered, Bleazard hurried down the mountain to where he found Stevenson lying dead along the trail. A later autopsy indicated that he died of a heart attack, but his niece said she believed her uncle became so excited at being so close to the Rhoads gold that his heart simply couldn't stand the strain. Did Stevenson's death add still another score for the Killer Mountains?

One of the reasons the daughter of Nettie Miles wanted the author to have her mother's map was so Bill Bleazard might share it. When that original tracing was shown to Bleazard, he immediately recognized it as the same map which had been sent to Hap Stevenson, but the original was somewhat different from Stevenson's traced copy, but the reason for that difference was quickly explained by Bleazard. He said that every time he and Stevenson used that tracing during their prospecting trips, Stevenson would write in things or make changes which he thought made the map correspond more closely to the actual topography, so that in time his tracing was no longer exactly as he had received it. There is a story often heard that Hap Stevenson's tracing of the Tunnel in Timber Map was buried with him, but that is not true. That tracing is still in his family's possession, but it was so altered by Stevenson that it is no longer as accurate or correct as it was when he received it.

Not long after Nettie Miles made her original copy from the Rhoads buckskin map, an altered version was somehow obtained by Robert Snyder. Anyone familiar with that country would have quickly noticed that some landmarks had been purposely changed. Snyder wasted a great deal of time searching for the Tunnel in Timber along Blind Stream instead of in Farm Creek Canyon. After he despaired of ever locating the lost tunnel, his son and grandson decided to make a search for the mine, but they correctly decided to make their quest along Farm Creek. With two saddle horses and a pack

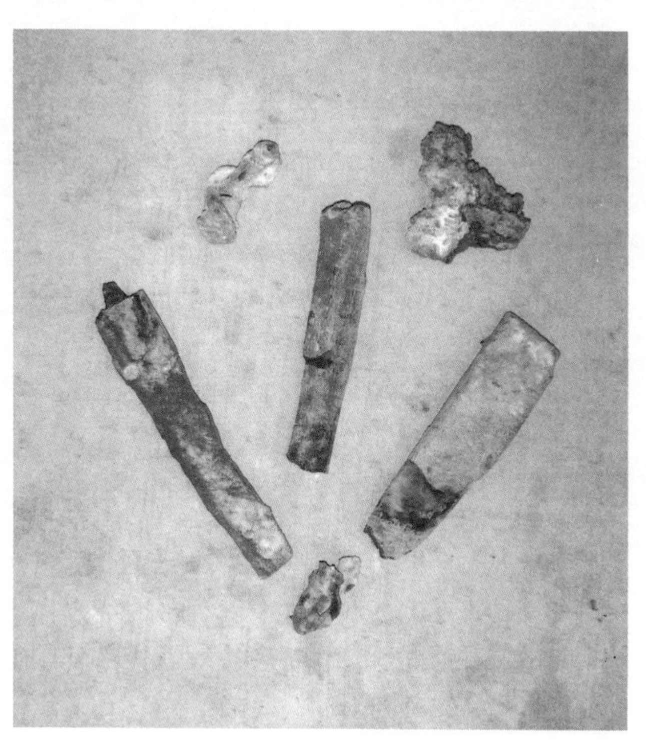

Heavy bars of slag, encrusted with silver, recovered from the Farm Creek smelter.

animal, they went into the mountains above Hanna. Weeks passed and the first snow of winter began falling, but they never returned nor were they ever heard from again. Several searches were made the following spring and summer, but no trace of the missing men was found. Then several years later, the skeletons of their horses were found in a dense pine thicket, where they had been shot. To this day no trace of either man has been found. The Killer Mountains keep their secrets well!

Hap Stevenson's death may have been due to natural causes, and we may never know what happened to the Snyders, but if you don't think there are still murders in those mountains, then think again! A funeral director from a local mortuary talked about several unexplained deaths in those mountains. One incident in the Farm Creek area north of Hanna clearly demonstrates that there is still danger in the Killer Mountains. A young male prospector had been shot in the back of the head at close range. Even though no weapon was found at the scene, the coroner's office ruled the death to be a suicide. But when the body was prepared for burial, the mortician discovered that there were two bullet holes in the skull, and when those bullets were recovered, they were of different calibers. They had been fired from two different weapons, hardly an indication of suicide, nevertheless, the coroner's verdict remained unchanged. If you plan on prospecting in the Farm Creek country, it might be a good idea to keep a close watch on your back-trail!

Not very far up Sink Canyon from where Hap Stevenson died, there is a vein of gray-colored rock which assays high in gold. It matches closely the description of a ledge which Caleb Rhoads once told Abraham Powell he dug gold from. He said that vein crossed a small meadow and passed under some ledge rock at the foot of a ridge. He told Powell that all of that vein carried gold values, but that it was richest right where it dipped under that ledge. Recently a deer hunter came upon the letter "C" cut into that ledge. He thinks it was put there by Caleb Rhoads. It's a hard place to get to now, for much of that land is owned by the Red Rock Corporation, and is closely watched over by the half-blood Utes who run livestock there.

In 1954, the Ute Termination Act separated half-blood and mixed-blood Utes from the full-bloods. That act resulted in the half-bloods being given their choice of a tract of reservation land of their own, separate and apart from the Ute Nation. After long and thorough deliberation, the half-bloods chose the Farm Creek country between Hanna and Rock Creek, and west to the North Fork of the Duchesne. That long, narrow strip of land includes 14,000 acres. Because it appeared that other areas might have been better farm land, livestock range or otherwise more productive, some have wondered why those half-bloods chose Farm Creek. One of the half-blood leaders was asked that very question, it being pointed out that most of Farm Creek couldn't be cultivated, nor was it the best grazing land or endowed with good stands of timber to be harvested. He replied, "There are things about Farm Creek which you don't know!"

The Tunnel in Timber Map copied by Nettie Miles shows a gold placer along Farm Creek, as well as a "gold lead," galena ore and the tunnel itself. Up stream from the Red Narrows and Round Grove there is a fence which separates Red Rock land from Forest Service land, but don't try to figure your location by where that fence crosses the mountains, for just about everyone in that area knows that fence was placed far enough up country to ensure that the Tunnel in Timber and those gold placers appear to be on Indian land, even though many believe they are actually on forest land. That there is placer gold in Farm Creek Canyon is certain, but it may not be in the creek, but in some centuries-old river bed, now high and dry above the present stream.

There was a story told around Hanna a few years back that Lot Powell and Rob Snyder hiked into Farm Creek Canyon just below the Red Narrows. At a place on the hillside where they stopped to rest, Snyder began scraping at the loose soil with a walking stick he was carrying. Right on the surface, barely covered with red sand, he dug loose a gold nugget as large as a robin's egg. He and Powell lost no time digging in all of the top soil and sand there, but nothing more was found. Powell later said that he thought that lone nugget probably washed down the slope from the gold placers marked on the Tunnel in Timber Map.

Early in September 1992, a long-time acquaintance was spending part of his vacation camping with a sheepherder friend high on Farm Creek. He had his metal detector with him, mostly just to keep in practice with it. Close to where the sheep wagon was parked he began looking for rifle cartridges or other "junk," but suddenly his detector sounded off, indicating something much larger than a brass shell case. Digging into what appeared to be undisturbed earth, he soon exposed some closely-fitted stones which appeared to be the rounded top of a dome shaped lime kiln. After considerable digging, he uncovered an ancient Spanish ore smelter, and when he tried his detector again, he quickly located a pile of metal-like bars. Most were about three inches in width, an inch or two thick and from twelve to twenty inches in length. Their exact identity is still somewhat of a mystery, but there is no doubt they are bars of smelter slag, what the Spanish called "escoria" and the Mexicans "grasero." Their centers are metallic while the outer surface is a dull-gray one molten hard dross or slag material. Metal fragments throughout the bars are obviously silver. They look very much as if a sword or similar metal instrument was used to skim dross from a pot of molten metal; causing a heavy coating of that dross to adhere to the sword blade.

Only a few weeks after that discovery, a half-blood Ute informed the author that he had discovered where someone had been digging in an old Spanish smelter at Farm Creek. When asked about that smelter site, the Ute volunteered that it was a very old place where silver bars were once made; and added that if one looked closely he would see where a ditch or canal once brought water to that site, although it is now nearly covered with brush and aspen trees. He also said that not far from that smelter there is a mine, covered over so long ago that it now looks like just part of the mountain side. That half-blood also said that he intends to keep a close watch at that smelter site, should anyone else attempt to dig there. The man who dug up those bars of silver slag told me he has all the slag he needs; he won't be going there again!

In my library I have an early tape recording made by a man who was the first abstractor for Duchesne County. It was part of his job to spend time at Fort Duchesne researching old mining claims on Indian land so they might be correctly plotted on maps and added to courthouse records. During that interview he talked about mining claims in the Farm Creek area.

> Old Rhoads wanted to transfer some of his claims from Indian land to Wasatch County, when the reservation was still part of that county, but the Indians wouldn't agree to the trade. He and Hathenbruck then tried to lease some Indian land, but there was no survey to plot a legal lease on, so Hathenbruck drew up an agreement designating the area he and Rhoads wanted in degrees of latitude and longitude. If you trace the area they wanted, it began at the mouth of Rock Creek and followed up Farm Creek to the mountains on the Wasatch County line, but nothing ever came of that lease because the Indians wanted to know exactly where they intended to mine before the lease was granted.
>
> I talked to this old Indian, Louie Apprats, and he told me about Rhoads and Hathenbruck being seen at Farm Creek, so I looked that area over pretty good. I got to using a dowsing rod, and got some good indications there. I know there's something up on Farm Creek; that's the hottest place I've ever dowsed. There's a place that would just pull those rods! I wanted to dig there and the tribe said they would give me a prospecting permit, but they wanted ninety percent of everything I found. I figured that state and federal taxes would take the rest so I never took them up on it. You know, if that wasn't Indian land, I'd probably be a rich man now!

Many prospectors and weekend explorers have had frightening experiences in the Farm Creek country. There is a Ute Ranger Station on the Rock Creek side of Farm Creek Pass, so that the steep road which passes by it is closely watched by Indian Police on duty there. That road has been open to the public since pioneer days and cannot legally be closed, but the adjoining lands are Ute property and are closed to trespass. During the fall a man and his wife drove up to Farm Creek Pass to see and photograph the autumn leaves. They noticed that there was no one on duty as they passed the Ranger Station, nor were there any tire tracks on the dusty road, other than their own.

They parked their car at the side of the road near the pass and walked a short way to where they planned to take a photo. Just then they heard the sharp crack of a high-power rifle and a split second later the whine of a bullet screaming just over their heads. Both quickly dropped to the ground and were beginning to think the shot had been only a random one, when another bullet whistled close overhead. Realizing that those shots weren't accidental and that they were the target, they ran back to their car and left the area. Since then they have tried to rationalize the attack on them, but several questions are hard to answer. Since there were no vehicle tracks on the road other than their own, and because they saw no footprints or tracks along the road or side hill at the pass, they had to conclude that whoever shot at them had to have been stationed somewhere near the summit just waiting for them, or for someone, perhaps anyone. They are quite sure that the shooter could have hit them had he chose to do so instead of just firing warning shots. They can't help but wonder what is so valuable at Farm Creek Pass that they were shot at, and they also wonder that if they go back there, will the next shots be only warning shots?

Foster Rhoades was a nephew of Caleb Rhoads, and he lived all his life near Hanna, close to Farm Creek. Back around 1925 an elderly blind man named Sharp came to Hanna, and began asking people who lived there about certain landmarks in the nearby mountains. In a town as small as Hanna was then, his path eventually led to the door of Foster Rhoades. Sharp told Rhoades an intriguing tale. He said that many years ago, long before he became blind, he used to go to a place along Farm Creek, above the Red Narrows but well below Round Grove, where after rain storms he could pick gold nuggets right off the ground. They could be seen at no other time, only after a rain storm washed them loose from the red sand soil. He told Rhoades that if he would help him find that place again, they would share equally in the gold.

Rhoades agreed to guide the old man, and following his instructions they went east from Hanna to Farm Creek, where they turned northward and continued past the Red Narrows. As they approached Round Grove, Sharp directed Rhoades to watch for a place where a stand of cedar trees came to a point on the mountain, across the creek on a north-facing slope. Sharp said that was where the nuggets were. It didn't take Rhoades long to spot just such a place, and he helped the old man across the shallow creek and began climbing the hillside. It was autumn and the hills were knee-deep in yellow straw grass. Sharp asked Rhoades to pick up some of the nuggets for him, but Rhoades replied that he couldn't see any nuggets, only a dry hillside covered with dead grass and weeds. Sharp immediately became very angry and accused Rhoades of keeping all the gold for himself. When Rhoades tried to explain that there weren't any nuggets, the old man yelled that he was a liar. Rhoades got Sharp back across the creek to his car and they returned to Hanna. He never heard from Sharp again, or did he ever go back to that dry hill side above Farm Creek. It wasn't until many years later when he was telling about Sharp that he suddenly remembered that the nuggets could only be seen after a rainstorm. Maybe if someone checked that hill side right after next spring's first rain, before the summer heat turns the bunch grass to yellow, maybe . .

"Snooks" Roberts of Hanna has run livestock in the Farm Creek area for well over a half-century, so there is little about that country that he doesn't know. Not long after 1900, a cowboy stopped to water his horse at a small tributary stream along Farm Creek. When he knelt to drink, he saw a narrow seam of yellow gold crossing the creek bottom. Realizing what he had found, he decided not to tell anyone, not even his employer. After breaking several small pieces from the vein, he carefully covered it with heavy rocks so no one else would find it. Not wanting to arouse anyone's suspicion, he continued working until the fall roundup was over. The following spring that cowboy returned to Farm Creek, but he couldn't locate the small side stream where the vein of gold was hidden under a few rocks. He spent every daylight hour searching every side gulch and draw, hardly taking time to eat or sleep, so by the time snow began to fall he was gaunt as a ghost and half-crazy over his lost fortune.

Several years passed before Roberts saw his old cowboy friend again, when he came to Robert's house to tell him about the vein of gold in the stream-bed along Farm Creek. He admitted that he had been confined to an asylum and then asked Roberts if he knew of any place in the mountains where a small stream might have gone dry. Roberts told him he knew of no place like that. Several times that summer Roberts saw the old cowboy riding the hills along Farm Creek, carefully checking every side canyon and gulch. He never returned the following year, nor has Roberts seen him since, but Roberts has kept an eye peeled for that lost creek. He hasn't found it or the narrow seam of gold hidden under a few rocks on its bed, but he's still looking.

A man I know named Gudmunson tells an interesting tale about early days on Farm Creek. A government surveyor was employed locating and marking section corners. He was an older man who was married and had a family of nine children. He had a young and well educated Indian youth as an assistant. While surveying in the foothills along Farm Creek, the older man became too ill to continue working, and his young assistant replaced him. That Indian youth had come to like the older man and he felt sorry for his impoverished family, so one day he took him to a place where there was a cache of Spanish treasure. He told the surveyor to take from the cache whatever he needed to keep his family in food and necessities, but he cautioned him to take no more than was needed; nor was he ever to reveal the location of that cache to anyone. For some time the old surveyor took enough treasure from that cache so that his family could live in comfort, but the time soon came when he was no longer able to go into those mountains anymore. He carefully marked the location of the cache site on one of his survey maps and gave it to a close friend; that friend being Gudmunson's father.

On that survey map he marked a green-colored ledge and a rock monument atop a promontory directly above the cache site. The elder Gudmunson followed that map into the foothills above Farm Creek, to where he was stopped and turned back by an Indian Ranger. He made several other attempts to find the surveyors cache, but every time he went into those hills he was stopped by an Indian who kept watch over that area. It was his hope that one day he would uncover that cache, but passing years, family responsibilities and eventual old age kept him from doing so. He intended to pass the surveyor's map on to his only son, but before he could do so that son was called to serve with the military and spent several years overseas. While the son was gone from home, his father died and other family members disposed of his personal possessions, including the surveyor's map with the cache site marked on it.

Young Gudmunson had been shown that map when he was a young man, so after his return from the military he tried to locate the cache site from his memory of that map. He couldn't find the green-colored ledge nor the old rock monument which stood on the point of land above it; but he did find a lot of trouble. One day while he was in the mountains above Farm Creek, he met a young half-blood Indian who showed him a bag of gold nuggets he had found. That Indian told Gudmunson that he had dug into an old grave which he had found at the base of a green-colored ledge of rock. He dug up three skeletons, buried on top of each other, all in a single grave. He uncovered fragments of European type clothing and some glass buttons in that grave, while the small leather bag of nuggets had been beneath the bottom skeleton. As Gudmunson was talking to the Indian youth, both were apprehended by Ute Police who took the half-blood into custody. They then told Gudmunson in no uncertain terms to get off their land and never to come back. He never returned, for he still remembers their warning: "Come back and there will be four white men in that grave!"

A few years ago I interviewed an old man who was well over ninety years of age, a man who had spent most of his life close to Farm Creek. He told me a story which I paid little heed to at the time, but later I came across almost certain proof that his story was true. He told me how in the early days there was an Indian whose name was Nephi Winchester, but who was better known to white men as Bill Pritchett. He was very rich and influential, and he was a brother-in-law of David Copperfield, a Ute tribal leader. Pritchett claimed that he owned a forty acre piece of land which actually belonged to a poor Indian woman named Alice One-Leg. Her real name was Itch-A-Boom Cuch, but because she had been born with only one leg, she was always called Alice One-Leg. Being her neighbor, the white man knew that the land belonged to her, so he agreed to go with her to the Indian court at Fort Duchesne and testify in her behalf. But it so happened that Pritchett had more influence with the court than either he or the old woman had, so she lost her land. Still, she was grateful for her white friend's help. During the next few years before she died, he helped her in other small ways, truly being a good neighbor.

Spanish treasure still lures men to search for the Gudmunson cache near Hanna.

During that interview, the old man told me that shortly before her death, Alice One-Leg called him to her cabin where she told him a very strange story. She said that before the Ute Reservation was opened to settlement by white men, the government had made a treaty agreeing to pay the Utes a settlement of $1.25 per acre for certain lands. Washington also agreed to pay $25,000 in gold coin at that time and an equal amount each year thereafter until the debt was paid. She said that the payment was made to the tribe in twenty-dollar gold pieces, twenty-five coins each in fifty leather bags. But the Old Ones, tribal elders, feared dividing so much money among the young bucks, afraid that they would spend it foolishly on the white man's fire-water, so it was decided to hide most of it until the land settlement was completed and tensions lessened. Alice One-Leg said that the gold was taken to a place along Farm Creek where it was cached in a location known to only a few of the Old Ones. She described that place to her white friend as her way of thanking him for his past kindness to her.

A close examination of treaties made between the Ute Tribe and the government reveals that what Alice One-Leg said was true. In a treaty negotiated by Col. H.O. Irish on January 8, 1865, the Ute Indians agreed to move to the Uinta Basin, relinquishing their claims to all other lands in the Utah Territory. In return the government promised to pay the tribe $25,000 in gold each year for ten years, $20,000 annually for the next twenty years and $15,000 for each of the following thirty years. During that same year, 1865, Brigham Young advised the Utes through Chief Sowiette that they should accept the government's offer, promising that the Utes would receive $900,000 over a sixty-year period, actually $150,000 less that Col. Irish had negotiated. That discrepancy wasn't explained, but it mattered little to Chief Sowiette who told Young, "We do not want to sell our land, we want to live by the graves of our fathers." Because the government disagreed with Brigham Young's Indian policies, it took congress nearly forty years to approve that payment, a clause in the agreement stating: "The Indians are to be paid in gold coin, they will not accept paper currency."

The payment made to the Ute Tribe may have in fact been even larger, for in 1902, just before much of the reservation was declared to be public domain, the same agreement which granted the Raven and Florence mining companies special advantages over other settlers, also provided for payment to the tribe for claims made against the government for lands already taken from them. That section of the agreement stated in part: "The sum of seventy-thousand and sixty-four dollars and forty-eight cents is hereby appropriated; ten-thousand dollars of that sum being compensation for deletion of certain lands on the reservation's east boundary, those proceeds to be applied for the benefit of the Indians."

By order of Col. Irish on March 2, 1868, all white men were to be moved from the reservation, the boundaries of which were described as being: "The entire valley of the Uintah River, extending on both sides to the crest of the mountains." That area was interpreted to include all lands from the Strawberry Valley to the Green River and from the crest of the Uinta Mountains to the Book Cliffs. After the Uinta Basin was designated as a reservation, the first Indian Agency was established at the head of Daniels Canyon, during the early spring of 1868. That location proved to be impractical because even then snow covered the summit six feet deep, so the agency was moved to the foot of Tabby Mountain near Hanna. Shortly afterwards it was moved again, to near present Utahn at the forks of the Duchesne River an Rock Creek; but most business was conducted at the Indian village at Sadies Flat, a few miles further up Rock Creek. It was decided that location was too remote so another and final move was made to the Whiterocks Indian Village, on Christmas Day, 1868. Shortly afterwards a trading post and sutler's store were established at Whiterocks.

Several dozen agents and traders have been assigned to the Indian Agency at Whiterocks over the years, but the one who served the longest was Robert Marimon, who came from Kentucky to be Post Trader in October, 1902, and remained in that position until 1928, when he sold out to Oscar Lyman. Marimon had also worked at the Ouray Trading Post on the Green River as early as 1886. Marimon's son continued to work at the Whiterocks post until it burned in 1930. From Marimon's post records as well as from the reminiscences of his daughter, Sarah, we know that the Ute Tribe was in fact paid $25,000 in gold coin just as Alice One-Leg said, and as the treaty stipulated when the reservation was opened to settlement.

"Each Indian, even the smallest baby, is to be given forty acres of land, and each family at least eighty acres; grazing lands to belong to the tribe in general. Rights to hunt and fish are guaranteed." So proclaimed the opening section of the treaty of 1902, when the Utes lost their ancestral lands. Prior to the land rush of 1905, Post Trader Marimon was advised to stock up on all manner of goods, since it was anticipated that when the Indians were paid for their lands, they would buy everything in his store. In correspondence with the author, Sarah Marimon wrote that the gold payment was made just as ordered, with at least part of it if not all being given to designated Indians. She wrote: "They bought everything we had in the store!" She also described the method in which payment was made to the tribe. "A military escort brought the gold to Whiterocks and remained to ensure that no white man came in to gamble with the Indians. But they gambled among themselves, playing the stick game, using twenty-dollar gold coins stacked like poker chips. I never saw so much money in my life!"

That a lot of those gold coins were lost by Indians who gambled with them is shown in a letter received from Anthony Colunga in September 1982. Colunga described how after months of negotiation, he finally obtained permission to use a metal detector at Whiterocks. Although he was allowed only two days to search, during all of which time it rained constantly, he recovered more than 200 coins, including many gold pieces and silver dollars. During that short search he was not allowed to use his detector at the old Indian gambling ground north of the village, where no doubt even more valuable coins were lost. It is also known that an Indian who lives at Whiterocks has recovered many valuable gold coins, many in new mint condition, which indicates they were lost soon after their receipt. Also found was a gold medallion with the likeness of President Buchanan engraved on it.

There is still more evidence to substantiate the story told by Alice One Leg. Don Carlos Foote, a part-blood Ute, told Stan Sharkey that he knew of several gold coins being found in Rough Canyon, just off Farm Creek. Perhaps connected in some way with those coins were two very old muskets found at nearby Winchester Flat, and just off that flat, two cannons which were pushed off a ledge into the canyon below, where they still remain, covered with dirt and brush. A lot of strange things have happened in that country, from Rough Canyon and Winchester Flat to the Lower Stillwater along Rock Creek.

Several years ago two men were camped in that area. While one spent the day fishing, his companion took his back-pack and a rifle and began to climb up to the ridge which separates Rock Creek from Farm Creek. Nightfall came, but he never returned to camp. His partner was agonizing over what to do, when just after midnight he came running and stumbling into camp. It was obvious that he had been subjected to a trying ordeal, for his clothes were ripped and torn, his hands and face were blood-smeared and he no longer had his back-pack and rifle. He immediately began throwing camp gear into their pickup truck and insisted they get away from that place immediately. His friend tried to calm him so they could wait until morning to break camp, but he was so frightened that they left in the middle of the night. Wild-eyed and hysterical, he refused to talk about what he had seen, other than to say he had been in Hell Hole.

Later, during the summer of 1991, that fisherman returned alone to that same area, his companion refusing to return to those mountains where he had been so frightened by someone, or something. He made his camp at Rock Springs, near the head of Farm Creek. One day he decided to hike over into the area where his companion had gone, although by a different route. Maybe if he was lucky, he might find his friend's back-pack and rifle. About two miles along the ridge leading to Hell Hole, he was stopped by two Indians. When he asked if he was on the right trail, they told him to get back to Rock Springs and nor to come back. He was surprised that they knew where he was camped, for he had seen no one while there, but one of the Indians told him that he had been watched every time he had been in those mountains. They even told him what kind of vehicle he was driving. Believing that he knew that area well, he started to explain that Hell Hole was not on Indian ground, but was on forest land. He said those two Indians suddenly became very mean and gave him "some extremely blatant warnings" that he better get out of that country and be quick about it. One of them added: "If you come back, your chances of leaving will be slim to none!"

After he broke camp he drove to Duchesne City, where he stopped at the Forest Ranger's office to make sure he had been on forest land and also to ask what was so special about Hell Hole that he had been prevented from going there. The answer he received was, in his own words, "That Forest Ranger acted as if I had said something bad about his mother! He got really hot! I thought he was going to lock me up!" He then asked the ranger if he could get a map of the Hell Hole area, to which request the ranger "became extremely offensive," and warned him not to go near that area. His friend had been frightened out of his wits at Hell Hole, the Indians had threatened him with dire consequences if he ever came back and the Forest Ranger ordered him not to go there. He then said he had only one question for me: "What the heck is so special about Hell Hole?"

I have an idea about what is so "special" about the Hell Hole area. Gold coins! There is no doubt that Alice One-Leg told the truth when she said the Ute Tribe was paid $25,000 in gold coin. Very likely the tribe also received other annual payments in gold, payments which few Utes except for a small group of elders knew about. None of those coins were ever returned to circulation, and few have been seen since, except for a few found in Rough Canyon. Did the Old Ones cache those coins near Farm Creek, perhaps at Hell Hole? With each of those coins worth at least one-thousand dollars now, what would that cache be worth? I know that Alice One-Leg's white friend never recovered that cache, and the fisherman and his friend never even got close to it, so those gold coins are probably still there just as Alice One-Leg said, somewhere near Farm Creek. If you don't frighten easily, the best place to start your search might be at Hell Hole!

CHAPTER 9

MISSION IN THE MOUNTAINS
Lost Treasures of the Black Robes

Rock Creek is a major tributary of the Duchesne River, their confluence being near the old ghost town of Utahn, a few miles north of Duchesne City. There is no doubt that the Rock Creek area was a center of Spanish activity more than 200 years ago. When the first Americans made their way into that country they discovered evidence of earlier occupants. There were strange signs and symbols cut into rocks and trees, many containing words of Spanish origin, and those cryptic marks were very old even then, covered with a patina of desert varnish or having moss and lichen growing on them. Desert varnish is a thin coating of manganese oxide which often takes centuries to form, while lichen is a parasitic growth which takes a lifetime to grow even a tiny fraction of an inch in depth. That glaze of desert varnish and the thin covering of lichen which coated those inscriptions after they were made is ample proof that they were inscribed very long ago. There is no way those inscriptions could have been faked or made as a hoax as some skeptics have suggested.

Who could have been in that area so long ago to perpetrate a hoax, and equally important, who would have been the victim of such a hoax? A hoax would have had to be perpetrated at a time when there were no Americans closer to Rock Creek that the British colonies on the far away Atlantic coast. The very first Americans to penetrate that wilderness were Mountain Men and fur trappers. They noted seeing those intriguing inscriptions, containing Spanish wording and measurements, with no similarity whatever to the Indian petroglyphs they were accustomed to finding. They also discovered ruins of stone structures; foundations and rock walls which subsequent research has proven were the remnants of an early Catholic mission. At the edge of the river close by those stone ruins was the skeletal remains of an ancient water wheel, placed to furnish power to turn an arrastra, a primitive ore grinding mill. It was obvious even then that those structures had not been built by native Indians. Research by the late Dr. Donald Moorman, Professor of History at Weber State University, reveals that a Catholic mission had been build in Rock Creek Canyon as early as 1700. Every year new evidence is uncovered to confirm his findings.

There is a great deal of physical evidence, as well as both oral and written history, to prove the Spanish presence at Rock Creek; more than is found at any other area in the Uinta Mountains. Many artifacts of ancient origin have been found there: Spanish swords, bronze cannon, match-lock muskets and pieces of body armor. Enigmatic and enticing waybills along with intricate and puzzling maps depicting that area have been found in the Spanish archives at Santa Fe, Mexico City and Seville, Spain. When the Ute Reservation was opened to homesteading at the turn of the century, Rock Creek was known to the Utes as Rocky Creek, hence its name, but white men called it the East Fork of the Duchesne River. The only areas suitable for farming were along the river bottom, which were the areas white men chose for their homesteads. Those pioneers soon discovered that others had lived on those same lands long before them. They noted in their journals or wrote to their families of the artifacts they found; old trails, stone foundations and long unused irrigation ditches and canals.

We know little of exactly when the Rock Creek Mission was built, but there are better records regarding its destruction. We have eye-witness accounts and historical documentation of the fate of nearby Fort Robidoux, the first permanent white settlement in Utah, other than that mission. That fort was located less than a mile from the Whiterocks Indian Village, and it was destroyed by Indians during the Ute uprising of 1844, only three years before Mormon pioneers entered the territory. At that time, Mexico was at war with the Navajos, and during a battle Mexican troops mistakenly killed a small band of Utes. The Utes sent a delegation to Santa Fe to confront Governor Manuel Armijo to protest those killings, but a misunderstanding and subsequent argument flared, during which soldiers opened fire and killed eight of the Utes. The entire Ute Nation went on the warpath, killing Mexicans wherever they found them. In October they attacked Fort Robidoux, killing everyone there and setting fire to the fort.

Fort Robidoux was a large structure; its foundation can still be measured as being two-hundred by three-hundred feet. Historian J.R. Kirkpatrick noted that its foundation was of heavy stones while the walls were built of stout cedar posts. It was located just above the forks of the Uintah and Whiterocks rivers, three quarters of a mile east and one-half mile south of the present Whiterocks store, on what is known locally as Big Tom's Allotment. Established in 1831 by Antoine Robidoux, it soon became a trade center for French fur trappers and a supply point for Mexican miners. Ute Indians brought their pelts there to trade for cloth, gun-powder and other supplies. Two other forts in that same area were also destroyed during the 1844 uprising; Fort Kit Carson at the junction of the Duchesne and Green rivers and Fort Davy Crockett, on the Green River at Brown's Park. A growing body of evidence gives proof that the Catholic Mission at Rock Creek met the same fate, although perhaps at an earlier date.

The Rock Creek Mission was not an elaborate structure like those along the long trail north to Santa Fe, for it was only an outpost in the wilderness, still it was significant, having all the trappings peculiar to a Catholic church. In the Rudo Ensayo (Rough Essays), an early Spanish record translated in 1762 by Father Juan Nentvig, it was written that all missions owned the appropriate vestments, although at remote sites like the one at Rock Creek, they may have been less ornate than those found at more populated places. Father Nentvig wrote: "All missions have altars, chalices (some only of silver, but in other places of gold), ciboriums, monstrances, both large and small candlesticks and crosses. All have silver statues of the Virgin, not only at principal missions, but at dependent ones as well." By church law, chalices used for communion wine and patens used to hold Holy bread for the Eucharist had to be made of precious metal, with their interiors plated with gold.

There is evidence that there were canals and cultivated fields at the Rock Creek Mission, and deep cellars where crops and other foodstuffs were kept. Small stone ruins found in the surrounding area are believed to have been out-lying storage buildings and shelters.

An unusual double-line Catholic cross. Treasure is hidden close-by!

Some in the nearby mountains and along trails leading to distant mines were visitas, small places of rest or temporary places of prayer between the main mission and reals de minas, groups of mines in those mountains. One of those visitas may have been located near the junction of Rock Creek and the Duchesne River. Many old artifacts have been found there, and it is a place closely guarded by Indians, a place where white men are not welcome. On a bench-land overlooking that visita site there is a very old cemetery, a place sacred to the Utes. On the flat-top close to that cemetery there is a place where Ute Indians once raced their horses. Cut into ledge rock just below that race track are petroglyphs depicting those racers of old. But among those stone carvings of Indians and their horses there is another figure, strangely out of place. It is the exact likeness of a giraffe! The mystery is, how could Indians at the time those petroglyphs were made know what a giraffe looked like? There was never a circus with a giraffe anywhere near that country. Like the petroglyph of a hairy mammoth found near Moab, it is an enigma. Utes treat that place with great reverence; the race track, the cemetery and the old visita mission site.

Ute Indians have guarded the secret of Spanish gold for more than a century.

There are several Christian religious symbols cut into ledge rock near that old cemetery, one of them an unusual double-line Catholic cross, with several letters and a directional arrow beside it. During the summer of 1992, the direction that arrow pointed out was followed past other signs in stone to a place where an old mine tunnel was found. Strangely, the same sequence of letters found at that Catholic cross were also found cut into stone above that caved tunnel portal. When work began to open that abandoned adit, it led to a truly astonishing discovery. Under the protecting branches of a giant cedar tree not far from that tunnel, buried in a blanket of fallen leaves and bark, an old pack saddle was uncovered. It is in excellent condition, having been protected from the weather and storms of a century or more. Carved into the wooden frame of that old saddle are two initials: "C R." There can be little doubt that pack saddle was once owned by Caleb Rhoads.

Between the Rock Creek Mission and the small visita near Utahn, at landmark places like Sadie's Flat and Dry Canyon, there are other places which also bear the signs of antiquity. In a sheltered alcove above the river, at the south end of Sadie's Flat, there is a centuries-old panel of Spanish signs. During more recent times Ute petroglyphs and the doodlings of white men have been added to that panel, defacing and destroying part of its ancient message. Ed Twitchell traced sections of the old Black Robe Trail from that alcove panel up on to Dry Mountain. Along the way, names and dates left by Jesuit Priests can still be found. At one time, when Twitchell first traced that old trail, there were still three old Spanish cannons, each strategically placed overlooking Sadie's Flat below. There is a "sighting rock" along that trail, a large round-top boulder with a deep groove cut in its top. If you sight through that groove, you can see an old mine tunnel portal far across the canyon.

At the base of Dry Mountain, along the west side of Sadie's Flat, there are three mining claims once owned by Caleb Rhoads. There is no doubt about that, for those claim notices are still on file at the Heber City Courthouse. (Duchesne County wasn't separated from Wasatch County until 1915) Some of his mining tools have been found there. On a gray-colored rock ledge just below his claims there is an Indian ghost figure; a strange petroglyph depicting what some say is a sacred

Bear Sign. It shows a bear standing with part of one front leg missing, and two circles, one over its bowels and another on its shoulder. One of the old Utes told Wayne Nelson that those circles represent treasure dug from the bowels of the earth, while the jagged and torn stump of the bear's fore-leg warns that if you try to steal his treasure, the bear will chew your arm off!

Dale Slade used to work for the Ute Tribe, for a time with Rex Curry, who was the Tribal Director Of Resources. Slade also found something unusual along that section of Rock Creek. All across that area there are countless round cobble stones. In that arid region, where much of the year's moisture comes from seasonal rain and flash floods, those stones have over a long period of time came to rest in that sandy soil with their heavy side down. Their exposed upper surfaces have become encrusted with a film of white alkali. If one of those stones is found with its alkali side down and the dark colored heavy side up, it is almost certain that it has been purposely turned over. Slade discovered a place where a long straight line of those stones had been turned over a long time before. At first it only appeared unusual, something which few people would notice, but when Slade climbed to a higher elevation and looked back, that line of turned stones formed an arrow, with its tip pointing up canyon. He has no doubt those stones were purposely placed as they were. The question is, why?

Slade recalled that he was told about a very old Spanish document hidden away in files of the Ute Tribe at Fort Duchesne; that document telling of an early Spanish expedition into the Ute Reservation area. That expedition had come north from the Henry Mountains of southern Utah, where the famous Lost Josephine Mine is said to be located. Since leaving employment with the Ute Tribe, Slade has researched early Spanish mining activities, and he has discovered many ancient Spanish signs and symbols in the Henry Mountains. Among them is a name and date cut into solid rock: Adolfo Caiecez, 1735. That is the name of the expedition leader on that old document locked away at Fort Duchesne!

Since pioneer days there have been stories of a rich Spanish mine somewhere in the lower Rock Creek area. Some of the old-time blanket Indians sometimes let slip leads which settlers interpreted as being clues to the location of that mine. Caleb Rhoads also hinted at clues to that mine, perhaps to taunt those who tried to pry information from him. Once when he was asked if government paid geologists might not find that mine, he replied that they were looking too high, that his mine was down low. Could those early Utes and Caleb Rhoads both have been saying that one of his mines is in the lower Rock Creek area?

Caleb Rhoads' pack saddle, found at one of his old mines in 1992.

Charles Kenworthy is one of the foremost authorities when it comes to interpreting Spanish treasure signs, and he has learned that all Spanish mines of any consequence always had a Catholic shrine close to the mine portal. He has determined that shrine was always in a place where miners could say a prayer before descending into those dangerous diggings. That shrine could be as simple as a niche cut into a rock ledge and housing a plain silver cross. At larger or richer mines it might be as elaborate as a small alcove or room, with an altar stone and a gold figure of the Virgin Mary. By royal decree, all Spanish mines had to have such a shrine, and by law that shrine had to be located within two-hundred varas (five-hundred feet) of the mine entrance.

An aged woman who still lives on the Ute Reservation was picking berries along lower Rock Creek not long ago. When she pulled some branches away from a rock ledge in order to reach more berries, she uncovered one of those old Spanish mine shrines. In a niche cut about two-feet wide and high and about six-inches into the rock, there was a sort of altar. On it, covered with a halo of dust and cobwebs, was a small gold statue of the Virgin Mary. The old Indian woman had no idea what she had found and called it an idol, since it had no significance to her except as a curiosity. Of course, she had never heard of a Spanish mine shrine. Several reputable people have examined that golden statue, but none have learned exactly where it came from, other than that it was found where she was picking berries, somewhere along lower Rock Creek.

"In The Steps," a publication which reports treasure finds, there was a recent story of a similar idol being found in the Four-Corners country of south-eastern Utah, as was described in the March, 1994 issue.

> What has every appearance of a little man idol, made of gold with empty eye sockets which may have contained some type of gem, was found in a

Church bells at the old mission at Rock Creek warned Spanish miners of attacks by Ute Indians.

> dry wash. Other such objects have been found in the same area. Most interesting is the fact that all of the objects have been found along the Old North Trail, from Santa Fe north through Utah. Similar items have been found in all four of the Four-Corner states. This particular idol closely resembles artifacts made by the Aztecs. It weighs 4.9 pounds.

That mine shrines do exist and may yet be found is further proven by a news report from Kane County, Utah, on April 8, 1993:

> A four and a half pound golden Madonna has been excavated at a local area. A highly qualified archaeologist uncovered the icon in a ruin which has never before been investigated. That Spanish icon was found about three feet deep, below a flat stone which is thought to have been an altar. The location is believed to have been an open air mission site, used during the early 1700s.

That similar icons or statues still exist in the Uinta Mountains is almost certain. If you are fortunate enough to locate such a shrine, you will be very close to a Spanish mine; by law no further than two-hundred varas!

While hiking in Rock Creek Canyon, several people have noticed some old stone outlines on the ground in an area of sage, willows and cottonwood trees, which have the appearance of a building foundation. Members of the Rust and Bleazard families who homesteaded near Mountain Sheep Pass on a bench-land above Rock Creek also wondered about those foundation stones. Even then they observed that it appeared that sometime long before those stones had purposely been broken apart and scattered so their real purpose wouldn't be recognized. Others have discovered that just at ground level there are the tops of several rows of cedar posts which have been burned. Ed Twitchell was one who realized those stones were structure foundations and those burned posts a barricade of cedar posts placed upright side by side, as an outer wall or stockade around those foundations. Wayne Nelson was another who concluded those foundations and cedar stockade were part of the old Spanish mission. Many of those stones have been moved over the years, but if you brush away a few inches of sandy top-soil, you can still see the burned tops of those cedar posts, for cedar posts in sandy soil can last for a long time.

During the past few years, several new pieces of evidence have been uncovered to place the site of the Rock Creek Mission more exactly. There were structures on both sides of the river, from Utahn to Daniels Flat, but most of them were near what is now called Treasure Hill, a low cedar covered knob on the east side of the river, near the south end of Daniels Flat. It is now quite certain that the main structure, probably the mission church itself, was located on the west side of the river, just up stream from Treasure Hill, on an inside bend of the river. That is a level area with the river on three sides and a steep mountain slope behind it; an easy place to defend. The testimony of several early settlers and later prospectors tends to verify that location as being the correct one. There were other structures both large and small, a prominent one at the mouth of Pigeon Water just north of Mountain Sheep Pass. It may have been a lookout station of some sort.

Wayne Nelson was one of the first to make a careful examination of that area, and during a recent interview he stated that the water wheel used to power an ore grinding arrastra was on the west side of the river, across from and just up stream from the old Walt Daniels cabin which still stands on Daniels Flat. Nelson judged that wheel to be about twelve feet in diameter, and had been built of rough timbers, only crudely squared and fitted together. It had fallen or had been pushed over long before Nelson first saw it, and a thick growth of willows had grown up through its broken sections. Those wheel segments were then so old that Nelson at first thought they were only pieces of drift wood, and he burned several pieces in his campfire. It was only then that he noticed that they had once been cut and joined into sections.

That water wheel had been constructed to turn an arrastra which was used to grind and crush ore-bearing rock. That fact was verified when Nelson discovered broken pieces of several large stone vessels, which at that time he called "smudge pots." He pieced together at least six of those large vessels, and from those remnants he could tell that they had been about eighteen inches square and deep. Fragments of burned slag containing a high metal content could still be seen adhering to the bottom and sides of those "smudge pots." Only a few years later, Ed Twitchell saw those same pieces of fragmented pots, but he called them smelters. There were fewer pieces then remaining and even less of the water wheel when Twitchell was there, but he could still recognize them for what they were.

Both Nelson and Twitchell recognized something else which few others did; that there had once been a wagon or caretta trail on the west side of the river, which led directly to that old mission site. The present road crosses to the east side of the river just below Treasure Hill. Early settlers like Bill Bleazard also noticed that old caretta road turned west up Chokecherry Draw and then turned sharply downhill past a small spring to the mission site flat. The author has a complete set of early government survey maps of the Uinta Basin, most of them dating from 1869. It is significant that those old maps show an abandoned wagon road on the west side of the river. Who other than Spanish miners could have built such a road so long ago that it would have been marked "abandoned" in 1869, nearly forty years before that area would be opened to homesteading? Those maps reveal that the road crossed the river just upstream from the mission site, but apparently went no further.

In 1969 a fisherman made a startling discovery at what has since come to be called Treasure Hill. While fishing along the river he observed a hole behind some willows and brush close to the edge of the river, a hole which appeared to extend back into the hillside. After pulling brush and driftwood from that opening, he discovered a man-made passageway. Being curious as well as a part-time prospector, he called to his partner and they decided to explore that strange hole. Removing his heavy boots and jacket, he squeezed into that small, wet and muddy cavern. A short way into the mountain side that passageway dipped down into what he later called "a water-trap of ingenious design," where he had to swim under water for a short distance, with only a flashlite wrapped in a plastic bag for light. The floor of that "water-trap" rose sharply with the passageway continuing into what they later learned was a series of five chamber-like underground rooms. Each of those rooms were square in shape and had been made by man, tool marks evident on chamber walls and connecting passages.

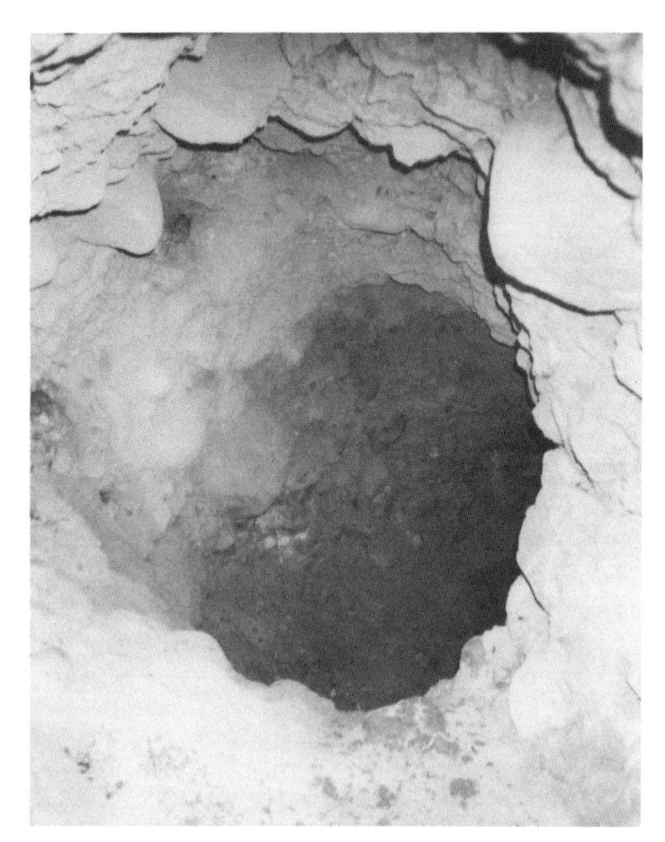

Descending into the ancient tunnel which leads to the "Water-Trap," deep inside Treasure Hill.

All of those chambers were empty except the second one, in which he discovered a square stack of heavy gold bars. For several hours both he and his partner took turns examining those silent chambers and counting those bars of gold. Both agreed there were eighty bars, each so heavy that a man alone couldn't lift one. Realizing the enormity of what they had found, they carefully covered the hole leading to the "water-trap" and made plans to remove those bars. The next several times they returned there were other fishermen or campers nearby, and once when they made a night time attempt, they had to flee to elude Indian game wardens who thought they were poaching. After a number of adventures they finally succeeded in dragging several bars out through the "water-trap" to the river. Even then their work was slow and dangerous, mostly because of the weight of the bars. It took them only a few trips to acquire enough gold to last them a lifetime. One of the partners was killed soon afterwards and the other had no reason to ever return to Rock Creek. The last I heard from him he was vacationing on the Mexican Riviera!

The only photo ever taken of the mysterious "Water-Trap," at the treasure chamber inside Treasure Hill.

So far as is known, the remainder of that cache of gold bars is just as the two fishermen left them. Because of the finder's religious persuasion, he called that cache the Jaredite Treasure, a name referred to in the Book Of Mormon. It was his belief that when the Nephites and Jaredites engaged in one last great battle, the Jaredites concealed their national treasury in the "water-trap" chamber, thus the name, the Jaredite Treasure. Of course, a lot of other people who saw parts of that gold believe it is Spanish gold, cached long ago by the Black Robes who manned the old mission. Since that discovery in 1969, the river has changed its channel, so that now the "water-trap" entry is now covered with piles of cobblestone, while a dense growth of willows and wild rose grows right to the water's edge. Today Treasure Hill is under almost constant surveillance, so that any attempt to dig there could be a dangerous undertaking. But at least one other person has profited from the same or a similar cache at Treasure Hill.

That there are adventurers willing to face those perils is proven by a recent find at Treasure Hill. A lone traveler, riding horseback, rode his mount into Rock Creek only a short way from where the two fishermen discovered the entry into the "water-trap." That rider, totally unaware of that earlier discovery, stopped his horse at mid-stream, where from his high vantage point he saw what he thought was a badger hole or opening, well above water level and in a place where a badger or coyote would have a hard time getting to it. It was in a place where a fisherman or someone on foot would be unable to see it. Thinking it strange that there should be a hole in such a place, he investigated, and discovered that hole opened up into a tunnel-like passage he could crawl into, although it was not large enough to stand erect in.

He crawled back into that hole for a distance he estimated was about sixty feet, lighting his way with matches. The tunnel opened up into what in the near darkness appeared to be an underground chamber, its floor under water of unknown depth. In that dripping water and mud his hands touched several stones which were different from the round cobblestones he had been crawling over. Those stones were sharp-edged with square ends. When he pulled one loose from the mud he knew it was heavy and was probably metal, perhaps lead. Backing out of that hole he managed to drag two of those heavy stones behind him. It wasn't until he was back in the daylight that he knew they were gold bars, each about eight inches in length and about one and a half inches square. He later found they weighed about twenty pounds each. By comparison they are not nearly as large or as heavy as the Jaredite Treasure bars found a few years earlier, but he says that's alright, they spend just the same!

Over the years several people have tried to obtain a prospecting permit to legally trespass on Indian land to investigate the mystery of Treasure Hill. Wayne Nelson had prospected that area, and the many old Spanish treasure signs he found there encouraged him to seek a permit. One of those signs was an almost life-size figure of a bear, cut deep into a giant pine. It was beautifully made as a photo he has shows. That bear stood facing Treasure Hill. Caleb Rhoads sometimes let slip clues to where some of his mines or caches were located, and on rare occasion some of the Ute elders did the same. Those slips weren't intentional, but still they were clues which someone familiar with the area might profit from. One of those clues was that a bear guarded a mine or cache. Another was that an eagle points the way to treasure. Some say a mine can be found in the eagle's nest. Not far downstream from the bear sign, there is a place on a rock-slide hill where sometime long ago stones were turned over so that they are of a different color that surrounding rocks. When viewed from across the canyon, those lighter colored stones look exactly like an eagle with spread wings! Like the bear sign on the pine, it is part of a waybill to a treasure cache. Unfortunately, that unique and important bear sign was destroyed when a greedy "crazy" chopped and slashed it with a pick so that no one else might see it. The woods are full of nuts!

Wayne Nelson and Orville Galloway finally convinced tribal leader Alonzo Jim to issue them a permit to prospect at Treasure Hill. Based on that permit, they brought in heavy equipment and began digging into the hill. Earl Garrison and Karns LaRose were hired to help with the work. They soon broke into an old tunnel and chamber, but found nothing there except some strange looking leather boots. Next they uncovered a vertical shaft which still had a wooden ladder in it. Alonzo Jim told them that shaft once contained a cache, but young Utes would steal from it to buy liquor, so the gold was removed and the shaft covered over.

The day following the discovery of that shaft, a member of the Ute Tribal Council came to Treasure Hill and ordered Nelson and Galloway off Indian land, but Nelson waved his permit at him and in turn told that tribal leader to get away from Treasure Hill. That was a mistake, for the council member hurried to Duchesne where he rounded up a mob of angry Utes, all of them heading for Rock Creek. But Karns LaRose had been in town and he hurried up Rock Creek ahead of the mob. When he arrived at Treasure Hill, he warned Nelson and Galloway to get away from there fast if they valued their scalps! Quickly loading their equipment, they fled up stream and over Mountain Sheep Pass. They had just crossed the boundary line leaving Indian land and were in sight of Mountain Home when the mob caught up with them. They had narrowly escaped with their lives, but they got the word that if they ever returned, they would be killed. Nelson says he hasn't returned yet, but he's been thinking about it.

Indian slaves descended into the black depths of Spanish mines on rickety "chicken-ladders."

Bill Pritchett (Nephi Winchester), one of the old Utes long since deceased, used to tell that some of the Old Ones like Red Cap and Unca Sam had told him the mission site had been a favorite Ute camping place long before the first Spanish miners came into the northern mountains. They told how soldiers drove the Indians away and built a settlement which had a church, irrigated crop lands and many small buildings. On the west side of the river just down stream from where the chapel stood, the Spaniards dug a large underground room where food supplies were stored. Today there is a large sunken depression in a grove of cottonwood trees where that cellar was located.

There is another underground chamber close by, one which served as a shelter for Spanish pack animals. Anyone old enough to remember when mining companies used horses or mules in the mines of the West knows that such underground barns were once commonplace. Several such shelters have been found along Rock Creek. At one a number of burro or mule hooves were found. According to a Ute Indian legend, after a battle during which many Spanish miners were killed, the hooves of their pack animals were cut off to keep the Spaniards from using their animals in the hereafter.

Many years ago, Ross Iorg, a half-blood Ute, found the opening of what appeared to be a cavern on a low bench-land overlooking Rock Creek, and thinking that he may have discovered an old mine portal, he asked Lloyd Roberts of Hanna to help him uncover it. Roberts took a large tractor to the site and dug out the entrance. Instead of a mine, they uncovered a place where in the long ago Spanish miners had stabled their pack animals, out of the weather and safe from theft by Indians. Roberts recalled that there was more than a foot of manure on that cavern floor, along with bits and pieces of harness leather and some rusted iron buckles.

The purpose of the mission and Spanish settlement on Rock Creek was to serve as a center for five reals (groups) of mines in the mountains beyond. There will be some who will say that Jesuit priests never mined or hoarded gold and silver at that mission; that the clergy's only concern was saving souls, but if they did not work mines of precious metals, why did the Spanish king issue decrees in 1592, 1621, and 1702 ordering those same priests to cease and desist mining gold and silver? And why did the king order all Jesuits arrested and returned to Spain in chains in 1767 if they hadn't

Only ruins remain of this old stone visita site in the Pigeon Water country.

A Spanish "midden," used to melt gold, found by the author at Pigeon Water.

obeyed his decrees? Those priests oversaw the work of packing ores from out-lying reals of mines to the mission settlement at Rock Creek, where its water-wheel driven arrastra milled those ores. Gold and silver ore came from places like Squaw Basin, the Brown Duck country, Blind Stream and Farm Creek and others as far away as Moon Lake and the Granddaddy Lake Basin. After that ore was ground to powder and amalgamated with quicksilver, it was melted in a smelter and reduced to bars of bullion; the same bullion which Thomas and Caleb Rhoads packed to the Mormon Mint.

Many Ute slaves died at those mines and mill before they rose up in rebellion and drove the hated Spaniards from those mountains. The Old Ones told Pritchett that when the Spaniards were killed and their mission chapel burned, such distant tribes as the Navajo, PiUtes and Northern Shoshones came to the aid of the Utes. Trail mission visitas were destroyed at the same time, at places like Pigeon Water, Sadie's Flat, Utahn and the Yellowstone River. The water wheel was toppled into the river while pack animals were killed and the carettas used to haul bullion bars were burned or covered with stones. Because forced labor in Spanish mines was the cause of the Ute revolt, they hid all evidence of those mines. It was their hope that if they concealed every source of the hated yellow metal, the Spaniards might never return. When the first miners began returning to those mountains, no trace of those mines could be found, so effectively had the Indians concealed them.

Ute Indians had been used as slave labor at many places in the Uinta Mountains, at Spanish and later Mexican mines and mills. In the Rudo Ensayo cited earlier, Father Nentvig wrote: "Indians have been trained from childhood to remember and describe places by the direction of the winds. Take an Indian into any depth blind-folded, whether in mines, dungeons or intricate labyrinths, and turn him around a thousand times; he will still be able to tell you what direction he is facing, as if he were a living compass." Another priest, Bishop Tamaron Romeral, recorded: "Indians are very ambulatory, and are very useful in all aspects of mining. They are very strong in their tasks. They know the veins in the mines and discover them. For these reasons miners seek them out; because they are hard workers and are able-bodied."

But although they labored hard, many Spanish miners did not consider Indians to be human beings, but only property to be used and worked until they fell. One early priest made the following report: "Missionaries forced them to plant fields of maize and beans, and perform all necessary tasks, as slaves. Moreover, they inflicted harsh punishment on those who did not do their duties. The Jesuits, under pretext that the Indians are incapable, made them work without any payment, but gave them only rations as if they were servants." Father Gumilla reluctantly conceded that Indians might be human: "Generally speaking, the Indian is a man, but one so uncivilized that I dare say he is a never seen monster with a wild brain, an ungrateful heart, an inconsistent soul, a lazy backbone, leaden feet and a thirsty belly. He is inclined to imbide until he is drunk. All these things transform him into an irrational being."

That Ute Indians at places like the Rock Creek Mission were abused and subjected to the harshest treatment is well documented. At the Archives of The Indies at Seville, there are countless millions of documents pertaining to the mines of New Spain, few of them indexed or cataloged. Many of those documents pertain to isolated mountain missions, and their contents tell a sordid story of Indian slavery. One such case involved Don Antonio de Mendoza, Governor of New Spain. Alarmed at reports of Indians being mistreated at those mission mines, the Spanish king dispatched Viceroy Tello de Sandoval to investigate allegations made against the governor. Viceroy Sandoval eventually brought forty-four charges against Governor Mendoza. Item 38 of those charges reads in part: "Many of the Indians taken in conquest were put to death in his presence, and by his orders. Some were placed in a line and blown to bits by cannon fire; others were torn to pieces by the dogs, some were killed by knife thrusts, while others were hung. He also left the branding iron of His Majesty with others so that they might capture and brand slaves." Is it any wonder that the Utes rebelled against their Spanish task-masters?

The late Mark Bleazard, an elder son of one of the first homesteaders along Rock Creek, was among the first to herd cattle at Daniels Flat and past Treasure Hill. He could still remember the massacre site near the old Spanish ruins, and even several of the carettas burned or covered by Indians. When he was still a young man, the ground was strewn with reminders of the battle between Indians and Spaniards. In a small grove of aspens below Mountain Sheep Pass, Bleazard found the remnants of three carettas. He recalled that they were crudely built and not very large, about the size of a small buckboard wagon. Because that area is so rough and broken, he often wondered how they were brought into the canyon. He thought it likely they were built at the mission site. He also found where six or eight holes had been dug near those carettas, which looked very much like rifle-pits where Spanish miners might have defended themselves. Bleazard later discovered a second massacre site further down country, just off the Blue Bench at Zimmerman Wash. It might have been where a last stand was made by fleeing miners and soldiers. He recalled that early settlers at nearby Utahn found old caretta wheels, sun-warped leather shoes and pieces of match-lock muskets. That is close by the old burial ground atop the ledges by the Indian race track.

There's a story which one keeps hearing about those old carettas found near the mission and massacre site. When homesteaders from Towanta Flat, which is on a bench-land between Rock Creek and Mountain Home, first saw those broken and burned carettas, the ground around them still looked like a battle field. Pieces of bone, both animal and human, protruded above the ground, where coyotes had dug into gravesites many years before. Metal buttons, sun-blackened

Ghosts of murdered miners still stand guard over the hidden treasures of the Rock Creek mission.

soles from leather boots and parts of brass tools could still be seen. Willie Davies said that he picked up several skulls, one of which still had an arrow embedded in it, and anyone who knew Davies knew he never lied. A few of the old Utes like Towanta and Bridger Jim said that when the massacre was over, all of the gold and silver which the miners and soldiers had, along with the church fineries, gold and silver goblets, chalices and urns were heaped in a wagon box taken from one of those carettas. That box of treasure was buried close by. Tales of that wagon box don't die easily.

Many years after victorious Utes heaped that wagon box full of treasure and buried it, the discovery of part of that cache came back to haunt them. When pickup trucks and sheep camps began to replace pack trains and tents, the Ute Tribe found it necessary to build a road across Towanta Flat to Rock Creek. Men still living remember the excitement when Indian workers cutting that road grade below Mountain Sheep Pass dug into a cache of ninety-one bars of silver! That bonanza find was obviously of Spanish origin judging from the Quinto or Royal Fifth marks stamped on them. A cross indicated they came from a church mine while the letter "V" proved the kings tax had been paid.

Because that cache was uncovered so close to the massacre site, tribal leaders at Fort Duchesne were quickly notified. It was their order that because of the curse on Spanish treasure, those heavy bars were to be re-buried at a place known to only a few select elders. An appropriate place was chosen close by, but when that site was prepared and ready for the bars to be unloaded, it was discovered that ten (by some accounts nineteen) of the bars were missing. Suspicion immediately fell on Wabun Wanzitz, a young college educated Ute. It was said that he hid those stolen bars near his cabin on Sadie's Flat, not far down country from where the

Wabun Wanzitz left this peculiar spelling of his name and strange symbol on a stone ledge above Rock Creek.

remainder were buried. If those suspicions were correct, then Wabun Wanzitz may have become the first Indian victim of the Spanish curse, for ever afterwards he was ostracized by the tribe and no longer accepted as a Ute, and he was shunned by white men who tried to learn the secret of those hidden bars. He became an alcoholic who lived in the worst of both worlds. Although his real name was Wabun, most now remember him as "Wobbin."

Wabun's mother was Sadie Wanzitz, for whom Sadie's Flat is named, and he was only one of a large number of brothers and sisters, few of whom lived beyond childhood. Because his father, Jimmie Wanzitz, had lost most of his children to sickness and disease, he somehow got the idea that Wabun, who grew up strong and healthy was some sort of witch, so he disowned and abandoned him. While he was still only a child, old Chief Dick Wanrodes took custody of young Wabun and sent him to an eastern boarding school, Carlisle College, where he became an accomplished musician. His father died in 1912 while Wabun was away at college, and he was adopted by Chief Wanrodes. He returned from college more white man than Indian, unable even to speak the Ute language. But Chief Wanrodes treated him as a son, and according to some, told him things about Spanish mines and treasures which few Indians knew.

Keith Rowley of Duchesne was probably Wabun's closest friend, and spent days with him roaming through places like McAfee Basin, Dick Hollow and Miners Gulch. Rowley recalled that Wabun always seemed to be searching for something he never found. He went to California, and while he was away from Rock Creek, another Indian died in his cabin. After the custom of the Utes, his cabin was burned to rid it of the dead man's spirit. When Wabun returned from California, he was heart-broken at his loss, for he had left all of his personal papers, an English-Ute dictionary he had been working on and dozens of old Spanish maps, charts, waybills and documents which had been given to him by Chief Wanrodes. He was told that all had been lost when his cabin burned. He sought comfort in whiskey, the white man's poison. Many thought that Wabun remembered the location of the mines and caches shown on those lost documents, and remembered that he had buried the silver bars from the Mountain Sheep Pass cache near his cabin. If he did, he never profited from them, for he died a pauper. The bitter irony of it all is that none of his personal papers, maps and documents were lost in that cabin fire. One of the old Utes removed them and only recently passed them on to his son, a fact known to few. You could say that Wabun drank himself to death for nothing.

Strangely, history sometimes does repeat itself. When the new Stillwater Dam was being built just upstream from the Upper Stillwater on Rock Creek, the old road which had been built below Mountain Sheep Pass was straightened to accommodate large trucks and construction equipment. A story then heard on the Ute Reservation told how a heavy equipment operator uncovered a cache of silver bars, very likely the same hoard which Ute elders had moved when the original road was built. That operator acted quickly to cover those bars with dirt and rocks before anyone else saw them. Well after dark he returned to recover them, but because of their weight he moved only a few before daylight. He covered the remaining bars with a covering of soil, but when he returned that night he found only an empty money pit. You can still see that hole near the road below the pass. No one seems to know who took the remainder of those bars. Were they the same bars which had been moved once before by Ute elders? We'll probably never know.

During March 1989, Duchesne County law officers received a report of a body in a shallow grave near the head of Tommy Hollow, below Mountain Sheep Pass and at almost the exact place the construction worker uncovered those silver bars. Subsequent reports from the agencies investigating that report were conflicting. All agreed there was a freshly dug hole, but agreement differed on whether it was a grave-site. One agency reported that a cache had been removed from that hole, marks indicating that something very heavy had been dragged away from it. Newspapers failed to report anything conclusive, but insiders around Duchesne claim that the FBI was on scene, since it was a crime on a federal reserve, and that agents found the body of a young Ute Indian in that hole. Some say that it was a case of two thieves discovering the construction worker's cache and then having an argument over division of the spoils. Strange things can happen when thieves fall-out, but anyone knows that a whole loaf is better than a half loaf!

There may still be some who doubt there was a mission along Rock Creek, or that an Indian uprising and massacre led to the caching of silver bars and golden church artifacts. If all we had to rely on was twice-told tales of early settlers, those skeptics might have a point, but we also have an account from an Indian who participated in that massacre, an account passed on from father to son in the Ute tradition. That oral history tells of the battle between Spanish miners and a combined force of Indians, including some of the Navajos who overthrew their masters at Lightning Ridge. This is the account of that battle, passed on to Dale Stevens, the man some call Gray Wolf, and recorded for posterity as it was written by him for the public record many years ago.

> TRIBUTE TO UNSUNG HEROES: Many years ago my grandfather took me high into the mountains, where I was shown a battle ground where a great number of people fought and died. If those who were enslaved had lost the battle, the whole Uinta Basin, and perhaps the whole United States would have been under jurisdiction of the Spanish. The straw which broke the camel's back was the brutal murder of fifty Navajo braves and their Chief. As the Spaniards were ruthless and blood-thirsty if anyone opposed them, the Indians killed them. There was a council and Indians came from all the surrounding area to answer the call to freedom. The battle commenced one afternoon in a long valley with a creek running down one side. Now picture in your mind, over three thousand Indians, armed only with bows and arrows and some very crude weapons, facing an enemy armed with cannons, muskets, rifles and much more modern weapons than the Indians had. How brave and courageous those Indians would have to be to face that blood-thirsty lot, considering that they were outnumbered by the difference in weapons. The positions which the Spaniards took were almost impervious. They had at least three large cannons and how many rifles and other weapons is unknown. The Indians had to approach and climb hills to reach the enemies positions. The bravery of those Indians is unsurpassed, and yet there are no markers or monuments erected to honor those fallen heroes. The people of the Uinta Basin have no remorse or feelings for those forgotten victors!
>
> Siguri Sioavi (Gray Wolf)

An ancient Spanish sword and double-barrel pistol, dating from the early 1700s. (Courtesy: Mr. Steve Malnar)

That massacre of Spanish miners has no date in the Indian's mind, for those Indians of old had little conception of time; only that it took place during the time of their grandfathers, which places it in the same time frame when Fort Robidoux was burned. It is important that Ute tradition and oral history records that the battle did take place, just as the white man's journals and written histories tell. Relics from that battle used to be commonplace, and even today it is not unusual for some new artifact to be uncovered. Ed Larsen found where dozens of hooves had been cut from pack animals just as Indian legend claims happened. One Ute elder said those hooves were severed and heads cut from Spanish miners so that their spirits would be condemned to roam that canyon forever. Some treasure hunters who have spent long, sleepless nights in that canyon swear those spirits still stand guard over the miner's caches hidden there.

Gray Wolf isn't a spirit, but it is said that he is one of those who stands guard over those old Spanish mines and caches. Those who know say that he has covered or concealed fifty or more old mines all across the Uintas, and that he will cover still more to keep the white-eyes from finding them. He told one prospector that a white man had stolen some bars from a cache and that he was just waiting for him to return. Just recently he approached one prospector and warned him against trying to go to any of the mines he was guarding. A chill ran down that man's back as Gray Wolf looked him straight in the eye and said: "There is plenty of gold east of the Green River. Do not look for it on my land!" The problem with that is, his land includes all of the Uinta Mountains!

A list of artifacts recovered from the massacre site

in Rock Creek Canyon would be a long one. Reid Lyons picked up part of a broken Spanish sword from where it had lain for countless years in the red sand of Daniels Flat. Only two years ago, two oil company employees discovered a complete sword and an ancient looking matchlock type pistol in that same area. The authenticity of that find is not in doubt. Both the Spanish museum at Seville and the British museum at London have examined that sword in great detail, and both have pronounced it to be a type made in Spain during the early 1700s. Known as the Pioneer Model of 1740, it was issued to civilian pioneers who accompanied priests, soldiers and miners who were sent to establish new missions in New Spain's northern borderlands.

In his notes on file at Weber State University, the late Dr. Donald Moorman described the type of sword and armament Spanish soldiers and civilian pioneers used:

> Spanish troops were armed with a broad-blade sword, a smooth bore carbine and a short barreled Migueler Lock pistol. Those weapons were considered to be nearly worthless even then, since each soldier was issued only three pounds of black powder each year, hardly enough to practice with. They wore a leather jacket and leggings, both made of four to six thickness' of bull-hide; very hot and heavy. That type of equipment continued to be issued until the time of Mexican independence.

Another account tells how the Spanish soldier must have presented a strange and fearsome sight to the Indians, arrayed as they were in metal armor and helmets, with a musket on his shoulder and a long lance of sword at his side. But by the early 1700s, their heavy armor had been replaced by buckskin shirts and trousers, with a heavy leather vest or jacket, which had a bull-hide shield (adarga) on the left arm to ward off arrows. Heavy coats of mail and metal helmets were conveniently lost or thrown away; bits and pieces of which are sometimes found even today.

Catholic crosses and rosaries have also been found at the massacre site, one particularly fine gold cross being discovered by a western history writer who wishes to remain anonymous. While exploring in that area, Earl Garrison came upon a sun-blackened leather pouch filled with gold nuggets, probably some miner's personal treasure. In 1966 an Indian found an unusual bronze bell, about four inches in height. Many similar priests bells have been found, but that particular bell is peculiar in that it bears two Catholic crosses and the words "Saigneleceur Chlantel Fondeur" inscribed on it. Other bells have since been found near the ruins of old Fort Kit Carson, another near the Ouray Trading Post and another near the North and West forks of the

A Catholic priest's brass bell, found along the old Black Robe Trail at lower Rock Creek.

Duchesne River. Still another was found recently in an old Indian ruin near Blanding, close to the Arizona border. Tests indicate that bell is more than 400 years old!

There are tales still told that when the Rock Creek Mission was abandoned by the clergy, they had no time to take anything except a few cherished items. The great bell which called miners to worship was hastily buried. The Old Ones say that bell was so heavy that it took six men to lift it. That might be dismissed as a tall tale if similar bells hadn't been found at other mission ruins. During the Pope Indian rebellion of 1680, the great bell at Mission San Miguel near Santa Fe was left behind by fleeing padres. That bell can still be seen at those old mission ruins. It weighs 780 pounds, and bears the date it was cast, 1356! One can only wonder how such a heavy bell was carried so far north across a thousand miles of barren wasteland. After the question "how" is resolved, there still remains the more difficult question, Why? If such a great bell was carried as far north as Santa Fe, why couldn't a similar but perhaps smaller bell have been brought to the old mission at Rock Creek?

Andy Nelson of Duchesne made a fascinating discovery. He was hiking at Pigeon Water, not far up country from the old mission and massacre site. He was hunting forest grouse and when he spotted one in some brush, he knelt to pick up a stone to throw to make the bird fly. The ground was covered with oak and aspen leafs, and while feeling in search of a small stone, his fingers touched something cold and metallic. To his surprise, he picked up a pair of ancient looking binoculars from where they had been lying on a rocky outcrop for probably a century or more. They are peculiar, being

Antique binoculars found at Pigeon Water, bearing the name, Seville, Spain. (Courtesy: Mr. Andy Nelson)

made completely of brass, with each of the four lenses being individually adjustable. They had been on that ledge rock exposed to the sun for so many years that their lenses had turned purple with age, while their surface is pitted from the abrasion of countless sand storms. They are a valuable artifact, but if their true story could be known, they would be doubly valuable. Nelson can't help but wonder who left them on that rocky ledge along Pigeon Water so long ago. Could it have been a Spanish soldier fleeing the Rock Creek massacre?

The artifact treasures found by Andy Nelson and others are legitimate finds, whether recovered through research or by accident; and in the author's opinion, but contrary to government antiquity regulations, should be recovered and displayed for the public's education and enjoyment; not left where no one will ever see them. But there is another kind of treasure which some seek, and which should be illegal. Those people are not relic hunters, they are ghouls, grave robbers. They sneak into old Indian burial grounds during the dark of night, to steal the trinkets and ornaments which were interred with the deceased. They are despised by the real treasure hunter or lost mine prospector, and they are fair game to the Utes whose ancestor's graves are being desecrated.

In the old days it was the custom to bury an Indian with all of his worldly wealth so that he might have his little treasures in the after-world. At my great grandfather's cemetery lot at Heber City, there is buried the body of Tom Tabby, son of Chief Tabby of the Utes. Recently the author was instrumental in having a commemorative marker placed at that grave-site in Tom Tabby's memory. Chief Tabby requested that his good friend, my great-grandfather, Bishop Joseph Stacy Murdock, bury his son in the manner of the white man. After the simple ceremony, the boy's favorite pony was led to his grave where it was killed, so that he might have a pony to ride in the hereafter. In that same manner, older Indians used to have their prized possessions buried with them: weapons, silver rings or bracelets and even gold coins. During the burial of one noted Chief, a long line of mourners passed by, each tossing a coin or piece of jewelry into the open grave. Those mementos are the treasures sought by grave robbers today. During the summer of 1986, six Indian graves were found open and desecrated, with pieces of bone thoughtlessly tossed aside in the ghoul's haste to find treasure. Each had been buried with his rifle, and rusted parts from them were also thrown aside. But without doubt, some articles of value were discovered and stolen. The Ute Tribe at Fort Duchesne acknowledged as much in a press release which stated, "There were many artifacts stolen from those graves." At that time a one-thousand dollar reward was issued for information leading to those grave robbers. In 1992 it was reported that other graves on Dry Mountain were dug into. Whatever was taken from those graves was little more than curiosities or trinkets to the thieves, but they were heirlooms of great value to those whose family members were buried there. So if the Indian game warden who checked your fishing license or the ranger who directed you back to the main road didn't seem very friendly, perhaps now you know why.

In the old days Ute Indians were often buried in a rock crevice or some shallow excavation, but rarely if ever in a tree as some plains Indians were. Squaws did most of the mourning, braves seldom attending burial rites. Most high ranking Indians had their valued personal items buried with them, and doubtless many things of value were buried with the deceased. Records at Fort Duchesne tell of one instance when some greedy white men who wanted to get their hands on some of those artifacts were frustrated. Tim Johnson, who had the reputation of being a mean renegade, told of those white men's efforts to located the burial place of Chief Ouray, allegedly to erect a monument there, even though Indians do not place such markers. Johnson showed the white men a place where an Indian had been buried, which the white men dug into to verify it was an actual grave-site. All they found for their efforts were a few bones, but a small marker was erected anyway. Later Johnson said: "Ouray, he alright. White mans never touch him. They make big talk and we shake hands. White mans happy. I no tell 'em Ouray not there. Ouray have fine ring on finger, gold watch, silver belt. They no find those things. But white mans is happy. Ouray, he happy too!"

It is commonplace to come upon old Indian grave-sites in the mountains. Many are lone burials, but sometimes there are several graves together. Time and erosion have exposed some of those sites. At one place north of Duchesne, several graves were dug into. The bodies there had been buried in blue uniforms, with heir rifles buried with them. Very likely they were guides or scouts assigned to the 9th Infantry Regiment stationed at Fort Duchesne, and as such they died honorably and were buried with military honors. But some ghoul desecrated their graves. Not only is digging into Indian graves a felony offense punishable by large fines and jail time; it is also immoral and indecent. And it could be very dangerous. I wouldn't place much value on the life of anyone caught digging into an Indian burial site. Would you want someone to dig up your grandfather's grave?

!!!UPDATE!!! Just as the final editing of this book was being completed, a man who for the present must remain unidentified discovered the long sought-after entry to the "Water-Trap" chamber at Treasure Hill! Twenty-five years after two fishermen discovered that ancient passageway with its ingenious barrier, that mysterious labyrinth was penetrated again. The finder described how only a few feet from the entry, the passageway become large enough to stand in. A few feet more it turns sharply to the right and then descends steeply into a well-like pit filled with icy water, no doubt at river level; the intriguing water-trap. The water was crystal clear, without any evidence that anyone had disturbed it in decades. Dripping water had coated the rock walls with a calcite covering and a musty odor of damp air and wet mud long undisturbed by man permeated the atmosphere. The finder learned that the tunnel-like passage turns to the left as it descends into the depths. Without proper lighting and scuba gear he was unable to trace its depths, but by the time you read these lines the first person to probe the depths of Treasure Hill in a quarter-century will have unlocked its secrets. The accompanying photos verify his fantastic discovery. Who knows what treasures will be found? Stay tuned!

CHAPTER 10

LOST BONANZAS
The Riddle of Rock Creek

Whenever I am asked about lost mines or treasure caches, the place most people ask about is Rock Creek. There is a sort of mystique about Rock Creek, a feeling that when one is there, he has stepped back a century in time. It can be a scary place. People tell me that when they are there, they have a feeling that they are being watched, that they keep looking back over their shoulder to see who is behind them. Many search the Rock Creek area for the Lost Rhoads Mine, but from what I've learned through search and research, neither Thomas or Caleb Rhoads ever said the mine was there. For that matter, neither actually said they mined gold, only that their gold came from a mine; which suggests that their gold may have come from a cache hidden in a mine. On several occasions Caleb Rhoads referred to "the place where I got my gold." Nor do most of the clues they left fit either the terrain or geology of Rock Creek.

Almost everyone who asks about Rock Creek has a genuine Rhoads map, but the ones I have seen, whether "genuine" or fake do not correspond well with the Rock creek area. I am not saying none of their gold came from Rock Creek; undoubtedly some of it did. But whether their gold came from a mine or a cache, and whether they dug for it or were shown or discovered a hoard of gold already dug, there is something about Rock Creek that is hard to explain; something mysterious, and at times a little scary.

That there are old mines in Rock Creek Canyon is beyond question; too many have been found for there to be any doubt about that. Legendary mines and caches like the Carre-Shin-Ob, Brigham Young or Church Mine may not be there, but plenty of others are. Something like forty years ago, Bob Burns, a Park City miner, told the author of seeing an old mine along the river while he was fishing there. Being a mining man from a mining town, I have no doubt that Burns knew what he saw, but being unaware of legends of a lost mine somewhere in that canyon, he paid little attention to his discovery. He said he saw a tunnel portal behind some heavy brush, almost at river level, its entry covered by a heavy wooden door. That door was so old and weathered that it was green with moss, but Burns had seen many similar mine portals in the hills around Park City, so he gave it no thought until some time later when he first heard stories of Spanish miners digging for gold along Rock Creek. Burns described his find to Fraser Buck, a mining developer and close friend of the author. Always interested in new mining ventures, Buck made the long trip to Rock Creek to investigate the Burns find, but even with the best of directions he failed to locate it.

That old mine seen by Burns is almost certainly the same one later found by Harry Young, a construction contractor. According to an account given to the author by Gale Rhoades, Young also saw that mine with the wooden door while on a fishing trip. The base of that heavy door was covered with cobblestones and driftwood which had been washed up against it by the river, so that he was making little progress trying to force it open when he was suddenly grabbed from behind by two Indians. Without further ceremony he was dragged to a fallen pine tree, where his hand and fingers were spread out atop that log. Before he realized what was happening, one of the Indians swung a small hatchet and chopped three fingers from his hand. Young was yanked to his feet and given a rough shove to start him on his way, one of the Indians saying, "You get going! Next time you come here, we cut off your head!"

Not everyone who has prospected along Rock Creek was as lucky as Young. Several have lost more than just their fingers. A mortician who operated a

A forgotten grave along the trail. The Killer Mountains keep their secrets well!

funeral parlor on the reservation for many years told the author that during many years of service he had been called upon to help investigate no less than two dozen suspicious deaths in the Killer Mountains, more than a few of them in the Rock Creek area. A young man was found hanging from a small cedar tree, but that tree was only five or six feet high, so that his knees rested on the ground. The sheriff pointed out that it was impossible for anyone to hang themselves from such a small tree, and also noted that the dry cedar needles and soil under his knees hadn't been disturbed as they would have been by a person thrashing about in his death struggles. It was obvious that the victim had been hanged elsewhere and was then brought to where the body was found, perhaps as a warning to others who might try to find the sacred gold; nevertheless, the coroner ruled the death to be a suicide.

That same mortician also investigated several deaths where victims were found with their necks broken, in each instance found in a place where they could not have broken their neck from a fall or accident. Investigating law officers were in agreement that they had been killed by persons unknown, but by whom was never determined. Several more victims have died from strangulation, and almost everyone would agree that it is difficult to strangle oneself. Retired Duchesne County Sheriff George Marrott investigated a case in which an old man who was a prospector had been reported missing in the mountains above Rock Creek; in the month of July. Several searches of that area were made to no avail. During the following October, deer hunters came upon his body lying between two large logs which had been rolled close together. His body could only be seen by someone standing on top of those logs and looking straight down. His death was attributed to natural causes.

Miner's Gulch is a steep canyon located just upstream from where Albert Peterson located his ranch homestead along the Lower Stillwater on Rock Creek. Peterson and others who then lived in that canyon came upon places in Miner's Gulch where Indians had dug into deposits of red-colored iron oxide ore, which they used as paint to cover their face and bodies during tribal rituals. William Bleazard, Ed Haddon and Eph Mecham sunk a shaft about twenty feet deep into one of those iron ore "blowouts," to where they encountered high-grade ore. They weren't interested in iron, but hoped to find other minerals there. That iron was so pure that dust from it mixed with sweat on their brows turned their faces blood-red. Bill Bleazard later recalled that a grain sack filled with that ore was so heavy two men couldn't lift it.

One day while hiking to their prospect in Miner's Gulch, Ed Haddon picked up and pocketed a small chunk of heavy black rock which he found at the mouth of a small draw along the trail. He carried that rock which he thought was hematite iron with him until its black surface patina was worn off and it became shiny. Then one day he noticed small specks of yellow color coming through that black varnish, and upon closer examination he discovered that it was really a gold nugget which had been coated with black manganese. He lost no time searching for more of those black nuggets, but he could never find where they came from. Jed Pulley was another who found some of those black nuggets coated with manganese, but that was a few years later and several miles downstream from Miner's Gulch. They looked exactly like the nugget Ed Haddon found. Pulley recorded a mining claim on his find, but it was on Indian ground so he never described its location in detail, since he knew he would probably never be allowed to work it. That location notice states simply, "Along Rock Creek, about fifteen miles from Duchesne."

Willie Davies was one of the early settlers in Rock Creek Canyon, and it was he who built the original resort lodge at the Upper Stillwater. He discovered a mine completely by accident. He was inching along a narrow ledge near the head of Miner's Gulch when he noticed a narrow slit-like opening in the cliff he was following. That crack was nearly hidden behind a large boulder where the ledge turned sharply to the left. Slipping sideways into that narrow opening, Davies discovered that he had entered a room-like chamber. A tunnel partly filled with fallen rock led from that chamber deeper into the mountain. Davies was close-mouthed about his find, but those who knew him best said that he later excavated that tunnel and uncovered a thin seam of gold. They say that is where he got the money to build the Rock Creek Resort.

During the summer of 1991, one of Willie Davies' descendants was attending a family reunion at the Yellow Pine Campground just below Miner's Gulch. He decided to hike into that steep, rocky defile to see if he could locate the mine Willie Davies had found. He came upon an old trail which passed under the limbs of a giant pine which had a Catholic cross cut deeply into its trunk. Following in the direction of the longest arm of that cross, to the northwest, he came to a place where there was a small flat knoll on that otherwise steep mountain front. Looking upwards he could see the cliffs which are just to the west of Bear Lake, still higher above him.

At the back of that knoll, up against the mountain front, he saw a low wall of closely fitted stones, with a tiny trickle of water coming from under the edge of a ledge rock. He noticed that the rocks above that low stone wall had been fire-blackened sometime in the past. He discovered an old ash pit behind that wall, where he dug out enough cinders and slag to fill a small gold pan he carried in his pack. Panning those ashes produced about a spoon-full of tiny brass-colored pellets. Those small globules proved to be mallable and easily flattened when he hammered them between two stones. A later test revealed that they would dissolve in nitric acid, and an assay proved they were nearly pure copper. The assayer volunteered his opinion that those copper pellets may have come from a primitive smelting operation. He explained that in ancient times, molten bullion metal would be poured into a container of cold water. Some of the gold could be separated from copper that way, with most of the copper splattering into tiny grains or pellets. It seems logical that the place where that copper-gold bullion came from can't be very far from that old smelter site.

Only a few months later, in August 1991, a prospector named Jim McGee reported finding a Spanish arrastra in Peterson Gulch, just over the ridge from Miner's Gulch. That old mill was unusually small, being only four feet in diameter, just about the size someone might build to work a small high-grade gold vein. It was expertly made of closely fitted stones, so old that those stones were covered with moss and lichen. One can't help but wonder if there might be some connection between that arrastra, the gold vein found by Willie Davies and the smelter ruins found on the mountain side above Yellow Pine Flat.

During the 1950s, Cecil Dalton discovered the portal of a long abandoned mine tunnel in Miner's Gulch. It had been cut through hard rock with the use of metal tools, and was large enough for a man to stand erect in.

Ruins of an old smelter near Bear lake. Can the gold ore smelted there have come from very far away?

Seventy feet into the mountain, that tunnel branched into two separate drifts. Careful sampling of the foot and hanging walls of the ore vein produced high value iron and copper, but Spanish miners wouldn't have dug a tunnel so far through solid rock for iron or copper. Could there have been a narrow seam of gold like that found by Willie Davies, or perhaps some manganese covered nuggets like those discovered by Ed Haddon and Jed Pulley? If so, those miners of old had picked that vein clean.

Still higher in Miner's Gulch there is a rock slide, and at its base there is a strange circle of trees; but they are too perfect in shape, too symmetrical to be just an oddity of nature. Just inside that circle of trees there is a circle of stones, one in front of each tree. There are similar stone circles in Pigeon Water and atop a ridge overlooking Sadie's Flat, which some say are council circles, but the one in Miner's Gulch is different. All of the stones there are odd-shaped and sharp edged or pointed on top, certainly not the sort of place a council member would chose for a seat. An aged Indian patriarch once said that a rock slide covers the entry to a Spanish mine. He may have been right, for one of those he confided his secret to says that old Indian still has a small box of gold ingots which came from that mine, ingots about the size and shape of a biscuit or muffin. If you get up that way, stop and take a closer look at those pine and stone circles at the foot of that rock slide. It could be worth your time!

A few years after Dalton discovered that old tunnel in Miner's Gulch, Jim McAfee told him of an un-nerving experience he had in Miner's Gulch. McAfee owned a small homestead, where one of his neighbors was Henry Wash, a Ute Indian. Over several years McAfee loaned Wash small amounts of money and helped him in other ways. When Wash became old and ill, he decided to repay McAfee for his past kindness in the only way he knew, by showing him the location of a mine in Miner's Gulch.

Traveling by night so they couldn't be seen by rangers stationed at the Farm Creek station, Wash led McAfee up Rock Creek Canyon past the Farm Creek road to the mouth of Miner's Gulch, where they arrived at the first light of dawn. The further they climbed up that steep gorge, the more nervous Wash became. McAfee noticed that he often stopped to look back down canyon, as if checking their back trail. As they rested, Wash told McAfee about the renegade Ute white men called Red Cap. When the reservation was opened to white settlement, Red Cap and a band of dissidents fled the reservation and went to South Dakota where they tried to persuade Sioux Indians to help drive the white men from Ute country. His mission failed and Red Cap and his followers were forcibly dragged back to the Ute Reservation by government troops. After his return, Red Cap vowed to kill every white man he found in the Uinta Mountains which the Utes still claimed as their own. According to Wash, Red Cap had killed several white prospectors he had found in Miner's Gulch. Wash's story did little to calm McAfee's rising feeling of apprehension.

When they came in sight of a ledge of rock high in the gulch, Wash suddenly stopped and said he would go no further. He told McAfee that if wanted to continue alone, he would find several graves under a large pine just below that ledge, a place where both white men and Indians were buried. He said their spirits still guarded the gold in the old mine, which McAfee would find just beyond that ledge. He pointed out a faint trail which led around that ledge, but he refused to go any closer to it. McAfee went on alone, but just as he started up that ledge trail, he heard a rifle shot echoing from somewhere along the canyon rim and turned to see Wash running down the gulch.

Not knowing what had happened, McAfee followed after Wash as fast as he could. Wash never stopped until he got back down to Rock Creek. Completely exhausted and obviously very much afraid, Wash insisted they were being followed and that they had to get out of that canyon fast if they wanted to get out alive. McAfee hoped to return one day but it wasn't to be, for Wash died soon afterwards and McAfee had no stomach for going back alone. He sold his homestead not long afterwards and left the area. He never tried to find that old mine again.

Mysterious stone circles are still found in the mountains. What strange tales might they tell?

Rarely did an Indian reveal the location of a Spanish mine to a white man as Wash tried to do for McAfee. Almost without exception, Indians have a religious fear of places where in the past their ancestors had labored and died for the white man's money-rock. Recently a white man personally experienced that taboo. Through his position in government service, he became closely acquainted with several high-ranking tribe members, so close that he was sometimes invited to go fishing with his Indian friends. One day when they were at Rock Creek near where a Spanish mine was said to be located, he asked if either of his two companions knew where that mine could be found. He received no reply, but he couldn't help but notice the change which came over his friends, as if they were suddenly afraid of something. When he asked again, one of the Indians replied: "You are my friend. I don't want you to die. If you get too close to those mines the spirits will get you. You will get sick and die!"

But that was then and this is now you say, and experiences like McAfee's or the death warning my friend received don't happen anymore. Don't you believe it. During the summer of 1991, two men were covertly making their way up a steep canyon just below Mount Albert and across the canyon from Miner's Gulch. They knew they were trespassing but thought they were alone until they heard someone on the ridge above call to someone else unseen in the canyon below. Watching the ridge-line, they spotted two Indians, one on each side of the canyon, both of them carrying rifles. It didn't take them long to figure out that those Indians were hunting, and that they were the hunted! They crept and slid down the rough canyon bottom on hands and knees until they got back down to Rock Creek, and there they swam and crawled along the river's edge until they were well out of sight of Mount Albert. Neither of them is much interested in prospecting there again!

Long after McAfee and Wash climbed into Miner's Gulch, a huge reclamation dam was built not far upstream, just above the Upper Stillwater. One of the men who worked on that construction project was named McKenzie, and during his off work hours he liked to spend spare time prospecting. One day when he was high in Miner's Gulch, he came upon what he thought might be an old mine at the foot of a ledge, below a large pine. He dug into a sunken area hoping to uncover a mine shaft, but instead he dug into a grave which had four skeletons in it. From scraps of clothing and small artifacts, he determined that two white men and two Indians were buried there. McKenzie told an Indian co-worker about his find and was told that if he wanted to learn the story of those old graves, he should go to an Indian woman, a mystic known as The Old Squaw Of The Spirits.

The old Indian woman warned McKenzie that his life would be in jeopardy if he continued to dig at that grave site. She said that a spirit more terrible to behold than he could imagine guarded those forsaken graves. That psychic also told him that there was an ancient treasure cached close to those graves, but not where he had been digging. McKenzie heeded the advice of the old woman and left the graves of the ancients alone, but he did pass that way often, and every time he felt as if he was being watched, even though he never saw anyone there. Then one day McKenzie saw something so frightening that he gave up his search for the treasure cache and quit his job. No one ever knew what he saw near those graves in Miner's Gulch, but he told his Indian friend that he was getting out of that area while he was still alive, which is more than can be said for those four unknowns buried beneath that pine. Score four more for the Killer Mountains?

Some have wondered about several Spanish treasure sign trees along Rock Creek at the edge of Treasure Hill on Daniels Flat. On three large pines, one of them now fallen and nearly hidden in thick willows, are what appear to be Catholic crosses and a square box figure; according to some an indication that there is church treasure buried nearby. But others say they are not treasure signs at all, but only a cattle brand, a T-Up and a T-Down, cut into those trees by Ed Rowley back around 1920. Bill Bleazard grew up on Daniels Flat and he is not so sure about that claim, for he has seen those same signs in the mountains a long way from Treasure Hill. One prospector who has covered a lot of that country has found what he believes was a small trail mission or visita in the mountains near Atwine Lake. He found those same kind of crosses on trees there, so we have a question: Are those tree markings at Treasure Hill and near Atwine Lake only cattle brands, or are they Catholic crosses which point the way to church treasure.

Nearly seventy years ago, Mark and Bill Bleazard worked on a leased section of Indian land just down stream from Miner's Gulch, the area now known as Daniels Flat. There was then, and still is, a log cabin there which had been built by Walt Daniels, an Indian ranger who kept a close watch on that area, including the old mine in nearby Dick Hollow. The Bleazard brothers herded cattle in the canyons above Daniels Flat, at places like Pigeon Water and Miner's Gulch.

Only ghosts frequent the last cabin still standing at the Sadie's Flat Indian settlement.

They would ride from their homestead on Towanta Flat down Twist Hollow to Sadie's Flat. At that time Sadie's Flat was a small Indian settlement, boasting about two dozen log cabins. Only one of those cabins remains today. There were also several cabins near Daniels Flat, three of them on the west side of the river. Except for the Daniels cabin, none remain today.

Bill Bleazard recently recalled that while herding cattle above Daniels Flat, neither he or his brother were ever bothered or followed by their Indian neighbors, that is not unless they entered one particular canyon. Bleazard said that every time either of them rode into Dick Hollow, an Indian would appear to ride with them; and quite often it would be Walt Daniels. Another old Ute who sometimes watched their movements was Ike Sauschnick, who most white men called Shotnick. Nearly everyone said that Shotnick was "a mean old man," whose job it was to watch over old Spanish mines and caches. He was a Ute elder of sufficient status that a trail and basin were named for him. The Shotnick Trail ran from the Uinta River near present day Neola across the bench-lands to Mountain Home. From there it angled up onto Towanta Flat and then wound down Twist Hollow to cross Rock Creek at Sadie's Flat. It continued up Sadie's Draw into Shotnick Basin and crossed the high ridge at Farm Creek Pass to descend down to the Duchesne River and on to Tabby Mountain.

More than a few of the mysterious deaths of prospectors and treasure hunters in the Killer Mountains have been attributed to Ike Shotnick. Two men named Snyder, father and son, disappeared in the Farm Creek country not long after being seen in Shotnick's company. He had a cabin along the Duchesne River near Tabiona, but most of his time was spent in the high country near Shotnick Basin, where he could keep a close watch on Rock Creek below, from Sadie's Flat to the Lower Stillwater. It is known that Shotnick went to prison allegedly for killing a prospector. Records reveal that he became so sick while in prison that it was thought he would soon die, so he was given an early parole. It was almost miraculous how quickly his health improved once he got back into the mountains at Shotnick Basin! One of the old Utes recently told me that it was Shotnick's son who killed that man, but that Shotnick took the blame to keep his son from going to prison. He said that young Shotnick actually killed three men in the Rock Creek country. He also said that old Shotnick killed his Mexican wife and then hung her body on a meat hook, like a deer carcass. Like some said, "he was a mean old man!" No one ever wanted to cross him a second time. LaVar Thompson could testify to that. He was prospecting near Dick Hollow when Shotnick stopped him, telling him that if he ever returned, he would kill him. Old Shotnick has been dead a long time now, but Thompson says he still hasn't gone back to Dick Hollow!

Some of the old-timers around Tabiona say that before he died, Shotnick cached $10,000 in gold coins somewhere near his cabin. They say those coins came from the cache on Farm Creek described by Alice One-Leg. If he had some of that gold to cache it may still be there, for no one has ever reported finding it. Ike Shotnick sometimes was seen at an old mine not far down stream from the old Heber and Jesse Moon homestead on Rock Creek. The Moon place was located just below Albert Peterson's ranch. The tunnel to that mine was covered over when the original Forest Service road was built, but there are a few old range riders who can point out an air shaft along its course just above the valley floor. If you hike into the canyon just above that sir shaft, you will see two Spanish signs cut into the rock of the canyon wall. Shotnick left a map which was found after his death, but it is an enigma. Whether it pertains to the old mine near the Moon Ranch is unknown. That map shows a ledge with the notation, "Dig Here," but the directions and compass headings are so reversed that it is unknown what part of the canyon the map refers to, or in which direction one should proceed, or how far. Early settlers like the Moons and Albert Peterson knew that Shotnick still brought gold from that mine even after its portal was covered over, but neither they or anyone else has been able to decipher that strange map he left. Ike Shotnick probably got the last laugh after all! (See Map M)

It is possible that the Shotnick Mine may have been found some years ago by an early resident of Roosevelt. Several years ago Jim Nebeker told Les Bush about a man who was as poor as anyone during homesteading days, just barely keeping body and soul together. But then he rustled a job working on the road then being built along Rock Creek. After only a few weeks on the job he suddenly quit work without notice, and began investing a lot of money in his homestead. Those who remember him say that he never worked much after that time, but that he always had plenty of money to spend. A nephew noticed something peculiar about his schedule. About once each year he would make a lone trip to Rock Creek, but only during the dead of winter when it was stormy, probably so that there would be fresh snow to cover his trail.

When he became old, the uncle summoned the nephew to his home and told him that for years he had been taking gold from an old mine he had discovered in Rock Creek Canyon. He showed his nephew some unusual looking pieces of gold, in wires and thin strips which were bent and twisted, as if they had been pried or cut from a narrow seam of ore with a knife or similar tool. The old man knew that his days were numbered, and said that when spring came he would take the nephew to his mine, but the old man's summer never came, for he dropped dead from a heart attack during the winter. The only clue he had given his nephew was that the mine was just a short distance below the old Heber Moon place, but on Indian land. That narrows the search area quite a bit.

There is a strange story about a cache site instead of a mine where that old man might have found his gold. Two different men have seen that place, but far separated in time so that there is no possibility of them knowing each other or fabricating a story. The tales they told are so similar that there seems to be little doubt they were both telling the truth. In the mountain country above the old Moon Ranch, but still well below the high ridges leading to Monument Knoll, there are many small crawl-size adits, places a miner might call a coyote hole. For the most part they are natural openings, and most are shallow, penetrating into the mountain for only a short distance. Many have colorful calcite crystals in them.

One of the pioneer Davies family said that when he was young, an Indian friend pointed out one of those coyote holes and said: "Plenty money rock there!" When Davies had the opportunity he crawled back into that small hole, and after he had squirmed his way for about thirty feet, that hole opened up into a large underground chamber. Both the hole and the cavern appeared to be natural formations, but inside that chamber he found a man-made rock wall, where chunks of stone obviously broken and worked with tools were piled up. Behind those rocks rocks there was a shaft, inclined downward at a steep angle. Curious why someone would have dug a shaft there, he began moving fallen rocks from it. He continued that hard work on and off for several months, until he could no longer lift those heavy rocks up to the collar of the shaft. His Indian friend had warned him of the consequences he might suffer if he was found there, so he only worked when he knew he would be unseen. Finally, after a summer of back-breaking work in that dusty cavern, he gave up and never returned. But he would often remember the Indian's words: "Plenty money rock there!"

In December 1992, just as the first heavy snow of winter was burying the Rock Creek county, the author received an anonymous telephone call. The caller asked for information about how to sell a large quantity of gold with no questions asked. When asked what form that gold was in, its weight, purity, etc, the caller related a strange story. He had been hiking with two friends in the rough country between the river and Monument Knoll when they came upon several coyote holes, one of which was big enough to crawl back into. The caller said that he volunteered to investigate its depths and crawled back into the mountain for forty feet or more to where the hole opened up into a room-size cave. In the back of that room there was a near vertical shaft. At first he thought the small passageway and chamber were natural formations, but then he discovered a hand-forged chisel and noticed drill marks on some of the rocks.

Wondering why there would be a man-made shaft in that cavern, he agreed to be lowered into it by his companions, on a short length of light-duty rope one of them had. Moving broken chunks of rock aside at the shaft bottom, he uncovered a mound of rotted leather bags, each filled with what the caller described as being gold disks; larger than a dollar and weighing about one pound each. Those disks were thicker at the center than at their edges, and each had some sort of symbols cut or stamped into them, strange characters he described as hieroglyphics. With only a light rope to bear his weight, he took only a handful of disks which he could carry in his pockets. Before leaving the mountain they piled some rocks over the entry to that coyote hole and began making plans to return with a heavier rope and other equipment. The caller was advised of the proper way to dispose of those gold disks and the agencies he should report his find to, after which he agreed to tell where that coyote hole and shaft was located, since he expected to remove all of those gold disks within the next few days.

There is too much similarity between the two accounts of the coyote hole and chamber with a shaft in it to be a coincidence, especially since the two incidents were well separated in time and were told by persons who not have known each other. The story differences are minor; the length of the coyote hole and the depth and angle of the shaft. Both holes were described as being above the old Moon Ranch, but not as high as Monument Knoll. If you get up that way next summer, check that coyote hole out. No doubt the gold disks will be gone by then, but at least you'll know where they were. There's always the chance they missed a few!

Not many years ago, Charley Iorg, a part German, part Cherokee Indian located a mine only a short way up-canyon from the Ike Shotnick diggings. Iorg called his mine the Wild Bull. It was located just above Yellow Pine Flat near what old timers call Limestone Point. That is exactly where the Hoyt Peak Fault crosses Rock Creek. Iorg built a small cabin on the west side of the river, overlooking Yellow Pine. For several years he dug out high-assay silver ore, which was said to carry good gold values also. Once claim jumpers tried to move in on Iorg, but he ran them off. His mine tunnel was about seventy feet deep at that time, a depth no sane man would have dug single handed through solid rock if there wasn't something of significant value there. Then for reasons which only the Forest Service in its infinite wisdom knows, he was ordered to quit mining. Over the years since then talus rock has slid down the mountain and covered his tunnel, while his cabin fell into ruin and rotted away. Only a few logs remain to mark its site.

Many years after Iorg left his mine and cabin, a man named Dick Park began taking groups of Brigham Young University students to an ore outcrop close by the caved Iorg Mine. Their training included work on an ore vein between an iron cap and limestone. Each student would carry buckets filled with heavy ore down from the mountain to Park's truck. Brett Sutherland was one of those students, and he recently told me that was about the hardest work he ever did, balancing two heavy ore buckets and crossing Rock Creek on a slippery log. After the ore was trucked to Provo, Park would crush and then concentrate it on a wilfley table. One of those students told me those concentrates would assay more than fifty ounces of silver. Sutherland verified those figures, stating that as he recalled, they tested sixty-three ounces of silver and four ounces of gold. He also said that Park knew of an old mine about a half-mile north of where the students worked. That old mine had some peculiar and ancient looking smelting pots in it, but the students never worked there because it was caving and extremely dangerous to enter.

Several prospectors have found gold nuggets near the old Iorg Mine. One recently told me that he pocketed some real good gold there, but he's not going back for more. He said he hadn't seen anyone all day, yet he had a feeling that he was being followed and he was sure he was being watched. As darkness fell, he noticed that night birds seemed to be calling all around him. To keep off the evening chill, he built a small fire. When he walked outside its circle of light to pick up a few more sticks of wood, he spotted another campfire only a few hundred feet away. Then he saw two more campfires, three camps in a circle all around him, where he hadn't seen another person all day. He hurried to each camp, expecting to encounter someone, but he found no one, not even a footprint in the sand, only three campfires. He called out, but there was no answer, only the cries of some night birds, in a circle all around him! He said it took him less than two minutes to throw his bedroll and tools into the back of his truck. He hasn't been back since, and he says he isn't going back. If you're up around Yellow Pine you might want to check out the old Iorg Mine; but don't stay after dark, not if you're afraid of night birds!

Strange things have happened at Bear Wallow, north of Pigeon Water and east of the long ridge separating it from Dry Canyon and Miner's Gulch. Several hikers have seen a well hidden mine portal there. The late John Sprecker of Duchesne was hunting deer when he noticed a peculiar slab of rock, obviously not native to that area. That rock was quite large and triangular in shape, but what set it apart from the common red sandstone was its white color. Sprecker wasn't aware that Spanish miners often placed such stones as trail markers or to draw attention to a particular area. He tried to lift that rock and in so doing discovered that it was balanced so that it could easily be turned sideways. When that white stone was rotated to the side a small vertical shaft below it was uncovered. That stone had been purposely placed to conceal the entry to that shaft. A rotted chicken pole ladder proved that men had once descended into that pit. Sprecker planned to have someone help him investigate that shaft, but unfortunately, he died not long after his discovery, so that its location was lost.

Curiosity sometimes tempts fate, for just last summer that same out-of-place white marker stone was found again. Mark Mason was hiking in the Bear Wallow area when he stopped to look more closely at a symbol cut into a tree; a triangle with a dot in its center. Looking beyond that tree, he was surprised to see a triangular shaped white stone shaped just like the triangle on the tree. He said that he wouldn't have seen the stone if he hadn't sighted past that sign tree. Mason wondered why there should be a large white stone where all of the rock was red sandstone, but like Sprecker before him, he too was unaware of Spanish marker stones and continued on his way. It wasn't until I told him of John Sprecker's discovery of a shaft beneath that stone that he decided to return to Bear Wallow and check out that marker rock, that is if he can find that Spanish tree sign again!

Stan Brennan, a gray-bearded Mountain Man type, has spent a lot of time at Bear Wallow. About four miles north of Mountain Sheep Pass, Brennan discovered a place where there are many old Spanish signs cut into large and obviously very old trees. By following those signs he came to an old log cabin. In the pine forest close to that rotted old cabin he found the remains of a corral, so dilapidated that it was hardly recognizable. That corral and cabin are not very far away from the 1856 Pine Mine. Wayne Nelson also found that old corral many years ago, but at that time there were still some old-fashioned leather ore sacks there, what Spanish miners called "zurrons" or "cueros." According to Nelson, the Jeep road through Bear Wallow and to the 1856 Pine Mine and Monument Knoll beyond was built by Les and Dan Pope. Cecil Dalton had been working an old diggings near there and he told the Popes about his claim. When Dalton returned several weeks later he found the Popes in possession of his diggings. Angry words erupted into a brief shoot-out, with gun shots being exchanged by both parties, but with no one being hit or injured.

Nelson and his Indian partner, Alonzo Jim, prospected well beyond Monument Knoll, into the cliffs and ledges near the headwaters of Brown Duck Creek. Along an old pack train trail they picked up several pieces of black colored rock, heavy with visible gold. Alonzo Jim told Nelson that ore came from the Red Man Mine, and was probably dropped from pack animals carrying gold ore from that fabulous diggings. He never told Nelson exactly where the Red Man Mine was located, other than that it was further around the head of the canyon, well hidden along a line of dangerous vertical cliffs. In later years Nelson made a few attempts to locate that fabled digging without any luck, and he thinks it would be harder to find now, more than forty years later.

Old-timers around the Duchesne area tell of a Spanish mine at Bear Wallow and a small cabin close to it. The portal of that mine is concealed by stones piled over it, but even though it is well concealed, it is still guarded by an Indian whose only job is to keep "white-eyes" away from the hoard of gold said to be cached there. Two prospectors told of accidentally stumbling onto that old cabin site, and they said that although its roof is sagging, its walls are still standing. Their tale of seeing a human skeleton sitting on a chair inside that cabin might be attributed to too active an imagination, or perhaps too many drinks of Old Blabber Mouth; that is if their story wasn't corroborated by others who claim to have seen the same thing, when they were perfectly sober. A man of some stature and civic responsibility at Roosevelt City, a man who is a teetotaler, told of seeing that same cabin while deer hunting with his father-in-law, who is a Ute Indian. He said the body inside that cabin is still dressed in leather boots and shreds of what were once trousers, but his shirt has completely rotted and fallen away. But most un-nerving to look at is that skeleton's long, stringy red hair, all covered with dust and cobwebs! But steer clear of Bear Wallow, he cautions, for it is certain death to be caught near that old cabin and mine tunnel!

Pigeon Water, Dry Canyon, Peterson Gulch and

Miner's Gulch all drain the high country between Monument Knoll and Rock Creek. That is the area north of the old Catholic Mission. Early trappers and explorers discovered several old mines along the Pigeon Water drainage. Aaron Daniels told about an old smelter he found there, only one of several known in that area. There is good reason why those smelters are there. Until recently, most geologists have thought that area was barren of minerals, but a heretofore unknown mineralized fault has been found as the result of aerial photography. That fault's location was verified by the Utah Geological Society's recent treatise on the Uinta Mountains, which states in part: "Somewhat surprising is the occurrence of a large, well defined fault, only recently discovered at Towanta Flat, west of Mountain Home. It is clearly visible from the air, and shows well in photographs." Recent infra-red studies suggest some very exciting possibilities in that area.

Even before modern technology began to draw attention to Towanta Flat and Pigeon Water, an Indian legend told of a gold mine there. William Peterson, the same man who many years later discovered Spanish mines near the head of the Weber River, was a game warden for the Ute Tribe, not long after the reservation was thrown open to settlement. During an interview Peterson told of a time when he rode his horse down out of Pigeon Water and was surprised to see a Model T Ford auto parked just about where Mountain Sheep Pass is now located. It was the first auto he had seen in that country, and since a young Indian girl was all alone in the car, he stopped to ask if she needed help. The young woman apparently mistook Peterson for an Indian, since he was riding an Indian branded pony and was wearing an Indian Police badge. She told him an interesting story.

The Indian woman told Peterson she was waiting for her grandmother, who had left the car and climbed out of sight over a ridge leading towards Pigeon Water. The grandmother told her that she had not been in that area for many years, and had returned only to investigate something her father had told her long before. She said there was a stone chair somewhere along that ridge, and if a person sat on it exactly at sunset on a certain day of the year, the last rays of the sun would shine directly into a mine tunnel. It could not be seen at any other time or on any other day. Her father told her that something of great value was hidden in that mine tunnel. Peterson never learned if the grandmother found that stone chair, or if she saw the sun shining into the mine tunnel. If you're up Pigeon Water way, you might want to keep a close watch for that stone chair. It could be a good place to rest awhile!

There is another story about Pigeon Water which might be worth looking into. A few years back a small crew of Indians from Ridleyville, a small group of Indian houses located just north of Whiterocks, were building a new road up Pigeon Water. While grading with a tractor, one of that crew cut into a vein of gold ore, only a foot below the surface. He later said that yellow wire gold was plainly visible in the rock there. That vein was uncovered just north of Mountain Sheep Pass. As soon as the crew foreman saw that vein, he had it covered over and re-routed the road up another draw. You may not believe this happened, but you can still see scars on the land where those cuts were made, and if you examine the boundary line between Indian and Forest Service land at the mouth of Pigeon Water Hollow, you will see that it is a zig-zag border, checker-boarding back and forth across the foothills for no apparent reason. But to the Indian mind, there is logic in that crooked, wandering boundary line.

The place where the tractor operator dug into that ore vein lies on the Indian side of that boundary line, but by only a few yards. If you follow that line east and then north, and then east again, you will come upon the site of an old mine shaft. It was covered with logs and rocks about twenty years ago, but it is not hard to find. Just across the boundary line to the southwest, on forest land, is Gooseberry Springs. That shaft is also on Indian land, by only a few yards, so I guess you could say there is some sort of logic to that zig-zag boundary line. At about that same time, an Indian named Redhorse drew a sketch map which shows both the ore vein and the shaft. (See Map N)

The first homesteaders along Rock Creek found evidence of earlier habitation. After Don Rust's father established his claim on Towanta Flat, he and some of his neighbors decided to build a canal from near Mountain Sheep Pass eastward, to bring water from Pigeon Water to their farms. In 1908 an engineer was hired to survey the proposed route, but he made a startling discovery. There was already a canal there! He found an ancient water course winding along the foothills and across Towanta Flat to Mountain Home and the Altamont area beyond. After careful survey and measurement, that engineer told the settlers that he couldn't map out a better canal than the one they already had.

Although giant cedars and cottonwoods were growing along that old water course, all that was needed to put that ancient canal back in service was to remove some of those trees and repair places where the earthen bank had eroded. After repairs were made and water was turned into it, they discovered that old canal had been built at nearly a perfect water grade. It is now part of the Farnsworth Canal System. Ute Indians certainly never surveyed and built that canal a century or more before the homesteaders discovered it. What other conclusion can anyone arrive at except that it had once been part of an irrigation system built to serve fields cultivated by Jesuit priests from the old mission on Rock Creek?

Among the first families to homestead on Towanta Flat above Rock Creek were the Rust and Bleazard families. Anyone not believing that Spanish pioneers had lived there long before them would no longer be a doubter after talking to Don Rust or Bill Bleazard, both of whom grew up there. The author has spent many hours talking to them, as well as to others who settled in that area. Don Rust remembers well the Old Spanish Trail which followed the Duchesne River to its junction with the Strawberry River. From that place it turned up Benson Draw and followed Big Hollow to his homestead on Towanta Flat. He traced that old trail up Pigeon Water and past Bald Knoll and Mountain Sheep Pass to Monument Knoll and across the head of Brown Duck Creek to Kidney Lake, Tworoose Pass and the high country beyond. Unknown to most prospectors and hikers today, that old trail wasn't marked by secret signs or strange marks on trees. Rust said that at regular intervals along the trail, there were markers on the ground; four small square-shaped stones, each about hand-size, fitted together to form a square, about twelve inches in size. That old trail passed right by his cabin, but over the years livestock and the works of man have broken up most of those old markers.

The late Colin Murdock, a half-blood Ute who was the author's cousin, recalled that some years ago he was trailing sheep with Melvin Hatland, a Ute Indian, along that section of trail described by Don Rust. Murdock said that Hatland stopped and picked up one of those trail markers. There were two smaller rocks under the marker stones, and he picked them up also. He examined each of the stones carefully and then replaced them exactly as he had found them. Murdock said he never knew what Hatland learned from those stones, but he is certain they conveyed some sort of message to him.

Don Rust's homestead was the first one adjoining Indian land east of Mountain Sheep Pass. Old Towanta was one of his neighbors, and had his own cabin close by. Rust remembered the time when the old Ute named Nar-Var-Putz, better known to white men as Mountain Sheep, was shot by Jim Olsen, a nineteen year old youth who was herding livestock for the Redden family. Mountain Sheep claimed that Olsen's stock were fouling a spring near his cabin and ordered him to move them; threatening Olsen and a Mexican herder with a tent pole. He struck at the Mexican and turned to strike at Olsen, when Olsen fired several shots at him from an old percussion revolver he carried. One bullet struck Mountain Sheep in the chest while another went into his mount and out through his cheek. Fearing reprisal from a group of Indians who quickly gathered, Olsen fled to Fort Duchesne and surrendered to the Indian

Ute Chief Dick Wanrodes. He took the secret of Spanish gold to his grave.

Agent. There were such loud cries of vengeance from the Utes that Olsen had to be spirited out of the fort, hidden under a load of mail bags being sent to Price.

Jim Olsen was the grandfather of Mrs. RuJean Rowley and she has the old court records and newspaper accounts of his trial. They describe how Mountain Sheep survived his wounds, although his face was disfigured by the shot which was fired into his mouth. In the biased language typical of news accounts of that time, reporters wrote that Mountain Sheep didn't die as he was supposed to do! But of interest to treasure hunters is the real reason why he was shot. It was said that Olsen had discovered a cache of gold nuggets which Mountain Sheep had the responsibility for guarding, and the old Indian had become irate when he learned that Olsen had stolen some of those nuggets from their hiding place, which was said to be at the foot of a large pine tree which stood near a small mountain lake. Those who saw the nuggets said they were pea-size, with a few being as large as a robin's egg.

Bill Bleazard also told of the shooting of Mountain Sheep. He said that after Olsen was acquitted of shooting the Indian, he spent all of his time at his farm. So great was his fear that he would be killed by vengeful

Utes that he never went anywhere alone. He never attended any public functions such as a ball game or rodeo if Indians might be there. Bleazard was sure that Olsen wouldn't have risked his life by going back to that nugget cache under the pine in the mountains, especially since it was still guarded by old Mountain Sheep. Those nuggets are probably still hidden right where they were then, for both Olsen and Mountain Sheep have long sine moved on to greener pastures.

Bill Bleazard also remembered his father telling how old Chief Wanrodes would come to their cabin almost every evening just about dinner time. He would accept a plate of food, but wouldn't sit at the table, preferring instead to sit cross legged on the floor. One night he came to the Bleazard cabin when he was very sick. After a time he removed a small leather pouch of gold nuggets which he always carried around his neck, saying: "It is the white man's money rock that is making me sick!" He took that small poke of nuggets outside into the darkness and returned in a few minutes without them, saying they would bother him no more. Nevertheless, he died soon afterwards. No one ever knew what happened to his bag of nuggets. He had no time to bury of hide them, but no one ever saw them again.

The late Mark Bleazard remembered the death of Chief Wanrodes. He was very old and nearly blind when he died during the winter of 1918. He wanted to be buried along Rock Creek at the sacred place where his father and other chiefs had been buried before him, but deep snow made his request impossible, at least until spring. He was interred in a temporary site under deep snow on the "Bleazard dugway," but when spring came his body was exhumed. Bleazard with young Dick Wanrodes and several neighbors dressed the old Chief in Mormon temple garments, for he had been a friend of Brigham Young and had been baptized by the Mormon leader. As he wished, his coffin was loaded on a wagon which was driven by Clyde and Ralph Rowley, with Bleazard and Chief Wanrodes' son leading the way to the place where his forefathers had been buried, a place the author has promised not to reveal.

Don Rust had a hair-raising experience near his homestead on Towanta Flat. One evening just at dusk, he and a young cousin were riding up the steep grade from Rock Creek towards his cabin when they passed by a place in the cedars he hadn't noticed before. Protruding a few inches from the hillside were the ends of a row of cedar poles. Wondering what their purpose was, they dismounted and Rust pulled one of the poles loose, leaving an opening into the hillside where it had been. Thinking they had found an old mine tunnel, Rust began pulling another pole loose when his cousin whispered to him that they were being watched. Rust said that even before then he too had a feeling that someone was watching them. It was almost too dark to see anything, but he turned and peered into the cedars which surrounded them. He couldn't see anyone, but still he felt a cold chill and the hair raise on the back of his neck. He turned to his companion and said, "Let's get out of here quick, we'll come back in the morning at first light."

A Mexican miner displays ancient "zurrons," leather ore sacks carried by the Indian slave miners in the Uinta Mountains.

The following morning they got up early and hurried back to that mysterious hole in the cedar poles on the hillside. To their amazement, the entire front of that opening had been dug out, leaving only a gaping hole on the hillside. Right up against the entrance were tracks showing where a wagon had been backed up to that hole, and just inside the entry there were deep marks showing where something heavy had been dragged out of that hole and onto the wagon. There was nothing left in that hole but some bits of rotted leather. Rust never learned what had been taken from that hole on the hillside below his cabin, but he will never forget the cold chill he felt tingling along his spine when someone, or something, watched him from the darkness of the cedars. He still doesn't know if he was only inches away from being a rich man, or from being a dead man!

CHAPTER 11

ECHOES FROM THE PAST
Tales That Endure

Nearly all so-called Lost Rhoads Mine maps show Moon Lake or Kidney Lake as points of reference. Both lakes are shown on several of the maps used by Caleb Rhoads, including his Pine Mine map. One of his maps also shows Lake Fork and the Yellowstone River. Some of the Spanish mines shown on those maps have been found. The 1856 Pine Mine was found some years ago in the ledge rock country north of Pigeon Water and west of Moon Lake. Apparently its gold vein had been almost mined out during Spanish days, for that vein now runs into an iron ore body. The Paint Mine is located just north of the Pine Mine on a rough side trail which follows Brown Duck Creek west from Moon Lake. Its ore vein has also run into iron. As related earlier, the Rhoads and Powell families were life-long friends. Once when Abraham Powell asked Caleb Rhoads why he no longer worked some of those mines, Rhoads replied: "The ore is running out and iron is coming through."

Pine Mine #2 was discovered west of the 1856 Pine Mine back around 1930 by Clarence Mecham, who learned of its location through an Indian acquaintance. It was an inclined shaft which had been concealed with a covering of pine poles and soil, so that a century or more only a shallow depression remained on the surface to revel its site. Mecham labored for many years to reopen that old diggings, but he had the shaft cleaned out to a depth of only eighty feet when the sidewall collapsed and filled it with rock and debris. Although his family has continued work as time and finances permit, that old shaft is still buried beneath tons of broken rock.

From atop the high ridge between the Paint Mine and Monument Knoll, one can look out across miles of wild country, from Moon Lake on the east to Brown Duck Basin and Kidney Lake to the west; and even further, to Squaw Basin and Dead Horse Pass. That there are old Spanish mines in that huge country is certain, for some have already been found. There are hidden caches of gold and silver there too, for a few of them have also been found. A lot more still remains, but like I said, it's a big country.

Among lost mine and treasure maps which have surfaced over the years, one is somewhat of a mystery. It shows several streams with some small mountain lakes at their headwaters, and several "Xs" which some believe mark the location of mines or caches. The only wording on that map is a name, Joe Sulser. Those who know about such things say that map was drawn by Sulser in a Mexican jail cell somewhere in Sonora, only minutes before he faced a firing squad. Who Sulser was or what his problems with the Mexican Rurales may have been is now forgotten, but sometime long after that map was drawn, it came into possession of Warren Sulser of Utah Valley. Other than that name, Sulser, which is not a common name, there is nothing to suggest that Joe Sulser and Warren Sulser were related, nevertheless, it does seem odd that Warren Sulser in Utah somehow acquired a map drawn by Joe Sulser in a Mexican jail cell. (See Map O)

On July 10, 1958, the Uinta Basin Standard at Roosevelt reported that the Lost Rhoads Mine had been found in the mountains west of Moon Lake by Ben Bullock from Utah Valley. Benjamin Hart Bullock was born October 27, 1878, at Provo of Mormon pioneer stock. He claimed to possess psychic powers which he said led him to "an ancient Nephite mine in the Uinta Mountains," which mine he obtained gold from over a period of years. It was while in those mountains that Bullock said he discovered the Lost Rhoads Mine through the use of a map owned by Mrs. Mary Sulser, the widow of Warren Sulser. That front page story fea-

Stan Brennan and a young prospector friend investigate a miner's cabin at Bear Wallow.

tured a map to the lost mine, which map was obviously a copy of the 1856 Pine Mine Map already well known to most treasure hunters. Those familiar with such maps couldn't help but notice the similarity between the Bullock Map, the Pine Mine Map and the 1851 Reinaldo Map. But strangely, the map printed by the Standard wasn't the Sulser Map which the newspaper article said led Bullock to the old mine west of Moon Lake. Why Bullock stated that he used the Sulser Map, but gave the Standard a copy of the common 1856 Pine Mine Map wasn't explained. Also, when referring to that map printed in the Standard, it was inferred that the three crosses below the pine tree symbol were mines, when in fact even then it was known that those crosses were graves of Spaniards killed by Indians.

According to that long and rambling newspaper account, When Enock Rhoads was killed by Indians near the narrows of Currant Creek, a map was found on his body, which map somehow came into possession of Warren Sulser. The only map which Enock Rhoads was known to possess was a small brown-paper sketch of mines located along lower Currant Creek. The story in the Standard did not explain that the body of Enock Rhoads was found by Mormon stockmen, who never mentioned any map being found on his body. But whether the map used by Bullock originated in a Mexican jail or was somehow recovered from Enock Rhoads' body, we know from Bullock's own statement that the map he used was the Sulser Map, and not the Pine Mine Map published by the Standard.

That Ben Bullock found an old mine is undisputed, but it was only that, an old mine. It may even have been a mine which Caleb Rhoads once worked, but it was not the famous mine which Thomas Rhoads brought gold from to mint coins for Brigham Young. Bullock's find was a very narrow shaft, dug in the manner of Spanish mines, in which only the ore vein and as little wall rock as necessary is removed from between the foot and hanging walls. Shafts or tunnels so dug are only as wide as the ore vein removed, and twist and turn as they follow the meandering of the vein. Little or no roof support was used in such workings, the roof being arched in a manner which is self-supporting. Bullock's shaft was hardly wide enough for him to enter, and was so cramped that he had little room to remove loose rock and debris which had fallen into it.

While Bullock was away from his find, someone, claim-jumpers according to Bullock, tried to open that shaft, but in so doing completely caved the old diggings so that the entry portal was buried beneath tons of loose rock. Before financing and heavy equipment could be obtained, Ben Bullock died, at Provo on July 12, 1962. After his death, his son, Vern Bullock, continued to keep that old mine under claim, but no meaningful work was ever accomplished. With its entry caved and buried, and with environmental regulations being what they are, it is unlikely that anyone will ever reopen those old diggings.

Kidney Lake is unique among Uinta Mountain lakes, not only because of its square shape but also because it is one of the few lakes in those mountains whose inlet stream is at the south end of the lake and its outlet to the north. Years ago, Henry Wash, the Ute Indian who led Jim McAfee to a mine in Miner's Gulch, recovered an eighty pound silver bar from Kidney Lake.

Ben Bullock. He used the Sulser Map to locate a Spanish mine above Bear Wallow.

Wash was at Kidney Lake at the same time Ed and Roy Boren were there. He was walking along the lake shore which was at its lowest level after several years of drought when he saw the sun's rays reflecting from some metal bars in water about six feet deep. He was able to hook onto one of those bars with a forked pine pole, but the others slid away into deeper water. There isn't any doubt about Wash's find, for both Ed and Roy Boren saw and handled that silver bar. There is some question of what became of that bar, for Wash never seemed to prosper because of its discovery.

Several fishermen have reported seeing a mine dump when the waters of Kidney lake are drawn down for irrigation in the fall of the year, just about where Wash found that silver bar. When the lake is at its lowest level, that dump is still under water, but it is possible to wade out to it. During the fall of 1974 when the lake was again at a low level, a fisherman recovered two more of those silver bars. Unsure of what he had found, he took them to the Ute Research Laboratory for analysis, which firm assayed those bars and reported they were 99.9% silver. The author has that original certificate of analysis.

Roy Snyder was another who prospected that same area, and just south of Kidney Lake he came upon a place where sometime in the long ago miners had built and used long-tom sluice boxes. Others also saw them and never doubted what they were, but in recent years hikers and fishermen have burned most of those old sluices as firewood. Snyder also found several chunks of gold-bearing ore in that same area, perhaps dropped from pack animals or washed there by the river. A few years ago a Forest Service employee discovered a fallen cabin and some lengths of rotted sluice box by it. In keeping with modern Forest Service policy and regulations, he burned those sluices and cabin to rid the wilderness of a man-made eyesore. It seems to be Forest Service policy that in order to protect the wilderness and its historical artifacts, we have to destroy them.

John Ofstad is a fisherman and backwoods explorer from California, whose enthusiasm and love of wilderness takes him into back country where others hesitate to follow. Using the Reinaldo and Pine Mine maps as his guide, he hiked from Moon Lake into the Brown Duck Basin and then scaled the high cliffs below the Paint Mine. From there he located the Old Spanish Trail just as it is shown on those maps. By following that trail, he located Mine #1 near the head of Slate Creek. He made a notation on his map that is exactly where the Hoyt Peak Fault allows igneous rock to break through the earth's surface. He found some magnificent crystals there, some more than a foot in length. Ofstad found Mine #2 on a ledge in a stand of pines near the head of Dry Canyon, not far from the old Mecham Mine. Mine #3 was located along that old trail where it crosses a small creek which empties into Kidney Lakc.

Before he reached Kidney Lake, Ofstad met a "rough looking character" riding horseback and leading two pack animals. The "character" had a large revolver stuck under his belt. Ofstad called a greeting, but the "character" was definitely unfriendly, and never returned his call. Ofstad couldn't help but notice that those pack animals were loaded down with something very heavy, even though their packs weren't large. Those pack animals left a much deeper track in the soft ground than did the horse carrying the "character." Several days later Ofstad met another rider, who judging from his description, was probably the late Duane Meriwether, a horse wrangler from the Rock Creek Resort. That rider was friendly, and after they exchanged a few words, Ofstad told him of his encounter with the "character." Meriwether told him that he had rented some pack horses to several men who were scuba diving at Kidney Lake, and added that he was sorry he had done so. He told Ofstad that those divers had recovered something from the lake which they packed down to Moon Lake, something they never let him see. But whatever they found, it was very heavy, for Meriwether said that when they loaded it into his horse trailer, the weight buckled the trailer's floor and heavy channel-iron frame. Meriwether told Ofstad that he later heard that those divers had taken 900 bars of silver from Kidney Lake!

Ofstad remembered something else, there was a lot of shooting at Kidney Lake, but only whenever he started around the lake to where those scuba divers were working. He said it may only have been target practice, but somehow he couldn't get it out of his mind that he was the target! One might chalk up Ofstad's story to exaggeration, except that others have encountered the same rough characters in the Kidney Lake area. There is a mild-mannered and gentlemanly prospector from one who goes his own way and minds his own business. During August 1993, he was heading for one of his prospects in the mountains beyond Kidney lake. He was leading a pack mule up a steep trail just below Tworoose Pass when he was stopped by two tough looking characters. He said that one looked like a Mexican bandit, having a cartridge bandolier across his chest and a 30-30 rifle point right at him. His partner looked a little bit more civilized, even though he carried a telescope mounted rifle. The "civilized" partner asked where he was going, and when it was explained that he was heading for Tworoose Pass to prospect on the opposite side of the mountain, he was told it would be better for him if he turned around and looked for prospects somewhere else. Judging from the rifle pointing at him and the tone of voice warning him to go elsewhere, he decided that wasn't the time or place to argue the point, so he turned his pack animal and went another way. But his encounter wasn't entirely unpleasant, for as he started back down the mountain, the "civilized" character called after him, "Good Luck!" It could have been worse.

Among the best proofs of Spanish occupancy in the Uinta Mountains are the many artifacts which have been found there. Several cannon have been discovered from Rock Creek to Moon Lake and eastward to the Yellowstone and Uintah rivers. Before the dam which raised the water level of Moon Lake was built, there was a small resort on its shoreline. It boasted a dance pavilion which floated on the lake surface, giving dancers a sort of rolling sensation as waves rolled in under it. A few old-timers still recall a tragedy in which two dancers who were a little bit tipsy fell against a guard rail which broke, letting them fall into the icy waters of the lake. Because it was late at night and confusion reigned, the victims sank into the depths before a rescue could be initiated. A diver was summoned the following day, and after several attempts he recovered the bodies. Later while talking about that dive, he mentioned that the bodies were found near the edge of a ledge where the lake bottom dropped off into greater depths. On that ledge he saw a small cannon which appeared to be made of brass. No one knew how that cannon came to be in Moon Lake, but it is not unlikely that it was dumped there by Indians shortly after the 1844 massacre in those mountains. There is no record that anyone ever tried to salvage that cannon, and since the lake level has been raised, it's doubtful anyone will try.

As described earlier, there were several out-lying visita type trail missions in the mountains beyond the central mission at Rock Creek. For some time now there have been stories placing one of those small missions near Atwine Lake, high in the Brown Duck drainage east of Kidney lake. One old cowboy noted that the same Spanish tree signs found near the Rock Creek Mission can also be found near Atwine Lake. A fisherman told of finding what appeared to be foundation stones in that area, while a man and wife team discovered a row of flat stones which may have been part of a wall.

Late in the summer of 1991, a lone traveler made an overnight camp near Atwine Lake, far enough back in the trees so that mosquitoes wouldn't be a problem. After he got his bedroll placed and a campfire burning, he noticed that a nearby clay bank looked as if someone had once dug into it sometime long before. Examining that bank more closely, he saw what appeared to be the square end of a brick or timber protruding several inches from the loose soil. He dug around it with a jackknife and was amazed to see that it was a metal bar, approximately three inches in width, two inches thick and about sixteen inches long. He estimated that it weighed in excess of one-hundred pounds!

Be Warned! Ute Indians keep a close watch on this old mine in the Yellowstone country!

That heavy bar was not badly tarnished or corroded, although pit marks on its surface had turned black, indicating that it might be bullion metal containing some silver. He had to struggle to lift that heavy bar onto his saddle horse and tie it in place. A later assay made by a reputable firm disclosed that bar was nearly pure silver, but it also contained some copper and a few ounces of gold. A few weeks later he returned to that clay bank with a new metal detector, but he found nothing else. Why that lone bar was found there remains a mystery, but a clue might be that it was found not far from those old foundation stones near Atwine Lake.

There may be even more to the Atwine Lake story. Recently there was a lot of excitement in the Duchesne area when it was reported that men representing Mel Fisher, the treasure hunter who recovered the gold of the Spanish Galleon Atocha off the coast of Florida, had been seen searching for a treasure cache in the Brown Duck Basin. The Ute Indian Tribe complained that

A centuries-old map carved in stone. It's your waybill to lost Spanish treasure.

A Catholic cross and a trail leading across a ridge. Church treasure is hidden just beyond.

A man-figured etched in stone points out the trail ahead.

Gold! Only four varas further!

Fisher's men had trespassed on their lands, while environmentalists were angry when they learned that helicopters were being used in the primitive area. One fisherman watched men unloading drums of gasoline and other supplies from a copter which had landed near Island Lake. A court reporter later claimed that the copter crew had been fined $5,000 for leaving a drum of gasoline after they left the area. A hiker said that when he encountered two of the copter crew near Atwine Lake, he was warned to get away from there and not to come back.

Because helicopters and mining or treasure hunting are prohibited in the primitive area, several concerned people made inquiry at the Duchesne Ranger's Office. At first personnel there denied any knowledge of what had been reported in the Brown Duck area, but when proceedings were initiated to obtain Forest Service records under the Freedom Of Information Act, a different story was told. It was acknowledged that a special use permit had been issued to an eastern firm on orders received from Washington. That permit was not in Fisher's name, however, those receiving the permit were allegedly business partners of his. Local officials were given no explanation why that permit was issued, but an item in the Daily Press at Hampton, Virginia, might shed some light on that question. That article described how Mel Fisher and his partner, Chris Widener of Denver, Colorado, planned to use a new high-technology metal detector and ground penetrating radar to attempt recovery of the famous Beall Treasure which is believed to be buried in Bedford County, Virginia. Those state of the art detection instruments would be mounted in a helicopter.

It is also interesting that Fisher was reported to be at Salt Lake City, where he was trying to locate either Gale Rhoades or Kerry Boren, co-authors of *Footprints In The Wilderness.* When he learned that Rhoades was deceased and Boren unavailable for interview, he returned to Florida. Several persons reported seeing Fisher at Rock Creek at about that same time. A Heber City man who worked on a construction project there noticed someone familiar to him at the Rock Creek

Lodge. As he was watching an account of the recovery of the Atocha treasure a few nights later, he immediately recognized Fisher in that documentary as the man he saw at Rock Creek. Still another man talked to a crew who had an expensive motor home at Rock Creek. It was pulling an odd-looking long and narrow enclosed trailer. When he asked about its peculiar shape, he was shown that it contained a helicopter with its rotors removed, and was told it could quickly be assembled to fly into those mountains.

One well known treasure hunter who is plenty savvy thinks the Fisher crew is interested in the Uinta Mountains because of something discovered in the Spanish archives at Seville regarding the treasure lost when the galleon Atocha sunk off the coast of Florida. He believes those records reveal that part of that treasure came from mines located in the Uinta Mountains. Those same records indicate that Uintas gold was also aboard the galleon Maravilla, whose treasure hoard was recently located by Herbert Humphries' Marex Salvage Company. First estimates place the Atocha's value at four-hundred million dollars, while the Maravilla hoard is expected to top one-billion dollars. If Mel Fisher's helicopter mounted high-tech instruments work in the mountains of Virginia, it might be reasonable to think they will also work in the Uinta Mountains! The mystery of Mel Fisher's interest in the Uinta Mountains was explained in August, 1994, when he came to Utah and met with the author and his long-time prospecting partner, Bill Bleazard. Because their investigations are still on going, it is not prudent to reveal all of the places they examined and the finds they made, but it can be told that part of their search centered on Soapstone Basin, Lightning Ridge, Rhoads Canyon and the Rock Creek area. (If we had visited all of the places the "backroom boys" said we were seen at, we would have had to be twins!) From maps and waybills in Fisher's possession, unknown tree signs, trails and at least one mine shaft not previously known to local searchers were located. After five days in the mountains, Fisher returned to his Florida treasure headquarters, but investigation of finds made in the Uintas will continue pending his return in several months, when he will have perfected an entirely new concept three-million dollar long-distance deep-seeking detector. The Fisher crew were great people to work with, and one couldn't ask for a more congenial host than Mel Fisher. The author had only one minor complaint; it was awfully hard to climb those mountains with the twelve-pound gold chain Fisher hung around his neck! Lake Fork flows from Moon Lake and joins the Yellowstone River downstream not far from Altamont. On some maps the Yellowstone is called a creek, but anyone who has seen it during spring run-off when it is a wild, raging torrent

Jesuit priests marked the Yellowstone Trail with church crosses.

knows it is a river, one which can be dangerous. The Yellowstone drains snow-melt water from the highest peaks in the Uinta Mountains, some of them well above 13,000 feet. A Spanish cannon was found near where Swift Creek flows into the Yellowstone. Since that discovery, fishermen, hikers and hunters have found cannon balls in that same area. Some who prospect the back-reaches of that country believe there was a Spanish settlement near there, perhaps having a small visita mission or place of rest and worship. It's possible, for several old mines have been found close to where they think that mission was located.

Some have reported seeing one of those mines near the confluence of Swift Creek with the Yellowstone. They tell of seeing the mouth of a tunnel deep in a thicket of wild rose bushes, only a short way up Swift Creek. Those who have searched that area have seen a peculiar symbol cut into an old aspen tree where those two streams meet. That sign depicts a square box, often a symbol of buried treasure, with a Catholic cross above it. Below that box is what may be a directional marker; a coiled snake with the four cardinal points of the com-

A primitive Spanish sword, proof-positive that Spanish miners followed the Yellowstone Trail.

pass marked on its outer coil. A small segment of that coil is missing, perhaps an indication that one should proceed in that direction. If you do follow in that direction, you will come to a small cave. Two young fellows who found that cavern removed some ancient looking artifacts from it. Those who have examined them say they are Spanish, perhaps left there by Spanish miners. There could be a gold lode close by, for modern day prospectors using small portable dredges have recovered flakes of gold from Swift Creek.

There are many places hidden away in dark canyons or along forgotten ridge line trails where mysterious signs and symbols like those found along the Yellowstone may be found. One might think that all of those old secret markings would have been found by now, but there are many yet to be discovered. As recently as 1989, some of the most perfectly preserved and intriguing treasure symbols ever seen were discovered in a box canyon where few Indians have ever set foot, and even fewer white men. You may doubt that such places still exist, but they do, for I have been there. When I first learned of that place from one of the Old Ones I too was skeptical, but when I entered that hidden canyon I found a Shangri-La, a place where there wasn't even a footprint to suggest anyone else had ever been there. In that canyon were mysterious carvings in solid stone, more than two-hundred years old!

There is a long line of treasure symbols leading from the river and across a long open flat to that canyon, each symbol leading to the next. If each is interpreted correctly, he who follows them will take the correct forks of the trail, and choose the true branch of the canyon. He will know he is on a proper course when he comes to a large panel of signs cut deeply into a sandstone cliff; some of those cuts nearly two inches deep! That panel pictures an exact map of the canyon beyond, with all of its twists, turns and forks. On that panel there is carved a date, 1791! Those obscure messages in stone were made more than two hundred years ago!

Photos of some of those signs accompany this text, but for the present their location will not be given, to protect them from vandals and "crazies" like the man who slashed the Bear Sign on Rock Creek. For now some of their meaning must remain a mystery. It is unknown how they were made, for they are beautifully cut, but not with a chisel or similar instrument. They are as smooth as if cut by a modern cutting tool, and they are deep, most more than an inch into the hard rock surface. Except for the patina of manganese oxide which covers them and the moss and lichen in their crevices, they could have been made yesterday. But they weren't, for photos have been shown to outdoorsmen who have spent all their lives in that area; sheepherders, prospectors, trappers and anyone else who might have seen them. Only one or two pioneer settlers knew of them, and they remembered seeing a few of them as long as eighty years ago.

During the summer of 1993, a small article in the Roosevelt Standard reported that the Ute Tribe had issued a minerals exploration permit to a Texas group to search for lost mines and treasures on the Ute Reservation, those mines or caches to be discovered within ninety days. That group was supposed to possess the necessary expertise and equipment to do the job, but what the Standard did not report was that the only information those searchers actually had were some infra red photos, topo maps and photographs of those 1791 signs; all stolen from a firm which had been doing exploratory work in that area the previous year. The Texas group's plans were soon frustrated, for they had neither the ability or know-how to interpret the material they had stolen. Nor was the fact they were from Texas and knew nothing of the places they were searching helpful to them. They were guided by a man who knew less than they did; the blind leading the blind. Weeks and then months were spent digging empty holes and drilling a tunnel nearly forty feet into solid stone, all at a considerable cost to the Ute Tribe who ended up paying the costs, but all to no avail. The message those stolen map and photos and the silent stone carvings conveyed was lost on them. They are not a mystery, for they have been interpreted by one of the world's leading cryptographers. Hopefully, the time may come when the Ute Tribe will allow his interpretation to be verified, but not by a con-man. Time will tell.

In addition to specific signs and symbols used by Spanish miners, the king directed by royal decree that specified rock monuments or trail markers be used so that mine inspectors and tax collectors could easily find those mines. The manner in which those markers were erected and placed were the same everywhere in the New World and did not change over centuries of time. Those monuments had a pointer stone directing the way to the next marker, or there might be a lone pointer rock purposely placed or resting on top of several smaller stones, so that daylight could be seen below it, ensuring that they would not be confused with a natural formation or work of nature. Those pointer stones were always placed so they could easily be seen by someone following the trail, but still not be so conspicuous that they would be noticed by someone not watching for them.

An "X" was used only on an "in" trail, leading to a mine, cache or perhaps a mission or visita. A cross always marked an "out" trail, or a route leading away from treasure, on a home-bound trail. A cross was always made with double lines, so as not to confuse it with a Tobias symbol, a single line cross which is often mistaken for a Catholic cross. The Tobias symbol is taken from the Bible, where an angel was directed to accompany and guard Tobias so that he could collect riches for the king. Nearly all Spanish signs are derived from the Bible; the numbers or markings on them often referring to chapter and verse. When on the treasure trail, watch carefully for those old signs and monuments along the way. Wherever you find them, you can be certain that they were made by Spanish miners, for few were made after Mexican independence in 1821.

Mrs. Douglas Henderson told the author that for many years her father was a Forest Service employee in the Yellowstone River area. He operated heavy equipment building roads and campgrounds, or cutting fire breaks across the mountain front. While cutting new trails he broke into several old mine shafts which he never suspected were there, since they had been cleverly concealed, leaving nothing on the surface to indicate their presence. From those old diggings and occasional smelter sites he recovered many Spanish artifacts, including bronze tools, cannon balls, a small brass bell and other relics so rusted or corroded by time that he couldn't identify them. For many years he preserved those relics and kept a record of where each was found, but several years ago he suffered a debilitating stroke which left him so despondent that he destroyed that record and threw away his relics, all except the brass bell, which his family still has.

If you look closely, you might see an old mine on the west side of the canyon just before the old road which used to follow the Yellowstone River leaves Indian land to enter the national forest. It is often difficult to see when summer growth covers the mountain, for its opening was covered with stones and grassy top soil a long time ago. Just beyond that mine are the Yellow Ledges, shown on most forest maps. Recently some tests were made on surface ore which came from those Yellow Ledges, which revealed that there is a lot of silver there. One of the clues Caleb Rhoads mentioned was that one of his mines was located close to some yellow ledges. That ledge area along the Yellowstone is peculiar, which places which are barren, where nothing grows. An engineer may have learned why that is, for he tested water from springs below those ledges and discovered they have a Ph factor of 2.8, which he says accounts for the sterility of the soil and rock. A PH factor of 7 represents a neutral condition, while 2.8 indicates that water is highly acidic, a condition no doubt caused by those spring waters percolating through ore beds somewhere below the surface; according to that engineer, probably close to the surface. It is noteworthy that those springs are located right where the eastern end of the Hoyt Peak Fault breaks the surface. There is something else odd about those springs. Their high water flow occurs late in the summer, not during the spring, which might make one wonder what the source of that water is.

They found gold! Mike Abt, Bill Bleazard and Mel Fisher.

There is still another Spanish diggings in sight of those Yellow Ledges, but lower on the slope and nearer the river. It is more difficult to find, for some years back the Ute Tribe had the tunnel and waste dump leveled off, but if you climb atop the low ridge which runs between the present road and the river, you can see where it was located. Before it was leveled, there was a large tunnel there, big enough that horse drawn carettas could be taken into it. A wide vein of low grade ore can still be seen about a half-mile before entering forest lands, just after a cattle guard is crossed. That low-grade ore was carried to the river on carettas, where it was concentrated. You can still see where rock from that mine was dumped into and along the river. There is a story that some bars of silver are still buried near that mill site. There are a few old-timers in the Altamont area who say that a Forest Ranger recovered two of those bars. One of the Ute tribal council warned him not to go back there, and he didn't.

There is a giant yellow pine which stands alone on an open flat near the river and not far from that mine by the Yellow Ledges. Carved deeply into its bark is a large Catholic cross. That tree isn't hard to find, for it is the only large pine on that flat, but it may not stand much longer, for it is so old that many of its branches are dying. That mine may be almost as old as the tree. Some who fish along the river near that old pine have encountered "Susie," an old Indian woman who packs baskets of heavy yellow sand from the mountain front near the old mine. Those who have watched her closely say that she carries that dirt to the river, where she washes it in an old dish pan. They say she makes beautiful things from the gold she recovers from her crude panning operation.

Frank Silvey, the early day Utah historian, wrote about an Indian woman he met in the Yellowstone country, not far from Moon Lake. She had a gold nugget as large as a robin's egg, which she wore on a leather thong tied around her neck. She had other nuggets nearly as large in a basket she carried. Silvey asked her about that gold, and pointing off towards the Yellowstone, she said that she found those nuggets just over the ridge from Moon Lake, in a swampy area where there were a lot more of them, only a few inches below the grass roots. Silvey spent some time hiking that country, and he discovered that there are a lot of places which answer that description along the Yellowstone side of that ridge.

Pick Murdock prospected with a lot of the old-timers, some of them Ute Indians who knew a lot about Spanish mines and caches hidden on Indian ground. Some of those he traveled with found death in those mountains. One of his partners, Cump Murray, died of a heart attack. Murray was one of the few Old Ones who knew where the Carre-Shin-Ob or sacred mine was located. Cump Murray never talked to many white men, but Ed Twitchell was one of the few he trusted. Twitchell operated a small country store where Murray often stopped to visit and purchase a few supplies. Twitchell really liked the old man and would often give him candy or soda-pop, so that in time Murray came to like and trust him. In return for Twitchell's kindness, Murray offered to show him a place where he might become rich.

Cump Murray was very old, yet he told Twitchell that his father had told him that even before his time Spanish miners worked a mine near the Yellowstone River. His father had told him that in the long ago those miners forced Indians to work in their mines. One day a fight between the miners and Indians erupted into a massacre, during which most of the Spanish miners were killed. Murray's father told him that the bodies of those miners and all of their tools were thrown into a shaft, after which it was filed with logs, rocks and dirt. Cump Murray had been shown that place when he was only a boy. Some years later, after his father had died, he returned to that place where he picked up pieces of rich ore. Later he had a white man assay those ore samples, and as he told Twitchell, they contained ten ounces of gold and forty ounces of silver to the ton!

Twitchell and Jack Spencer helped the old Indian into Twitchell's truck and following his directions they drove up the Yellowstone River Canyon northward from Altonah. They continued to the Yellow Ledges where the road forked. Taking the less traveled road to the west, they followed up a narrow canyon to where the trail finally came to an end at a small swale or valley. Murray pointed the way up a ridge beyond that swale, but said he couldn't walk that far. There were too many large rocks along that ridge for Twitchell's old truck to climb that grade, but the Indian said they were almost there, so all three men worked hard in the hot sun rolling heavy stones out of the way. Twitchell coaxed his truck up that last steep ridge, where he asked Murray which way to go. Cump Murray lifted his arm to point the way when he suddenly slumped over and died. Exertion from moving heavy rocks combined with excitement caused a massive heart attack. The Killer Mountains claimed still another victim. Twitchell and Spencer never knew how close they came to finding the Spanish mine by the Yellow Ledges along the Yellowstone.

Some believe it may be hazardous to their health to search for Spanish mines on Indian ground. During the 1989 deer hunt, not far from that giant yellow pine with the ancient Catholic cross and the old mine covered with rocks, a hunter from California came upon a shallow grave. Wild animals had dug into it, exposing part of a human leg and a hand and forearm. The remains were not old and appeared to be those of a young white male. That gruesome find was made at Mud Springs Draw, almost in sight of the old mine along the Yellowstone. Not wanting to become involved in a long drawn-out investigation which would ruin his hunting trip, the hunter waited until after his return to California before notifying the Duchesne County Sheriff's Office by telephone. A search was made but to no avail, since fresh snow had fallen in the mountains. That report could have been written off as a hoax, but it wasn't.

Aware that the Killer Mountains have a long history of hiding lost graves and that many cases of missing persons in those mountains are never resolved, the sheriff's office contacted the California hunter, and reluctantly he agreed to return to Utah to lead officers to the scene. With his guidance a new search was made, but again that grave couldn't be found. The early winter storm had so changed the land's features that the hunter could no longer be sure of his landmarks. That might have ended the matter, but it didn't, for the hunter knew what he had seen. Several weeks later he once more traveled from California to guide searchers; as a matter of fact he returned to those mountains three times. Then to resolve any doubts which law officers might have, he consented to a polygraph test to prove he was telling the truth. He really didn't need to do that, for traveling from California three times should have convinced anyone that his story was true. To this day he knows that grave is there, even though he can't find it.

Aaron Daniels used to prospect all along the Uinta Mountains, often with partners like Pick Murdock, Happy Jack and Caleb Rhoads. All of them found old mines in the Yellowstone country, but all were close-mouthed and left few clues for today's treasure hunter. It was well known to early settlers that Daniels with one or more partners was often seen going into the Yellowstone River Canyon, but those who tried to follow his trail always lost it somewhere in the broken country beyond Swift Creek. On one occasion when Rhoads accompanied Daniels and Murdock, he wandered away for a short time while they stopped to rest. When he returned he had a small bag of "money-rock," more gold than quartz. Daniels later said that considering the short time Rhoads was gone, the source of that gold couldn't have been more than a half-mile from where they stopped to rest. He speculated that Rhoads may have obtained that gold from a mine somewhere close to the river. Judging from the many Spanish artifacts which have been found in that area, he may have been right.

Ute Indian oral histories and stories handed down from early white settlers tell of a fabulous gold mine and cache somewhere in the Yellowstone River Canyon. Recent studies based on high altitude infra-red photography seems to verify the truth of those tales. Muse Harris, a son of old Henry Harris, a Ute interpreter during pioneer days, once admitted that he had helped another Indian hide some gold disks or plates along the Yellowstone. He said those disks had some sort of strange figures on them, unlike any writing he had seen. According to his story, those disks were some sort of cursed treasure which Ute leaders wanted to be rid of. Harris told his eldest son that those disks were buried at the "horse's head." His son never gave much thought to his father's tale while he was still alive, but after the old man's death, he thought a great deal about it. He never learned whether the "horse's head" was a natural landscape feature or perhaps skyline formation which resembled a horse's head, or if it might be someplace where his father had cut a horse's head on a tree or ledge of rock.

Ruth Lowry may have seen one of those gold disks. She was living at Ioka at that time, and one day as she was driving into Roosevelt, she picked up a hitch hiker, an Indian boy she called Duchesne George. Lowry noticed that he was wearing a large and peculiar belt buckle she hadn't seen before, and she asked about it. Duchesne George told her he had found a place where many yellow-colored metal disks had been buried. He took one of those disks and fashioned it into a belt buckle. Lowry recalled that buckle was large, similar to rodeo type buckles, and that it had some odd letters or writing on it. She said that Duchesne George was struck and killed by a car while hitch-hiking not long after she talked to him. She thinks that gold belt buckle was buried with him.

Ed Twitchell used to prospect all across that country, from Ioka to Whiterocks, and he tells a tale which was told to him a long time ago by Verlun Labrum. It concerned a time when Roosevelt City was only a couple of cabins in the wilderness, and is a story of murder most foul; a killing to silence a homesteader who discovered a hidden cache. Labrum said that a settler had a shack northeast of Roosevelt, not far from Blackhawk Bench on the way to Whiterocks. It was close to one of the many gray-colored mesas which are found in that area. After a torrential rainstorm he noticed that a side of one of those mesas had slid, exposing a heavy wooden door which covered a cavern of some sort. He told Twitchell that the settler pried the door ajar and crept back into that black hole, where he found a cache of silver bars.

Excited over his find, he hurried into town where he bought drinks for everyone and spent most of the night bragging about his good fortune. But while boasting and drinking his fill of beer, he failed to notice that among his listeners were two Ute Indians. Long after dark as he staggered back to his homestead shack, he was jumped by two Indians who threw him to the ground, and with a razor-sharp knife, cut him open from throat to belly. They pulled the intestines from his body and filled the cavity with bran poured from a grain sack. Somehow he managed to crawl back to the one room saloon where he told his horrid tale, dying only a few minutes later. He had no chance to tell where that gray mesa with the wooden door was located, nor has anyone found it since. Chalk up one more for the Killer Mountains.

Not long ago a Spanish sword was found half-buried

The trail to the Blackhawk Mesa cache begins at the old Roosevelt townsite.

in a thick mat of cedar tree leafs and bark under a gnarled old tree near Talmadge, a small hamlet not far from the mouth of Yellowstone Canyon. It is a crudely made weapon which might be better described as a machete. It is made of old-style wrought iron, the kind blacksmiths worked in Europe before harder irons were developed. The blade, grip and hand-guard are all made in one piece. The blade is thick and heavy, and sometime long ago several inches were broken from its tip. It is an intriguing relic, which experts who have examined say is definitely not of American manufacture. Probably not by coincidence, another valuable Spanish relic was found in the same area. It is a miner's "barrena" or hand-steel drill, but a most unusual one. It has four cutting edges with grooves between each edge to channel cut rock away from the tip, not unlike a modern star-drill. It has a sword-like hand grip with two protruding "ears," the purpose of which appears to be a place to strike with a hammer should the drill become stuck in a hole being drilled. With its unique cutting edges, a miner could quickly drill his hole deep into solid rock.

In his description of the massacre of Spanish miners and soldiers near the mission at Rock Creek, Gray Wolf wrote: "The Spaniards had at least three large cannons." What became of those cannons and the mystery surrounding them makes a fascinating story. Those cannon were too large and heavy to be moved far, so they were buried or covered with stones to hide them forever. But after more than a century of time, erosion and the works of nature have begun to uncover them. If you are trying to find one of those old cannon, one sign which might reveal their hiding place is a barren area of little or no vegetation or an area of discolored soil. A large mass of brass or bronze poisons the soil and its surrounding vegetation, so that any such barren place warrants close examination. The number of people who have seen cannons or recovered them over a long period of time dispels any doubt that Spanish soldiers used them at places like Rock Creek and the Yellowstone Canyon. Descriptions given by those who have chanced upon a cannon while hiking in those mountains suggests that there were probably two different types; a small mountain howitzer weighing from about fifty to eighty pounds which was carried on a pack animal, and a larger and much heavier type which could be mounted on a wheeled carriage, similar to the one found on the Kamas Prairie during the 1850s.

Because of their great value as artifacts, those who have found and recovered Spanish cannons are usually reluctant to advertise or acknowledge their finds. A museum quality cannon may bring the finder fifty-thousand dollars or more. Others hesitate to make their finds public because they know the locations where they are found is an excellent clue to the mines or missions they guarded. Still others are concerned with antiquity laws which prohibit individuals from owning ancient artifacts. To the Forest Service or BLM, ancient can mean almost anything. Those agencies seem to think that it is better to let an artifact rust away into scrap metal rather than to preserve and display it.

A gentlemen who found one of those old cannons along the Spanish Trail has allowed photos of his find to be included in this volume, but he has requested that his identity not be disclosed. His find is one of the small mountain howitzers, and it is a magnificent thing, a work of art, made of brass and heavily engraved with ornate scroll work. It is about thirty inches long, weighs approximately fifty pounds and has a two inch bore. But of greatest interest in the Royal Shield of King Ferdinand of Spain on the barrel, and the date 1517 engraved in Roman numerals. 1517! That date precedes the conquest of Mexico! It could very well be one of the cannons brought to the New World by Hernan Cortes and later taken north beyond the Spanish borderlands by Francisco Coronado on his epic journey to the Four-Corners of southern Utah. Later expeditions may have taken it even further north. If so, its value in incalculable.

Albert Peterson settled a homestead along Rock Creek just upstream from lands reserved for the Utes. His property, now known as the Robbins Ranch, is located just below Miner's Gulch and Yellow Pine Flat. There was an old tunnel on his homestead, but Peterson built a small sawmill at its portal and it soon became covered with slabs and debris. He also found another mine, a tunnel, on the mountain just above his sawmill, which was very cold inside. Peterson found evidence that others long before him had used it to keep meat cool, and he used it for that same purpose.

Soon after he got his ranch established, Peterson improved on a rough wagon trail which led up into Miner's Gulch and past Bear Lake, a road which had been built by loggers Van Killian and Chet Lyman. Peterson built a small sawmill at the end of that road. On a high promontory overlooking Bear Lake, sawyer Tom Draper found a Spanish cannon. Draper said that it appeared to have been placed so that it could be fired down onto almost any point along Rock Creek, from Yellow Pine to Daniels Flat. A lot of old-timers who worked at Peterson's sawmill saw that cannon, and no doubt it made an interesting topic of conversation when they gathered together at meals or in the evening when the day's work was done, but in those days that's all it was, just a curiosity, nothing more. There were no markets for Spanish cannons then, and just knowing where one was did nothing to feed Peterson's growing family. It was something everyone seemed to know about, but no one really cared much about.

Several people still living recall seeing that old cannon, among them Gordon Taylor of Kamas Valley. As a young man, Taylor with three companions made the long trip to Rock Creek as a fishing and camping adventure. They camped along the river at the Lower Stillwater, close to Peterson's ranch. After tiring of fishing and while planning a hike into the mountains above Miner's Gulch, Peterson suggested that they might want to climb onto the promontory above Bear Lake to see the old Spanish cannon. The climb was a rugged one, but Taylor recalled that although the cannon was nearly hidden under thick brush, they had no trouble finding it.

Since it was made of brass the fishermen thought it might have some value, so they decided to drag it down from the mountain. They soon discovered that over the years it had settled into the ground, while roots and tree limbs had grown around it and into the barrel. It also proved to be much larger and heavier that it first appeared to be. Lifting and pushing as hard as they could, they were unable to move it from its rocky perch in that brush thicket, so they left it just as they found it. Taylor still remembers how impressed he was by that cannon's strategic location, placed high on a point overlooking the canyon below and commanding a range of fire over the Lower Stillwater country.

For several years a story has circulated about one of those cannons being salvaged and hidden in a garage at Duchesne City. One resident there told the author the identity of the man who brought it from the mountains with a helicopter, and that person is not one known for telling tall tales. It may have been the same cannon seen by Gordon Taylor on the ridge high above Peterson's ranch. During the early days along Rock Creek, Willie Davies built a log cabin resort just below the Upper Stillwater, a place where fishermen, hunters and prospectors could camp and purchase supplies. Not long ago, two of Davies' descendants, one from Utah and the other from California, dragged a cannon down from a steep, rocky gulch overlooking the old resort. It was taken to a home in Utah Valley where it was placed on display, but there were rumors that the Historical Society intended to obtain a court order to seize that cannon as an antiquity, so it was taken to California. Hopefully, it will be returned to Utah where it can be displayed as an artifact of early Spanish mining days in the Uinta Mountains.

George Stewart was a respected attorney at Roosevelt, where he was equally well regarded as a historian and collector of Utah lore. I have a taped interview made with him in which he describes investigating reports of cannons in the Uinta Mountains. Stewart learned of one cannon which several mixed-blood Utes attempted to salvage for its value as scrap metal. One of those Indians told him that they pulled it down from the high ridge where it had been found, but when they got it into the deep sand of the lower foothills, they were unable to drag it further. When they discovered that they could not get a wagon to where they left it, they pushed it in a hole and covered it with stones. They intended to return for it someday, but it always seemed easier to talk about than to do. Stewart said they are all gone to the Happy Hunting Ground now, so unless the works of nature uncovers it, that cannon will probably remain right where they buried it.

Frank Murdock was a half-blood Ute as well as a half-brother to the author's grandmother. He found a Spanish cannon in Pigeon Water Hollow, its barrel protruding about one foot above the ground. It appeared to be made of bronze, which Murdock recognized as being valuable. It may have been the same cannon described by George Stewart. Murdock and his son, Glen, planned to drag that cannon closer to the old wagon road, but to prevent anyone else from finding it before their return, they piled more rocks over its barrel and covered the rock pile with brush. After Murdock told one of the tribal elders about his find, he was told of the curse placed on artifacts from the Spanish massacre, and he changed his mind about salvaging it. One of Frank Murdock's grandsons has kept a close watch on that old cannon, sometimes replacing stones or brush to keep it covered. Only a few months ago he told me its barrel is turning green with verdigris, but it was still well hidden when he last checked on it, during the deer hunt of 1992.

Many treasure hunters have sought those old cannon of the Uinta Mountains using modern magnetometers, but with no success. Of course, the reason they failed is because bronze cannons are non-magnetic and do not activate a ferrous metal seeking instrument. Most early cannons were made from bronze metal rather than from iron simply because iron heated much more quickly during firing, often resulting in greater casualties among those firing them than among those being fired at; that when gunpowder ignited prematurely when it came in contact with the hot iron. Most Spanish cannons were actually cast in Holland. If you should come upon one, don't think of it in terms of scrap metal. If properly preserved and sold to a collector or museum, it could add substantially to your retirement fund.

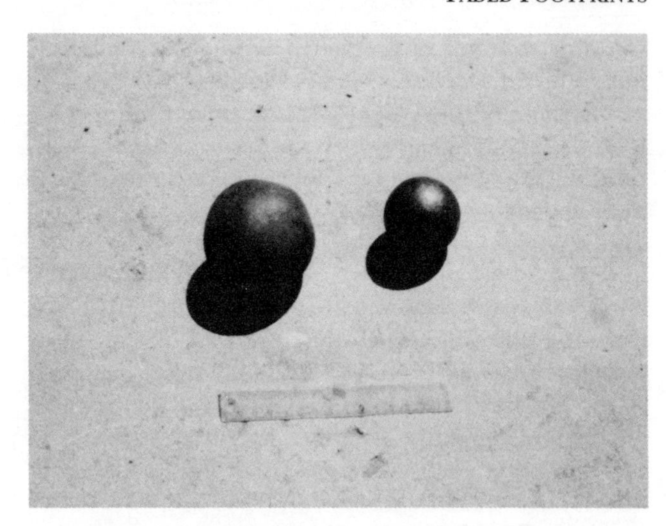

Tom Thacker found these old cannon balls at Dry Canyon, close to the old Rock Creek mission.

Many who prospect or otherwise follow the old trails have found cannon balls. Tom Thacker of Mt. Emmons found one while hunting deer in Dry Canyon, close to the old Rock Creek Mission. A light snow had fallen, and the sun's warmth had heated that metal ball so that the snow had melted from around it. It is about three inches in diameter and is flattened on one side, as if it had hit something quite solid. Skeptics have questioned whether it is a real cannon ball and have asked Thacker why it would be found in such an out of the way place. I heard him give a good answer to that question when he said, "If it's not real, then someone went to a heck of a lot of trouble to carry it so far up into the mountains, just to throw it away!"

No one can doubt the authenticity of a cannon ball found by Dudley Mallard. He operated a small sawmill north of Duchesne, cutting pine logs which were trucked in from the nearby mountains. While cross-cutting through a huge old log, the blade suddenly cut into a foreign object, and Mallard heard something being thrown through the air. Stopping the saw to investigate, he was amazed to discover that the blade had encountered a lead cannon ball, cutting a slice from one side of it. The remainder, more than three-quarters of that ball still remained embedded in the log. That cannon ball was near the center of an ancient pine. From counting that tree's annual growth rings, it was obvious that ball had been lodged in that tree for more than a century before Mallard's saw blade struck it! If that cut had been made only a few inches to either side, that ball would never have been found. More than a foot of living tree had grown around that ball, so that no sign of its entry could be seen before the log was cut.

Unlike many cannon balls which have been found

in the mountains, the ball found by Mallard deep inside an ancient pine was made of lead, with a hollow center which had held an explosive charge. For reasons which will never be known, that charge never exploded when that ball hit the tree. But one thing is certain; definite land and groove marks can still be clearly seen on that lead ball, leaving no doubt that it was fired from a rifled barrel. After seeing that cannon ball recovered by Dudley Mallard from the center of an ancient pine, no one could ever doubt that Spanish soldiers used cannons in the Uinta Mountains. Mallard isn't certain where that saw-log came from, but he has it narrowed down to two places. It either came from Hell Canyon on the Yellowstone River or from Wolf Creek on the West Fork of the Duchesne River. But wherever it came from, it is proof positive that Spanish adventurers used cannons in the Uinta Mountains.

With the congress readying the Ute Reservation for settlement, and with prospectors quietly slipping into the mountains and stockmen eyeing every meadow and river bottom for home-sites, Caleb Rhoads suffered an accident which hastened his death. While he was at his log cabin home at Price, he was suddenly crushed when the main support log of the roof collapsed and pinned him under its weight. Because of his advanced age he was slow to recover and made few trips into the high country after that time. He would send word to Happy Jack or Pick Murdock, asking them to meet him at some isolated spot so that neither Indians or Indian Agents would know he had returned to one of his mines. The places where they would meet would be at the forks of the West and North forks of the Duchesne or at the Hathenbruck Crossing on Rock Creek. After one of his last trips, Rhoads showed his half-brother, John, two sacks of ore he had brought back with him. Later John Rhoads said that ore was so nearly pure that it was more gold than quartz.

Pick Murdock once told how he and Rhoads would travel by night during the dark of the moon, and would keep to the roughest country to avoid detection. While waiting for the time when he could legally stake claim to his mines, Rhoads would say that if he could only live until that time, none of his large family, including his many half-brothers and sisters would ever want for anything. Once he told his wife, Sidsie, and his half-brother, John, of at least seven old mines where he could get gold, and that not including at least one place and perhaps more which were sacred to the Utes, and which he had promised Brigham Young and Chief Arropene that he would never reveal the locations of. During the last years of his life, Caleb Rhoads frequently urged Utah's two senators, Reed Smoot and George Q. Cannon to help him gain access to his mines, but in time he despaired of ever gaining their assistance. But in 1888,

Who can doubt the authenticity of this old cannon ball, recovered from the heart of an ancient pine? Note the land & groove marks left by the rifled barrel.

Utah Congressman George Sutherland did get a law enacted which eventually allowed white men to locate mining claims on the Ute Reservation.

There is a document which sheds a lot of light on the approximate location of what has come to be known as the Rhoads Mine; not to be confused with the so called sacred mine, the Church Mine, the Brigham Young Mine or Carre-Shin-Ob. Shortly before his death Caleb Rhoads signed a quit-claim deed granting a one half interest in the Rhoads Mine to John T. Clark of Provo, on August 8, 1904, in exchange for "good and valuable considerations." Although the exact location isn't spelled out on that document, it does state that Rhoads did locate the Rhoads Mine on the same day he located the Pine Mine; on July 10, 1859. It isn't certain which Pine Mine he referred to, but the location of each Pine Mine is well known. It is only logical to assume that if one of the Pine Mines and the Rhoads Mine were located on the same day, then the Rhoads Mine has to be in close proximity to that Pine Mine. If you goal is to find the Lost Rhoads Mine, that is a clue you shouldn't overlook.

Not long before Caleb Rhoads died on June 2, 1905, he and Pick Murdock made one last trip into the Uinta Mountains. Nearly seventy years of age and in poor health, it wouldn't have been unlikely that Rhoads told Murdock things he had told no one else, and took him to places where he had taken no other man. So it shouldn't have come as a surprise when Murdock staked his claim to a mine in the Uintah River Canyon in November of that year, only a few months after Rhoads' death. Those who were close to Murdock said that he brought nearly pure gold from that mine, the same kind of ore described by John Rhoads. The location of that mine is still on file at the Wasatch County

Courthouse, but like most claim notices of that period, the recorded description provides little information to revel the location, stating only that, "The mine is situated on the east side of the Uintah River, southwest of the sawmill." In Happy Jack's notebooks there is an entry mentioning Murdock's iron mine, but Pick Murdock never staked that mine east of the Uintah River for iron; he was too smart for that. Those who saw the ore he brought from it said it contained "stringers of gold in an iron-stained rock." Today that location is in Uintah County. Old maps show a dozen or more sawmills in the Uintah River Canyon, all of them long since gone. Pick Murdock's Uintah River Mine is now lost.

There are several clues in the Happy Jack notes which might pertain to Murdock's mine in the Uintah River Canyon, and one of them tells of a place where Caleb Rhoads found gold. East of Neola there is a large stone not far from the road, and on it there are inscribed several Spanish words and some numerals. Those signs point the way to a mine where Rhoads worked shortly before the reservation was settled. That mine was so close to a sawmill where logs were being cut for construction at Fort Duchesne that Rhoads once stated he had to take care not to be seen by mill workers. During the recent oil explorations in that area, a well was being drilled close to that place. One day work was suddenly stopped when a Ute medicine man arrived. All of the workers were told to leave the area and the road leading to the well was closed off. That night two of those employees returned to learn why they had been sent away. In a place thick with red willows they discovered an old wooden tripod and a directional arrow made of white cobblestones on the ground below it. That arrow pointed to the north, where that mine was later said to be. Several days passed before the workers were allowed to return to their drill rig. When they did, there was no sign of that old wooden tripod or the cobblestone arrow on the ground. Nothing remained to prove they ever existed.

Among the author's notes is a letter from Wallace Murdock, one of Pick Murdock's half-brothers. That letter tells how during the last years of his life, Murdock would disappear from his cabin at Whiterocks without notice, and would be gone for nearly a week. When he returned it would be during the night, with no explanation ever given to anyone about where he had been. But it was well known that when he returned, he would have enough gold to last him for another year. No one ever knew where he got that gold. Some said it came from his Uintah River Mine, while others were just as sure that it came from Little Farm Creek. Most likely it came from some place known only to himself. No one knows where his now faded footprints led, or to what secret places he went. That is part of the mystery of the Lost Rhoads mines.

CHAPTER 12

EMBERS FROM AN OLD CAMPFIRE
Pick Murdock, Prospector

The Whiterocks River Canyon is a frustrating place to search for lost treasure. There are dozens of rough, rocky side canyons, brushy gulches, numerous caverns, ice cold creeks and hundreds of mountain lakes; all hidden in a wilderness of thick pine forests and lofty peaks. Geologists recognize that the Whiterocks Canyon is a mineralized zone. A broken and fractured continuation of the South Flank Fault marks a huge uplift displacement near the mouth of the canyon, raised by a deep igneous intrusion. The Weber Quartzite is exposed in towering outcrops as much as 1,800-feet in height. Metaliferous veins may be found wherever volcanic dikes break through those outcropping formations.

From the Whiterocks Indian Village one can see more rough country in a day that he can explore in a year. David Copperfield, one of the early Ute tribal leaders, said that the mine where Thomas Rhoads obtained gold for Brigham Young could be seen from Whiterocks - - if one knew just where to look. But if someone wanted to prospect every part of those mountains he could see from Whiterocks, it might take him forever. And he might not find the mine even then, for Copperfield also said that if someone came too close to that hidden treasure, "The spirits would scatter his body all across those mountains." But who worries about an Indian curse when they're looking for gold!

All of the old-time prospectors passed through Whiterocks. Ute Indians like Happy Jack, Harris Murray, Bridger Jim and Pick Murdock lived at or near Whiterocks, while white men including Aaron Daniels, Pete Dillman, Jimmie Reed and Caleb Rhoads made the Murdock Trading Post there a jumping-off place for mysterious trips into that then unknown country beyond. Pick Murdock, Al Murdock's half-brother, spent a lot of time at that trading post, while Frank Murdock, a son of the Indian girl, Pernetta Murdock, was a blacksmith there and later operated a butcher shop. Art Murdock also worked there. When he was well past ninety years of age, Art told the author how he missed his chance to see the Lost Rhoads Mine.

Art Murdock a half-blood Ute, recalled a time long ago when he was a youth at Whiterocks. One day Caleb Rhoads came to the trading post where he met Pick Murdock. The two men busied themselves loading several pack animals for a trip to one of their mines. Pick asked Art if he wanted to go with them, but Art recalled that he was then still in his teens and he found Whiterocks too exciting a place to leave for what promised to be a long, hard pack trip into the Uinta Mountains. He recalled that there was a gambling grounds just north of the village, where he spent countless hours watching the old Indians play the stick game, using piles of gold and silver coins as chips. No, he didn't want to leave action like that to go on a rough trip into the mountains, so he declined Pick's offer.

About a week later Rhoads and Murdock returned from their trip. While they were unloading the pack string, Pick handed Art a chunk of rock "as large as a clenched fist." Art said that rock was mostly gold, a beautiful jewelry specimen. There was a jail and a jailer's cabin at Whiterocks, but they were seldom used, so Pick Murdock made his home at that cabin. Everyone who knew him said that Pick was very clean, and that he kept that jailer's cabin spotless. Art said that he placed that chunk of gold on display at the store and later put it on the window ledge at the jailer's cabin. That gold specimen was there for twenty years or more, during which time countless people handled it. Over the years pieces were broken from it until it was only half as large as it once had been; then one day Art

The Whiterocks Indian settlement. The trail to Pick Murdock's Uintah River Mine starts here.

noticed that it was missing. He never knew who took that chunk of gold, but he never forgot that he had a chance to go to the Rhoads Mine: "If only I had gone with Uncle Pick and Old Man Rhoads, I would probably be a rich man now!"

A lot of prospectors and treasure hunters have relied on stories told about Caleb Rhoads' friendship with an Indian Agent or post trader named Jimmie Reed. The author has compiled a list of agents and traders from 1868 to the present, which list reveals that Reed was never assigned as either agent or trader. The author has researched the so-called Reed Trading Post at great length, which has revealed that about 1828, by some accounts as early as 1823, a group of three men established a small fur trapper's post on the Whiterocks River. They were Agustus Archambeaux, Dennis Julien and William Reed. Julien may have trapped in those parts even before then. His name has been found cut into rock ledges from the canyon-lands of southern Utah to the site of Fort Robidoux, at which site his name and the date 1831 can still be seen.

Accompanying them was Reed's sixteen year old nephew, Jim Scott, who would later take his uncle's name and be called Jimmie Reed. The rough post they built was known as Reed's Trading Post. In 1831, Antoine Robidoux, one of a large family of frontiersmen from St. Louis, built a new trading post only a few hundred yards from the Reed Post. It was only one of a long line of posts extending from southern Colorado to the Uinta Basin. Because of his family's influence and contacts among Mountain Men and fur trappers, Robidoux soon put the Reed Post out of business. Robidoux had become a Mexican citizen and was married to the adopted daughter of New Mexico Governor Manuel Armijo. Governor Armijo issued him a permit granting an exclusive trading monopoly with the Utes. When the Reed Post closed, Archambeaux and Julien returned to trapping, while William Reed returned to his home in Kentucky. Young Scott, by then called Jimmie Reed, accompanied Dennis Julien to learn the trapping trade. It is known that in August 1826, Jedediah Smith led a small band of free trappers south through the Great Basin to southern California. Among them was a youth named James Reed. When Smith returned to the Utah mountains, two of his party deserted along the way in California; one of them was James Reed. Whether that was the Jimmie Reed who trapped with Julien is unknown, but it is known that both were in California during the 1830s. Archambeaux, called "Sambo" by the Utes, eventually returned to the east, where he was killed during the Civil War.

Jimmie Reed remained in California until the late 1870s, at which time he decided to return to the Uinta Basin. But first he traveled to Wyoming where he married a full-blood Shoshoni Indian woman he called Margaret. In 1878, Reed and his Indian wife moved to Ashley Valley, now Vernal. For the next four or five years

The Roubidoux inscription: "Antoine Roubidoux passed here 13 November 1831 to establish a fort on the river Wintey (Whiterocks)."

Dennis Julien left his name on a rock ledge near Fort Robidoux, 16 May 1836.

he ran cattle along the Green River north of Vernal. About 1883 he sold his cattle to Charley Crouse, one of the first settlers at Brown's Hole, after which he acquired a small farm consisting of a few acres near the mouth of Whiterocks Canyon, only a few miles up-country from where his uncle had built the original Reed Trading Post. He was then well over seventy years of age. At no time did he operate a trading post or was he ever an Indian Agent as some now claim; still, it is entirely possible that after his return to Whiterocks, he could have traded horses with Caleb Rhoads and others, and he probably did accompany him on prospecting trips. Jimmie Reed was born about 1811 and died in 1928, reportedly at the age of one-hundred and seventeen years During the period when he was in California, Alma (Al) Murdock established his trading post at Whiterocks, the first since Fort Robidoux was burned in 1844.

Interviews made with many of the old Ute Indians when the reservation was being settled by white men verify the foregoing. Jimmie Reed Jr. stated that his father traded with Shoshoni Indians in Wyoming before he moved to Vernal and later to Whiterocks. A daughter of Reed, Mary Harris, stated that the old Reed Trading Post was located only a few hundred-feet from the post built by Robidoux; which post was located just southwest of present day Whiterocks. Tribal Elder John Duncan may have said it best: "Trading Post at Whiterocks no built by Jim Reed. That post all gone before he returned to the Basin." Stories of Jimmie Reed trading with the Utes are mostly just that, stories, since Reed was a young man when he went to California with Dennis Julien and never operated a post after he returned to the Basin as an old man.

Pick Murdock, the Indian prospector partner of Caleb Rhoads, was known to frequent the high country between the Whiterocks and Uintah rivers east to the Yellowstone and Lake Fork. He probably knew that country as well as anyone, but he often traveled alone, so few today know the dim trails he followed. He also knew as much as anyone about Spanish gold in the Uintas, but in his own way he was as much an enigma as either Thomas or Caleb Rhoads. He was a prospector nearly all of his life, which fact is corroborated by a history of the Uinta Basin, which states that he was the first man to ship commercial grade ore from the Ute Reservation; that being from a place on the Uintah River where he mined ruby-silver. Little has been written about that mysterious Indian prospector and miner, and most of what has been written is wrong. Except for a single line in Footprints In The Wilderness, everything else in that book about Pick Murdock is wrong. That single exception states: "One who knew Caleb Rhoads better than most was Pick Murdock." It's time to get his story straight.

Pick Murdock was a full-blood Uintah Band Indian. That fact is reflected on the rolls of the Ute Tribe at Fort Duchesne. He was raised by the author's great-grandfather, Joseph Stacy Murdock, a pioneer of 1847 who was designated as the first Mormon Bishop of Heber Valley by Brigham Young. During an Indian battle near Utah Lake in 1849, Bishop Murdock and Porter Rockwell, the famed Mormon scout and frontier marshal, captured an Indian Chief of the Timpanogos Band. He had two small Indian children tied across his horse, both of them captives taken in battle with the Uintah Band. To save those children from a life of slavery with their captors, Bishop Murdock claimed and adopted them, raising both as his own flesh and blood. Judging from their baby teeth, he guessed the boy's age to be about six years, while his sister was about a year younger. Bishop Murdock, a polygamist, had six wives and thirty-two children, but those two Indian youths were educated and given the same advantages and opportunities as any of their half-brothers and sisters. That Indian boy's name was Spick-itta, as is shown on official Ute records, but that name was usually spelled Supicket by his family. He often signed his name Picket. He grew up to be the legendary Pick Murdock, a talented prospector and cohort of mining men like Aaron Daniels and Caleb Rhoads. He never married. His sister, Pernetta, grew up and did marry, having five children, and a posterity of prominent grand-children.

Pick Murdock was raised at Heber Valley. That he was well educated is shown by the fact that in later years he acted as an official interpreter during councils between the Ute Tribe and the U.S. Government. He accompanied several Indian delegations to Washington during negotiations with the Bureau of Indian Affairs. By the time he was a grown boy he was working with his white half-brothers at their livestock camps, from Kamas Valley to the Green River. Though he had been raised as a white man, he was a Ute at heart, so it wasn't unusual that he began to spend more time living the life he loved best, as an Indian; much of that time at Whiterocks. After his half-brother Al established his trading post at Whiterocks, Pick spent even more of his time there.

Having been raised in sight of the silver mines at Park City, Murdock took an early interest in prospecting, but he didn't get his name from that avocation as some have claimed; "Pick" simply being a contraction of his given name, Spick itta or Supicket. On mining claim notices still on file at the Wasatch County Courthouse, he signed his name as Picket Murdock. He was credited with discovering several rich lodes in the Park City Mining District, including the Glen Allen Claim, which later became the famous New Park Mine, the district's largest gold producer. He also prospected over the crest of the Uintas well into Wyoming Territory, for Rock Springs Attorney Pat Werner has discovered court documents which reveal that he located valuable mines on the north slope of those mountains. Several of those claims were recorded with Jesse Ewing as a partner, while others were located with Charley Crouse of Brown's park and with "Mexican Joe" Herrera, a local tough and notorious horse thief.

Pick Murdock may have acquired a taste for liquor while prospecting in such exalted company. Another of his prospecting partners, the Indian known as Happy Jack, kept a journal, and one entry tells of posting bail to get Murdock out of jail. Another who kept a journal

Pick Murdock, Indian prospector and partner of Caleb Rhoads. He knew the secret places of the Uinta Mountains.

record was William Boren, an old-timer who freighted goods from Heber Valley to the Ute Reservation. In his journal there is a note sent to him by Murdock:

> I am send this to you that mabe you will bring this things. One good hat, two wool shirts, butcher knife, mabe some rope. I am camp on Duchesne River. Be carefull. Indians not all happy, mabe fight soon. Better not come alone, all same with guns if things go bad. Mabe so bring whiskey. Murdock.

Several interviews with early travelers and settlers mention the Murdock Trading Post at Whiterocks during the 1870s. If that date is correct, then the post must have been established by Bishop Murdock or one of his older sons, and was probably nothing more than a one room building where only basic necessities were sold or traded. By the time he was a young man, Alva (Al) Murdock took over that store, no doubt with Pick to help him. Many pioneer accounts state that all of those old-time Utes thought highly of Al Murdock. They knew that his father had raised the Indian children, Pick and

Al Murdock, Uinta Basin pioneer and first trader with the Ute Indians.

Pernetta, and had always treated them fairly. To demonstrate their good feelings toward him, the Utes leased the entire western part of the reservation to the Murdocks to graze livestock on, for only one-thousand dollars per year. Other Heber Valley families included their stock with the Murdock herd and it became a "co-operative" venture. Co-Op Creek in Strawberry Valley received its name from the co-operative herd the Murdocks grazed there. In later years, Al Murdock said he never lost a single animal to the Utes by theft; but he would give any Indian family in need a beef animal whenever they needed one. Some years after he negotiated that land lease and was running sheep and cattle on Indian land, a newly assigned Indian Agent was sent from Washington, and one of his first acts was to order his Indian Police to remove Murdock from the reservation. But those Utes thought so highly of Al Murdock that they refused that order, and the Murdocks stayed.

Al Murdock later established a second store and trading post at the forks of the Duchesne and Strawberry rivers, only a few weeks before the reservation was opened to settlement on June 6, 1905. He erected what he called a "circus tent" where he stocked all manner of goods. When the land rush started, the entire population of what would become Duchesne City consisted of only fifty-two white men, an Indian named Sugoosie Jack and one white woman; Murdock's daughter, Dora. The new town was called Dora in her honor, which name it retained until 1911, when the name was changed to Duchesne. Murdock soon built a freight and stagecoach line connecting Price with Duchesne, Myton and Vernal. One day a shipment of goods came in on one of his freight wagons, and among the many sacks and boxes was a bag containing $60,000 in gold coin. After that wagon was unloaded, that bag of coins couldn't be found, and it was assumed that it had bounced off the wagon somewhere along the freight road. Two months later that bag of gold coins was found where it had been thrown into a corner along with bags of bolts, nuts and nails. An employee thought it contained washers!

In later years, Al Murdock sold his mercantile business at Duchesne. George Kohl, who owned a mercantile at Heber City, took over the Murdock store in 1916. Kohl suffered a personal tragedy related to that store and to the Lost Rhoads Mine, as was recalled by his close friend, Wilford Binggeli. Kohl employed a Ute Indian girl at his store, and during conversation with her one day, she told Kohl she had overheard some of the old men of the tribe talking about the Rhoads Mine. She told Kohl that she could lead him right to that mine, and although both he and his wife, Bessie, were reluctant to go with her, she continued to urge them to do so and they finally agreed; expecting it to be little more than an outing or holiday. The Kohl's were so crushed by the terrible thing which happened during that outing that they would never talk about it publicly, but they did confide in Binggeli. Kohl said they had ridden into the mountains and as they were passing through a narrow canyon, a shot suddenly rang out and the Indian girl fell from her horse, dead. They never knew how anyone could have known about their plans. Had their conversation been overheard by one of the old Indians who lounged around their store, or was it just the curse of the Killer Mountains?

Pick Murdock began spending most of his time around Whiterocks, often acting as an interpreter at nearby Fort Duchesne. Whiterocks was then a supply point for prospectors heading into the Uinta Mountains, so it was only natural that he would become acquainted with other mining men, men like Aaron Daniels, Caleb Rhoads and his Ute friend, Happy Jack. Like Pick Murdock, Happy Jack had also been raised in a white family, by Richard Jones of Heber Valley. He was raised as a half-brother to Jones' son Wallace, who Happy Jack always called Bill. He was sometimes seen at Bill Jones' cattle camp at Rock Creek. Even then Happy Jack was involved in prospecting with Murdock and Rhoads. One day he brought a beautiful chunk of native silver from one of their mountain trips and gave it to Pick's niece, Stella, who was the wife of Frank, his sister Pernetta's son. Stella Murdock kept and cherished that silver specimen for many years, often telling how it had been given to her by Happy Jack.

Al Murdock's Pioneer Store, located at the "Dora" Townsite, now Duchesne City.

Like Pick Murdock, Happy Jack was also educated in the white man's ways. He had a habit of keeping a record of things or events which were meaningful to him; things like horse trades, mines he had been to and prospecting trips with Murdock and others. His journals were nothing more than small leather-bound notebooks he could carry in a shirt pocket. In one entry he tells of going to "Supick's place at Whiterocks" where they put together a camp outfit for a two week trip. They made their camp at "South Fork" and brought back some ore samples which "Supick will have assayed." A later entry states, "Good ore, Supick all happy." That ore was described as "quartz with rose vein" and assayed "three parts gold, one part silver and ten parts copper." Still another entry in one of those notebooks tells of going to Murdock's store, where he met "Supick, David Copperfield, Bridger Jim and Kale Rhoads." On that trip they went to "the Yellowstone headwaters."

It is well known that Caleb Rhoads sometimes stopped at the Whiterocks trading post to purchase supplies and meet with Pick Murdock. It is very likely that Rhoads and Murdock were acquainted long before their meetings at Whiterocks, for while he was still a young man, Murdock helped his white brothers herd livestock at Pine Valley, between Beaver Creek and the Provo River, along the trail both Thomas and Caleb Rhoads followed into the mountains above Kamas Valley. After his half brother, Calvin, drowned in a mill pond at Pine Valley, the Murdock's sold that range for $300. Roy Boren, who also prospected with Murdock in the Whiterocks area, once said that Murdock told him he first met Rhoads near Kamas. Boren also said, "Pick Murdock knew things about the Rhoads Mine which no one else did; for example, that the Mormon Mine was near Whiterocks, not in Rock Creek Canyon as most men thought." There were probably few Spanish mines in the Uinta Mountains which weren't known to Caleb Rhoads and Pick Murdock.

Pick Murdock was sometimes seen leaving Whiterocks on the old Farm Creek Trail which led northward into the high peaks beyond, but from the head of Farm Creek he could range far, either east or west. One prospector told of finding one of Murdock's lone camps below Dead Horse Pass in the Squaw Basin country, many miles west of Whiterocks. There are several old Spanish maps which show the location of mines and caches in the Squaw Basin and Moon Lake area, and it is not unlikely that Murdock knew of them. But there was one map which he probably never saw, one which raises as many questions as it answers; the so-called Gringo Gordo Map. (See Map P)

Exactly where the Gringo Gordo Map originated will probably never be known for certain, but the story which Gene Keene of Kaysville tells is probably about as close to the truth as we'll ever come. Keene was raised on the Ute Reservation, where he was acquainted with a Catholic priest, a man who was of Italian descent. That priest made the mistake of falling in love and getting married. Back in those days the Church of Rome took a dim view of such happenings, so he was relieved of his parish and was excommunicated. It so happened that before being assigned to the Ute Reservation, that priest had labored in the faith at a centuries-old mission in the Mexican state of Sonora, where he discovered an ancient parchment map in the mission archives. He made a tracing of that map, and after his release from the ministry, that copy found its way into Keene's possession.

If one examines the Gringo Gordo Map carefully, he will easily recognize Kidney Lake and Moon Lake before either of them was enlarged by diking; as well as Rock Creek and the Squaw Basin country beyond. Not so easily understood is the priest's name for that map. Gringo Gordo appears to be Italian rather than Spanish. Perhaps it was as close as he could come to translating words from the original. Literally translated, that title seems to mean "fat white man," but if that is so, who was the fat white man? Perhaps some long forgotten Jesuit priest? Some recently made copies of that map have had several other strange words added to it, but by who and for what purpose is now unknown. "On Pasa Siempre" translates to something like "on foot forever," while the meaning of "Kaevin" is unknown. Some say it refers to a cavern, but near the border of the original tracing is the word "Cavernoso," which like "Cueva" is usually acknowledged to mean cave or cavern. That original tracing had the inscription "Ce Oro" along an unidentified river or stream, while later copies have the phrase, "The Gold Of Rock Creek." Every new copy which surfaces seems to have still more added words, which only confuses the real meaning of the original map. But if you can decipher that map's mixed Spanish and Italian wording, you could be on the trail to treasure!

The Gringo Gordo Map shows the San Pedro Trail leading into the high country at the head of Rock Creek, with an "X" to mark the site of an old Spanish mine. John Jones, an early settler of Tabiona, told of finding an old stone cabin just about where that "X" is on the map. He came upon that cabin near the crest of the mountains, in a desolate world of wind and snow high above Squaw Basin. It was apparent to him that the cabin was very old, everything in it being covered with the dust of centuries past. There were pieces of animal-hide pack sacks chewed to pieces by rodents, and in one corner there was a dusty pile of ore samples. Jones recalled that there were visible grains of gold in those rocks. He also found a faint trail leading even higher into the mountains, past the Rock Lakes and into a rocky box canyon. Jones was a rancher, not a prospector, so he spent little time there, for even though it was still September, it was beginning to snow and black clouds promised that a wild lightning storm was brewing, so he quickly retreated to lower country. Jones said he always regretted not returning to that old stone cabin, but time just seemed to slip away.

Gene Keene also saw that box canyon high above Squaw Basin. He hiked over Tworoose Pass and into the upper basin, well beyond Squaw Lake. Just outside the entrance to that box canyon he came upon an old grave, easily spotted since it was outlined with heaped up stones. Near the head of the canyon, where vertical cliffs rise to nearly 13,000-feet, he found the portals of two mine tunnels at the edge of a rock slide. That loose rock was so unstable that just walking across it sent huge boulders crashing down the mountain side. He could hear the sound of moving stones grinding and echoing from deep under that slide, so he made no attempt to pull loose rocks from those portals; it would have been suicide to do so. Across the canyon Keene saw the ruins of a stone cabin, no doubt the same one described by Jones. Keene also planned to return one day, but it is a long, hard hike into Squaw Basin. Those old mine tunnels and the stone cabin are still there, if they haven't been covered by slide rock. Just follow the Gringo Gordo Map to Squaw Basin. From there, you're on your own.

For many years stories have been told about a mine or cache near the crest of the Uintas, higher even that Squaw Basin. Several Indians have let slip hints about "plenty of money rock" beyond Dead Horse Pass; and it is known that Caleb Rhoads was seen bringing pack animals over that pass into the Black's Fork country. Willie Davies, who built the Rock Creek Resort, was also known to prospect in that 12,000-foot high country of wind-swept ridges and ever-moving rock slides. Just under the Rock Creek side of the pass there is a rough and broken line of cliffs. If you make your way along the base of those cliffs, taking care where you step above the abyss below, you can see places where Spanish treasure signs have been cut into solid rock. Some say that Willie Davies knew of an old mine there, so difficult to get to that he had to use a rope to climb down those cliffs. There are two ways to get to Dead Horse Pass, and neither is easy. From Continent Lake on the Rock Creek side of the pass, the trail switch-backs up a loose talus slope to the crest at 11,600-feet. From the north slope, following the West Fork of Black's Fork, you will cross Red Knob Pass at 12,000-feet and from there drop down to Dead Horse Lake, before ascending the final climb, also up a steep, rocky talus slope. Snow drifts often stay until September, when winter storms begin again. It's no place for a "paper-prospector." Unless you're on the trail of Spanish gold, leave it for the younger climbers.

During his many years of prospecting, Wayne Nelson discovered several very small log huts, too tiny to be called cabins. He has photos which show them to be no more than three feet high and wide and only long enough for a man to take shelter in. Without exception, every one of those huts is located in the center of a dense thicket where it is unlikely anyone would ever find them except by accident, the way Nelson found them. Alonzo Jim told Nelson those huts were built and used by Caleb Rhoads during covert trips into the mountains, after the time when Chief Tabby warned him not to return. One of those well concealed huts is just off the trail leading to the Dead Horse Pass mine found by Willie Davies.

Below Squaw Basin, but still high on Rock Creek, men still search for the Mine With The Iron Door. Almost every summer fishermen and hunters or others who have no interest in mines or prospecting report seeing that old mine, its portal covered with a rusted iron door. Nearly every report places it in the same area, about one mile upstream from Squaw Basin Creek, on the west side of the canyon and within sight of the creek. That waybill sounds easy to follow, but that is rough country, and it is heavily forested.

There is more to Squaw Basin than just the Iron Door Mine. A highly intelligent man who knows gold when he sees it has panned gold nuggets in lower Squaw Basin. While crossing a wild and broken country of rock slides and stunted jack-pines, in the jungle of loose rock and downfall timber below Cleveland Peak, he came upon a long abandoned cabin. It was obviously very old, its lower logs rotted to sawdust and its caved roof green with moss. Wondering why anyone would build a cabin in such a cheerless place so far from any trail, he stopped to investigate and ended up spending a week there. There are a number of small unnamed lakes in that boulder-strewn basin, a few having fish in them, but most only dead, stagnant water. In a tiny spring-fed brook tumbling out from under a rock slide above one of those lakes, he saw the unmistakable glint of gold reflecting a yellow flash in the early morning sun.

After making camp near that old cabin, he had no trouble panning enough BB size grains of gold to pay for his vacation fishing trip. A week of pleasant days were spent panning gold from that tiny brook, but before he left he noticed something which may be of interest, should you happen to be up that way. It didn't take long to recover most of the small grains of gold in that short length of stream between the edge of that rock slide and the lake, but even before he left, the gold he had panned was being replaced by other tiny grains, washing out from beneath that rock slide. He thinks that is why the cabin was built where it is. That prospector lives in another state now, but he doubts that anyone else has been to that cabin since he was there, for it is far off any trail. I wonder how much gold there is in that brook by now?

Many of the strange places in that high country may be connected in some way. That tiny stream of gold nuggets, the old cabin hidden in the jack-pines and the old stone cabin in the box canyon may all have something to do with a huge gold pan many have reported seeing along Cabin Creek, just below Squaw Basin. Those who have seen that old Spanish midden describe it as being too large and heavy to work with, but Cecil Dalton, one of the last to report seeing it, says the only reason no one has attempted to carry it out of those mountains is because of its size; about four-feet in diameter. It would be a difficult thing to carry it down the steep and stony trail down to Rock Creek. Dalton said it weighs about forty pounds, but it is so cumbersome that he wonders how it was used. He says it would take two men to work it, and even then it would be quite awkward. Perhaps in the old days there was some sort of water wheel or other mechanical means of operating it. Dalton placed that heavy midden atop a large flat rock along Cabin Creek, just above its junction with Rock Creek, along the old trail leading to Squaw Basin and the Mine With The Iron Door.

There is a place shown on the Gringo Gordo Map which was touched by death sometime in the distant past. Several years ago, Gene Keene and his son were hiking near Kidney Lake. There was nothing unusual about either the summer day or the alpine meadows they were crossing, except that young Keene had an odd feeling, as if someone was watching them, or that they were intruding in some sacred place. He told his father that he felt as if they were in a cemetery, and suggested that they go another way. Keene had experienced a similar feeling, but hadn't mentioned it. Stopping to look around, they discovered they were in the middle of a burial ground! They counted the outlines of eight separate graves. There were no tombstones, nor were any of those graves marked, still they knew they were graves. After they returned home, Keene noticed that those graves east of Kidney Lake are at exactly the same place where graves are marked on

One of a long line of stone markers below Mosby Mountain. What do they lead to?

the Gringo Gordo Map. Who is buried there? Could they be Spanish miners? Were they somehow victims of the Killer Mountains?

Old Bill Mosby was one of the Whiterocks prospecting club before he was killed by a falling log. He had a copper prospect near the head of Red Pine Canyon, a side branch off Whiterocks Canyon. He told one of his family that Caleb Rhoads had shown him a gold mine between the Whiterocks River and Mosby Creek, along the trail from Red Pine Canyon; a mine he promised Rhoads he would never take anyone to. Along that old trail, where it leads up to Mosby Mountain, there are some of the most elaborate Spanish treasure symbols cut into patriarch pines that one will ever see in the Uinta Mountains. Regrettably, during recent years persons not concerned with protecting the Spanish heritage of those mountains have cut down some of the large pines on which they are carved. Caleb Rhoads once said that near one of his richest mines there was a grove of large pines with Spanish signs cut into their bark. There are several ancient smelters in that same area close to Mosby Mountain; one along Mosby Creek and another on Crow Creek. Slag taken from either of those smelters assays high in both gold and silver. They are only two of many smelter sites which have been found in that area. Recent infra-red photography has identified still more sites nearby.

Near the head of Crow Creek, below Mosby Mountain, there are two wind-worn caves overlooking the valley below. On a high ridge just above those caves and leading northward towards Chen-Ob Spring, there is a long row of ancient looking stone monuments, each one progressively larger in size; the largest being on the end of that ridge above those two caverns. Those monuments have been there as long as the oldest Indian can remember, but their meaning has been lost in time. The Spaniards called those stone monuments "amojonar," distance landmarks or treasure markers. One of those caves is relatively easy to climb up to, but to gain access to the other requires a rope and some climbing skill. Inside that more difficult cave, on the north wall, there is an all-seeing eye, cut deeply into the stone wall. If you stand with your back to that eye and look out over the canyon below just as that all-seeing eye does, an interesting sight may be seen. Look to where the cedars stop growing and the sagebrush begins; look carefully and you will see an old mine shaft. If you look into it from a closer vantage point, you can see the end of a "muescas" or notched-pole chicken ladder sticking out of it.

A man who is part Indian, not a Ute but a Cherokee, was shown that shaft by a Ute he befriended. The Ute told him: "Climb down that ladder to the bottom. Take all the gold you need, but take no more than you need. If you take more than you need, or tell anyone where you got that gold, you might die!" That man, who was well over seventy years of age at the time, told me he took a long, hard look at that rickety notched-pole chicken ladder dropping down out of sight into the black pit below and decided he didn't need any of that gold. I'm not sure whether the Ute's offer applies to anyone else, but you might check it out!

There is something else of interest about those old stone monuments overlooking Crow Creek. Infra-red

Raymon Bascom examines an old Spanish tree symbol on Mosby Mountain. (Courtesy: Mr. Dale Bascom)

An expedition sign reveals that Spanish miners passed this way in the long-ago. (Courtesy: Mr. Dale Bascom)

surveys of that area have revealed an unusual trench-like cut on the mountain near one of those rock markers. There is about two feet of the top of a ladder sticking above the surface in that trench, but what it leads to no one knows, for if there is a shaft there, it was filled in long ago. Why else would there be a ladder sticking up out of the ground? Further down the ridge there are two old inclined shafts, but they are filled with water and look like two small ponds. Infra-red photography has proven they are in fact shafts, not springs or water holes as modern maps indicate. That Spanish miners searched for gold in that area is given further credance by the recent discovery of a sword found in a small rock cavern close-by. On its rusted blade are the words ART a FAB a DE TOLEDO, referring to its place of manufacture, Toledo, Spain. Only about a mile further up-country from those stone monuments there is something even stranger, a mine with an iron door!

There is no doubt that mine is very old, for there are large aspen trees growing on its waste dump. That dump is difficult to find unless one knows just where to look, for it is in a place which makes it look like a natural ridge. The portal of a tunnel against the mountain side is covered with a heavy iron plate. It would seem to be a simple thing to dig away the rock and brush which covers most of that iron door, but it isn't that easy. It would be difficult to get heavy equipment to that tunnel portal, and to dig it out by hand would not only be a difficult task, but a dangerous one as well. Several people have started to dig that rock away, but every time they get close to that iron door, someone unseen fires a rifle shot at them! That first shot always misses by a few inches and seems to be only a warning shot, but no one is real sure about that, for none have stayed around long enough to see where a second shot might hit! It all depends upon how badly you want to know what is behind that iron door.

There are a lot of old diggings along Crow Creek and Mosby Creek. There is an old trailer house near the smelter on Crow Creek, where the road forks leading to the Little Water Mines. On the hillside behind that trailer there is an old tunnel dug back into a gray-colored shale rock. Several Indian brothers have been seen digging there. Further along the road to the Little Water Mines there is a green and white house, and behind it there is another tunnel, driven along way into solid rock, with no supporting timber. Close by is still another old mine, which is recorded at the courthouse as the Nick-Shot #2 Lode. It is said to be a Spanish diggings. A Ute high in the tribal hierarchy has several claims on file at the base of the Chalk Cliffs, along the road to Paradise Park. There is a copper mine just above those cliffs. It would take a long time to investigate all the old diggings in that area.

North of Whiterocks, a stream called Farm Creek joins the Whiterocks River. Don't confuse it with the Farm Creek which flows into the Duchesne River near Hanna; they are not the same. All of the old Indians who lived at Whiterocks agreed that when Caleb Rhoads and Pick Murdock went to their mines, they always went northward, following Farm Creek. But there is no way now to know how far they followed Farm Creek, for in its upper reaches they could have turned eastward towards Mosby Mountain and Crow Creek or westward to the Uintah and Yellowstone river country. A few who knew him said that Pick Murdock sometimes went to an old mine located high on Little Farm Creek, what today is known as the West Fork. One of his half-brothers said that he sometimes went to that mine by crossing over from the headwaters of Pole Creek. If you check that area on the old maps used then, you will find that in those days the Whiterocks River was called the North Fork of the Uintah.

There is no doubt that Rhoads and Murdock went to mines which were not very far from Whiterocks, for Art Murdock and others said that sometimes they would return from their prospecting trips after only two or three days. On other occasions they might be gone a week or more. But no matter how long it took, they always returned with their pack animals heavily loaded

An ancient shaft on a quartz vein contact, Mosby Mountain.

with money rock. During recent years several old mine shafts and tunnels have been found within a day's ride of Whiterocks, but they have all been filled in or dynamited so that it would be a difficult task to dig them out. Only a short way up the Whiterocks River there is an old tunnel, on the right side of the canyon. It is nearly covered with brush, but if you watch carefully you can see it from the trail. Only last year a peculiar and very old Spanish spur was found along the trail close to that tunnel.

Some years ago Wayne Handy operated an auto agency on the Ute Reservation. One of his customers, a Ute Indian who he was friendly with, told him of a mine in Whiterocks Canyon where the skeletons of two Spanish miners still rested. Handy also knew a white woman who married an Indian. She told him of a cache which was on her property. She said that two men stole two bars of gold from that cache and sold them to a banker at Vernal, for $30,000. But she said the joke was on them, for everyone knew those bars were worth twice that much! That woman has seen some of those bars, and described the markings on them. Each had the numeral XV to show they were two-thirds gold content. A small "bite" had been taken from each bar which had been tested or assayed. The tax stamp proving the king's Quinto had been paid was indicated by a circular inscription bearing the king's name, Carlos. A few bars in that cache did not have the king's stamp, indicating they were church gold.

That woman's tale of the two bars taken from a cache near Whiterocks has more significance than appears at first glance. Many who have studied the Rhoads gold at great length have concluded that he brought bullion gold to the Mormon Mint, not gold ore from a mine. A strange sequence of events which had its start during the summer of 1992 adds a lot of credence to that theory and to the woman's story. A professional treasure hunter well known in the Arizona area found his fortune near Whiterocks, but then lost it forever. He located the cache he had been told about by an old woman, but ran into unexpected car troubles while enroute back to the Salt Lake City airport. He was forced to leave his rental car at Duchesne and call an acquaintance who agreed to take him to the airport. That man noticed that the only luggage he carried was a small handbag which he held tightly to at all times. During the long ride he told his benefactor that he had made a remarkable find near Whiterocks, and after some coaxing he agreed to show him the contents of that hand bag; three gold bars!

The man from Arizona told an exciting story. With cooperation of an old woman who lived near Whiterocks, he had found a cache of thirty-six bars of gold bullion. He said that cache site was in a very public place, in view of several Indian houses, so that he had to dig into it during the dark of night. Even then barking dogs alerted several residents who turned their house lights on. Those bars were all the same size, weighing about thirty-seven pounds each, heavy enough that he could carry only three of them from the cache site. He explained the danger of returning to that place, and then asked his friend if he would help him, for it would be far safer for two men to attempt the next recovery. An agreement was made that they would return together, after the Arizonan took the bars to Phoenix where he expected to sell them. He agreed to return to Salt Lake City in ten days.

The allotted time passed, but the Arizonan never returned. After several more days, inquiry at the airport revealed that he had in fact returned, but that he had been arrested by federal authorities at the terminal. It took several days more to learn that he had been charged with trespassing on Indian land, desecration of a grave and violation of the antiquities law. His Utah acquaintance learned through Indian contacts that many years before a load of gold bars had been moved from the Rock Creek Mission, and had been cached in one of the old Indian cemeteries near Whiterocks. After several days, the Utahn was allowed to talk to the prisoner, who told him that if would raise bail, he would take him to where the remaining thirty-three bars of gold remained hidden.

It never took the Utahn long to figure the odds against him being able to recover those bars, nor did he want to become involved in violating serious federal criminal charges, so he declined to pay the prisoner's bail. Soon afterwards the Arizonan was transferred to a federal prison which will be his home for a long time to come. But if you're still interested in getting rich fast

A Spanish sword, found hidden in a small cavern in the White Chalk Cliffs above Crow Creek.

and don't mind facing some very irate Ute Indians, as well as some very professional federal officers, there are thirty-three bars of Spanish gold waiting for you, in a Ute Indian Chief's grave not far from Whiterocks. You can have my share of it!

Wayne Nelson once found an old mine in the Whiterocks Canyon. He drove along an old logging road which led from Mosby Mountain westward towards the head of the Whiterocks River. He was walking along a rugged section of brush and rocks when he came to a place where someone had covered a tunnel portal long ago. He dug that tunnel out for twenty-five feet, finding pieces of ancient brass tools in the debris removed from it. When he came to a heavy log bulkhead which would require days of work to remove, he broke camp and returned to town for additional supplies. But when he returned to Mosby Mountain a few days later, he discovered that logging road had been closed with large boulders and a barb-wire fence. Nelson, who is disabled from a hip injury, couldn't walk back to that tunnel. The log bulkhead is probably still at that tunnel's end just as he left it, that is unless someone has covered that tunnel portal over again.

Just off the Farm Creek Road north of Whiterocks, near Snake John Spring, there is a curious treasure sign carved on an old aspen tree. It is the outline of a miner's shovel, and on that shovel is what appears to be a map of the canyon beyond, along with several numbers and directional markers. If you can decipher its strange message, there's no telling what it might lead you to. A clue might be in the ruins of a Spanish smelter along the creek bank just down-canyon from that tree sign. That silver ore was once smelted there seems a certainty, for during the summer of 1993, an old man offered several small silver bars for sale, bars which came from that smelter site. Many years ago an Indian came to his small store and offered those bars for sale. That was back during hard times, and as near as he could remember, he paid the Indian a dollar or two for each bar. They're worth a lot more than that now. As you can see in the accompanying photograph, they are curious little bars, crudely made with bits of sand and slag in the metal. Each has a Catholic cross and the letter "V" on it, indicating that the silver came from a church mine and that the king's tax had been paid on it. It might be reasonable to assume that the bars which came from that old smelter were made of silver from a mine not far away.

At many mines of gold or silver, many of them located far from the royal tax collector, many elaborate and devious schemes were devised to avoid paying the Royal Fifth. There were never enough inspectors to monitor every mine in the northern mountains, so that many illegal bars and ingots were smuggled to seaports along the gulf coast. Much of the treasure now being recovered from Spanish galleons like the Atocha and Margarita does not bear the official tax stamp indicating that the king's fifth had been paid. Many schemes were hatched to circumvent paying that tax. Personal jewelry was not taxed, so that many wealthy miners returning to Spain had long chains of gold or silver made which they could wear around their necks; some so heavy they could hardly stand under their weight. Nor were gold goblets or silver dishes taxed. One tax collector observed that a ship's anchor had been freshly painted, and discovered that the anchor was solid gold!

An old document tells of one of the means tax collectors used to ensure that the king's fifth would arrive safely at Seville. In the records of Gonalza de Aranda, one of the king's accountants, there is found a reference to "the chest of three keys." Gold and silver paid into the king's account was placed in a large metal chest having multiple locks, requiring three different keys to open them. Three trusted agents each had a key to one of the locks, but none could open all three locks. An accounting of the chest's contents was made only after they were loaded aboard a treasure galleon bound for Spain. In the document cited, when the ship reached the port of Veracruz, Aranda wrote: "What is being shipped to Your Majesty is gold of standard quality; 13,596 marcs, which amounts to one-hundred thousand castellanos largos. It is being shipped in ten ships conveyed by one of the fleet." Those old records reveal that during only a single fifty year period, two-million pounds of gold was shipped to the Spanish king from the mines of New Spain, no doubt part of it from the mines of the Uinta Mountains, shipped in "chests of three keys."

During the late 1880s, the Killer Mountains claimed two more victims in the rough country near the head of the Whiterocks River. An old miner had a good paying ledge where he dug out lode gold. For several years he packed his summer's find to Ashley Center, now Vernal, where he would trade that raw gold for minted coins. He would pay his bill at the mercantile, set aside enough money to last him through the winter, and send the remainder to his daughter who lived in the east.

There wasn't any secret about the gold he had; many people knew of it. When he failed to return to town during the fall of 1877, the store merchant became alarmed and organized a search for the old man. Pick Murdock, Pete Dillman, Aaron Daniels and Bill Preece headed into the mountains above Whiterocks to look for him. Dillman knew that area well, having been a law officer as well as the district's first Forest Ranger. The story of their search is told in Builders Of Uintah, a history of Uintah County.

Those men all later recalled that it was the hardest trip any of them had ever made. Snow had drifted ten feet deep and all of the search party nearly perished in a blinding blizzard which raged across the mountains for two days. They didn't know exactly where to look, but about a week later they discovered the frozen bodies of the old miner and his pack mules where they had become bogged down in deep drifts. The place where that miner died is still known as Deadman Basin. His body was packed out to Whiterocks where he was buried. Dillman cashed in the gold found on his pack mules and sent the money to the miner's daughter.

In the spring, Pick Murdock, Happy Jack, Henry Harris and John Murray, all Ute Indians, decided to search for the dead miner's diggings, but all they ever found were some pieces of canvas from his tent and a pine-pole corral hidden back in a stand of timber. The old man's ledge must have been somewhere close by, but they never found it. But they did find something else. In some notes which Happy Jack scribbled in his notebook, he told of a chilling discovery. Near what is now called Deadman's Lake, they came upon a crudely made log and stone shack, and inside they discovered the body of a prospector who had been reported missing several years before. He had been shot in the back and his tiny cabin had been ransacked. It was speculated that he had been murdered for his gold. He was buried by the side of the cabin, and from an address on a faded envelope, his family was notified. Several months later the mail brought Pick Murdock a crudely sketched map which had been sent to him by the man's family. It is unknown whether he and Happy Jack ever tried to follow that map, or what became of it, but the following year two brothers named M.W. Warner and E.P. Warner did find and began working an old mine they found about two miles west of Deadman's Lake. Frank Warner told the editor of the Vernal Express, "I think this will be a genuine bonanza!" There is no way to know whether it was the murdered miner's mine, nor do we know if it turned out to be a bonanza.

During his long life, Pick Murdock saw many men killed because of the lust for gold, or for gold their killers thought they possessed. Tom Rhoads was one of those prospectors who died a violent death. He was an old fellow who was in some way related to Caleb Rhoads, and he lived in a little cabin north of LaPoint, on the old Indian Trail to Mosby Mountain. Some thought that because of his connection with the Rhoads family, he probably knew something about the Rhoads Mine; a belief he did little to dispel. On occasion he would show someone an old map or perhaps some age-yellowed document; actions which would cost him his life. Some thought he had a cache of gold buried somewhere near his cabin, but that is highly unlikely, for those who were his neighbors said he was poor as a church mouse. But whether or not his maps and documents were valuable, or his gold cache authentic matters not, for just talking about them proved to be a deadly mistake.

Indian slave miners worked deep in Spanish mines, with only torches for light.

On September 25, 1924, two young Ute Indians went to Tom Rhoads' cabin to rob the old man of his gold. I have a statement written by Harris Murray, then age 18, in which he confesses that he and a companion murdered Rhoads. That document goes to great length describing the killer's movements before and after the murder, but in its pertinent parts it relates the following:

A tree symbol near Snake John Spring. It tells of an old Spanish smelter close-by.

> We rode down to Rhoads' cabin. Ernest got off his horse and broke two clubs from a tree. We went inside, where Rhoads was lying on his bed. Ernest raised his club to hit him, but he couldn't do it so I took the club and struck him several times. We both tied his hands and mouth. Ernest began searching the cabin for his gold, and I found his purse, which contained two twenty-dollar bills and about three and a half dollars in silver. We got a shovel and went down to the wash where I dug a grave, about three feet deep. We covered the body with dirt and used some brush to wipe out our tracks.

After the murder of Rhoads, the two Indian youths stole the old man's Model T Ford and fled to Nevada, where they were arrested near Ely. They killed Tom Rhoads for forty-three dollars and fifty cents. If he had any gold cached anywhere near his cabin, his killers never found it. His old maps and documents were never seen again. It is unknown what happened to them.

There are few places where there are more Spanish mines, both lost and found, than there are in Pole Creek Canyon, a tributary of the Uintah River, across the dividing ridge west of Whiterocks. Aaron Daniels described a Spanish mine in Pole Creek Canyon in his journal. That entry reads in part as follows:

> I was examining an iron lead along Pole Creek when Caleb Rhoads came by. He showed me an old smelter on the creek bank which had been built by Spaniards. He then took me to several old mine shafts, one of them about a half-mile from the Pole Creek Sink. He told me that he had dug gold from that shaft. It is hard to find unless you know just where to look. I went down into it for only a short distance because the sides were caving in. I think it could be one of his best mines.

Ute Indians recovered silver bars from this smelter not far from Snake John Spring.

In his journal, Daniels drew a sketch map of that mine. (See Map Q) Fred Cudney, a noted geologist who is very familiar with the Uinta Mountains and with Pole Creek in particular, said that he considers that area to be a most promising place to prospect for minerals.

Not long after Caleb Rhoads' death, a pick and shovel were found hidden in the branches of a large cedar tree on the divide between Whiterocks and Pole Creek. Pick Murdock identified them as belonging to Rhoads because of the peculiar way he had carved their handles. One of the old Indians at Whiterocks used to tell how after Rhoads' death, other white men tried to work a mine located in the bottom of the Pole Creek Sink; a huge open-pit glory hole where the waters of Pole Creek disappear into a cavern, only to reappear from the Pole Creek Cave, about seven miles down canyon. That tunnel in the sink bottom broke through into the caverns not far into the mountain. Those caverns are a series of limestone caves which honey-comb those mountains. No one knows how extensive they are. One old Indian liked to tell how the walls of one of those cavern rooms was encrusted with tiny wires of pure gold. He said that gold was so pure that it could be cut from the cavern walls with a pocket knife.

Early in the spring that tunnel in the bottom of the sink is covered when snow-melt run-off is high, which forms a small lake in that pit. An Indian from Whiterocks said that Utes have a special reverence for gold brought from that sink-cavern, and for centuries they have used it to make arm bands and other orna-

Men have died searching for this old mine near Whiterocks. Note the heavy mineralization and water.

The old mine in the Pole Creek Sink. Men have died there looking for gold.

ments which are used during sacred ceremonies. Other than Caleb Rhoads who had a special understanding with Chief Arropene, all other white men have been driven away from that place, or they have been killed. It has been said that more than a few of the prospectors who have mysteriously disappeared in those mountains have found their grave sites in those caverns. As recently as 1939 a prospector named Babcock built a cabin at the edge of the sink and began mining. He was found dead by the side of his cabin, the ruins of which can still be seen. Not long after Babcock's death, the portal of that mine was sealed shut to keep white men out of it. Some Utes believe that the spirits of murdered miners still stand guard over that strange sink hole, warning those who would steal the sacred gold.

During September 1977, Gale Rhoades located a mining claim at the Pole Creek Sink, which location was recorded at the courthouse and duly noted in the Roosevelt Standard. Rhoades failed to uncover the great hoard of gold he had hoped to find, according to him because of a lack of financing. Work progressed slowly until 1980 when development funds were obtained from Gunsite Oil Company, in exchange for a majority interest in that claim. After favorable reports were made by the Forest Service and an independent geologist, work began in earnest. But because of its isolated location it proved difficult to keep qualified miners on the job, and a number of men came and went until 1986 when Wallace Muir agreed to take on the job of opening that caved tunnel. Muir ran into many unexpected problems, not the least of which were caused by the claim owners.

September 5, 1986, seemed to be a work day just like any other at the Pole Creek Mine. Rock from the previous day's blasting was removed from the drift and a new series of holes were drilled into the face rock. Fuses of the same length used every day were cut and blasting caps were crimped onto them. Dynamite charges were tamped tightly into each drill hole. Muir prepared to light those fuses just as he did every day. An experienced miner, he knew that Bickford fuse burns at a rate of 42 seconds to the foot, so he cut each length of fuse long enough so that he could light every charge and still have plenty of time to be out of that tunnel before those charges started exploding. But that day something went terribly wrong. When he lit the first fuse, instead of it spitting and beginning to burn at its normal rate, the flame raced along the fuse in only seconds. Before he could even jump back, the flame ignited the blasting cap and the entire dynamite charge exploded in his face. Wallace Muir was killed instantly, while another miner further back in the drift was seriously injured.

Was Muir's death the result of some Indian curse or the wrath of spirits of Spanish miners whose ghosts guarded that gold? Those who knew Muir and knew how careful he was didn't think so, and neither did Forest Service and law enforcement officers who investigated at the scene. The coroner's death certificate concurred, listing the cause of death as "pending police investigation." We may never know why Wallace Muir died at the Pole Creek Mine, but a few old-timers have their own ideas about what happened; but their opinions are only an educated guess. They tell of similar accidents which weren't accidents. One old-timer recalled a claim-jumper's trick, in which a fuse used for blasting was carefully cut along its length with a razor-blade, a cut so fine it can't easily be seen, especially if that fuse is covered with mud and dirt as it usually is in a mine tunnel. When the fuse is ignited, it burns just like a line of black powder poured along the ground. Its speed is so fast the victim has no chance to escape. One might say that Muir was a victim of the Killer Mountains, but many don't believe his killer was a spirit or ghost.

Some Indians who have been seen in Pole Creek Canyon are definitely not spirits. During the summer of 1991 a prospector well known in the Uintas hiked into Dark Canyon, a seldom visited gulch not far down country from the Pole Creek Sink. He discovered an old Spanish smelter in that canyon, but as he approached more closely to inspect it, a rifle shot echoed from a ridge and a bullet kicked up dust just over his head. He quickly jumped backwards and began crawling into some thick brush. Crossing the canyon bottom unseen, he came to an open area where he could see that ridge above. He saw two men making their way down from the ridge to that old smelter. Although he was armed with a large caliber revolver, he wisely decided that brushy canyon bottom was no place for a shoot-out with two men armed with long range rifles, so he ran to his truck and got out of there. He is sure the two men he watched were Indians, and he is just as certain they were live Indians, not spirits!

CHAPTER 13

DIG HERE!
The Mine of Lost Souls

If the mysteries of Squaw Basin or the spirits guarding the gold of Pole Creek seem to be a little spectral or unreal, perhaps some of the mines and caches across the crest of the Uintas to the northeast may prove to be more substantial. Back around 1868, a half-mad prospector named Jesse Ewing discovered a silver lode which many have sought over the years. Most seem to think it is in Ewing Canyon, south of the Green River near Brown's Park. Ewing was a cantankerous old codger with few friends. He would take in a partner to help work his mine, usually some lone traveler without family or friends, or a derelict picked up in one of the saloons at Vernal. He preferred some husky young fellow who could work hard, but one not bright enough to ask questions about the value of the ore they were digging. Strangely, after Ewing took them to his mine, none were ever seen alive again! Ewing would promise them a half-share of the ore they dug, but whenever one of his partners began to suspect how rich that ore really was and asked for an accounting, he would permanently terminate their partnership and throw the body into the Green River.

One of Ewing's partners, a man remembered only as Coulter, was killed with rat poison, while another named Robinson was found in a dry gulch with a knife in his back. Others were shot or simply disappeared. More than a few of Ewing's partners were fished out of the Green River far downstream from Ewing Canyon, but none of their deaths could be traced back to that wily old rascal. Pick Murdock prospected in Ewing Canyon and one day while there he stopped to talk to Ewing and his latest partner, a mean looking hombre named Tap Duncan. Ewing didn't know it then, but Duncan would be his last partner. Murdock got Duncan aside and warned him that all of Ewing's partners seemed to mysteriously disappear when it came time to divvy up the profits. Duncan kept a close watch on Ewing after that, so when a few days later he tried to bushwhack him, Duncan got off the first shot, ending their partnership. No one knew much about Duncan, but several years later when Harvey Logan, a top-gun who rode with Butch Cassidy's Wild Bunch was arrested, one of the alias' he used was Tap Duncan.

Duncan helped himself to Ewing's money purse and before shaking the dust of Brown's Park, he stopped long enough to tell John Jarvie, a homesteader, where Ewing's body could be found. He said that since he was already wanted in Wyoming, he didn't care to stick around and explain what happened to Sheriff Pope from Vernal. When news of Ewing's death became known, dozens of prospectors headed for the hills, all in search of secret mines he was supposed to have had, but no one ever found them. Searchers did discover a copper prospect where Ewing had dug a tunnel nearly five-hundred feet into the mountain, but those who knew him said his money came from silver, not copper.

Some years later a Mexican who herded sheep in that area claimed that Ewing had a small vertical shaft which led to his silver diggings. Crazy like a fox, he built a cabin over that shaft and hid its entry under a heavy wooden trap-door. Ewing had cabins all through those hills at his different prospects, so it is unknown which cabin the Mexican referred to, but it is a moot question now, for all of them are gone, most burned in brush fires. Still, if you've got a good metal detector, you should be able to find some old-fashioned square nails or an iron hinge where that trap-door was located. If you should uncover some heavy, hand-hewn pine planks under those charred ashes, you've hit the jackpot!

The canyon where Ewing built his tiny one-room

Old cabins at the Dyer Mine. Millions in rich ore was dug here, proof of mineral riches in the Uinta Mountains.

cabins is still pock-marked with prospect holes; some he dug but many dug by others. If you have time and don't mind a long, hard hike, go down Red Canyon, route of the original wagon trail before the present road was built. There are several old cabins and mine shafts down there, so old no one knows who built them. That Mexican sheepherder told of seeing an old mine dump on the south side of the Green River, not far from the old Allen homestead. Any of the old-timers at Brown's Park can head you in the right direction. He said there were some really old signs cut into a ledge rock near that dump. He didn't know how to read their lost meaning, but he could tell they were Spanish markings. Take your metal detector and some good hiking shoes; like I said, it's a long walk!

During the late 1880s, Pick Murdock was a partner in a discovery which became one of the richest mines in the Uinta Mountains. In Kane Hollow, near the head of Brush Creek north of Vernal, he discovered an outcrop of extremely rich copper ore, much of it averaging fifty percent pure copper. With several partners he staked the Ace, Copper, Calumet and Hecla claims, all properly recorded at Vernal. In order to raise enough operating capital to start work, one of the partners, F.M. Dyer, went to Salt Lake City where he borrowed $10,000 (some accounts say $30,000) from a banker named Gates. But with cash in hand, Dyer apparently had second thoughts about how hard it would be to dig into that ledge of copper high in the Uintas, so he went south with the money. Banker Gates held a mortgage on the property and foreclosed when the partners failed to pay the first payment on their loan. Courts upheld his title to the property, and with Dyer nowhere to be found, Murdock was left holding a losing hand.

Gates hired a crew which at one time numbered 100 miners and built a smelter near the mine. Ore containing less than forty percent copper was refined and melted into ingots. Ore assaying more than forty percent was shipped directly to an eastern foundry as native copper metal. An average of ore mined at the Dyer Mine ran thirty percent copper, twenty-six ounces of silver and enough gold to pay expenses. Remember, that was in the Uinta Mountains which some die-hards still say aren't mineralized! Nearly all of the ore from the Dyer Mine was hauled north over the crest of the Uintas on the Carter Road, a corduroy wagon route to Carter Station on the railroad, but for a time some Dyer Mine ore was hauled by wagon and team two-hundred miles across the Ute Reservation to the Marsac Mill at Park City, and it still returned a healthy profit. Ore from Kane Hollow was so rich that Banker Gates realized a three-million dollar profit on his ten-thousand dollar investment! But not one penny found its way into Pick Murdock's pockets.

There was an earlier mineral discovery only a few miles down country from the Dyer Mine, where both gold in quartz and gold placers were found. On June 4, 1880, a group of prospectors met at the Vortex Cafe at Vernal, where they organized the Carbonate Mining District. In their by-laws, one article stated: "No Chinamen shall be tolerated in this district." Within a month a rough camp of log cabins and slab shanties blossomed forth on the mountain side, and somehow they were given the grandiose name of Bullionville. Presumably, no Chinamen lived there. Its site was near the head of Brush Creek, twenty-seven miles north of Vernal. For a short time Bullionville grew like Topsy, with new claims staked daily, but its placers were soon worked out while its quartz veins pinched tighter than a miser's purse. But during its short heyday, Bullionville attracted more prospectors to the east end of the Uintas, many of them sure they were closing in on the Lost Rhoads Mine at last.

During the summer of 1894 prospectors discovered gold placers at the foot of 13,449-foot Gilbert Peak. It was said that the first strike was made when a miner saw flakes of gold in a mound of dirt thrown up by badgers. Apparently, he worked his find in secret until he was killed by Indians. When two men came upon his remains, they discovered a poke of gold dust in his packsack. Within weeks men were staking claims, including the Victoria, Bromide and Legal Tender. One mining company, the Jeanette, was incorporated at Rock Springs, Wyoming. The Vernal Express enthusiastically reported there was $40,000 worth of ore in bins at the Victoria and $100,000 in sight at the Bromide. Every company promised rich rewards for their stock-holders, but except for the Bromide which shipped thirty tons of ore, there was more gold on their gilt-edged stock certificates than there was in the ground.

From the time that De Re Metallica was published in 1556, it was the Spanish miner's Bible. They used that book to become masters at mining and at smelting and refining gold and silver ores. That ancient volume, written in Germany by Georgius Agricola, described in detail every phase of prospecting, mining, milling and smelting which would be used during the next three-hundred years. As a matter of fact, most of the methods described in that book haven't changed much today. Spanish prospectors located mineral veins with the use of dowsing rods, dip needles and pendulums. Many of the ore bodies they uncovered using those ancient methods were low-grade, veins which a modern-day prospector might pass by, but the Spanish miner had two factors in his favor; lots of time of plenty of free labor, the latter often being the same Indians who had guided him to those ore veins.

Dowsing rods are well known to most prospectors, though few place much credence in their use. Many will acknowledge that a "witch" might locate water with a forked stick, but few believe the "enchanted twig" can lead the way to mineral veins. But strangely, if you add a battery to those rods and give them some fancy name like an Electro-Rod, those same skeptics will stand in line to shell out hundreds or even thousands of dollars to buy one! Someone who would laugh at a water witch will gladly give a month's pay for some chrome-plated rod fitted out with all of the latest electronic bells and whistles. They would be better off using the witch's forked stick!

Almost every treasure hunter has met someone using a "doodlebug" dowsing instrument. Usually they are some sort of long distance device guaranteed to locate buried wealth or mineral veins from several miles away. No matter what they might be called, doodlebugs are all about the same, having some sort of handle and a pointer rod which swings to and fro to point the way to treasure. The real fancy ones can be "loaded" with a sample of gold or silver, and have almost any number of antennas and radio-wave aerials sticking out of them. The more bells and whistles, the more they cost. Recently, I saw one advertised for $85,000! None of their owners can tell you how they are supposed to work, except that you have to have faith in them. It would be a lot cheaper to have faith in a forked stick! Nor will their manufacturers explain their mysterious inner workings, except to warn you that if you attempt to dismantle their instruments to learn what makes them tick, their secret circuitry will be rendered inoperative and your warranty will be voided.

No doubt every prospector and treasure hunter would like to own an electronic dowsing rod, one which is guaranteed to lead them straight to a concealed gold mine or some hidden cache, but before you rush out to lay down your life savings, stop for a moment and ask yourself one simple question: If someone has such a wonderful machine, one guaranteed to lead him to instant wealth, why does he want to sell it? Think about it. If that electronic gizmo really works, why isn't he rich? If I had a real genuine guaranteed gold finder, I wouldn't sell it for anything. I'd hide it away in some dark corner and never take it out until I needed more gold. Remember P.T. Barnum. He said there's a sucker born every minute. The con-men who build doodlebugs say that's not enough; they build two doodlebugs a minute!

Miners of old left tantalizing signs for today's treasure hunter.

There are treasure finding instruments of value. They're called metal detectors, and they have been around since 1930 when Gerhardt Fisher began building them. Those first detectors weren't very effective, but modern instruments can detect a single coin at a foot or more in depth. The Fisher Gold Bug can locate a pinhead size gold nugget in placer sands or in hardrock. A magnetometer is another instrument a prospector might find useful. It can detect differences in the earth's magnetic field, such as a large underground iron deposit, or a ship's cannon on the ocean bottom. Although it responds only to ferrous metals, it can detect iron cannon balls which can lead to the recovery of a bronze cannon, or perhaps an iron chest full of gold coins. Infra-red photography has already been mentioned. Used properly, it can reveal old trails or places where the earth has been disturbed, such as mine workings or building foundations, even though they may be purposely concealed or hidden under a growth of brush and trees.

Still another valuable tool for the treasure hunter is ground-penetrating radar, usually called GPR. It can be used to detect voids in the earth, such as an unsuspected underground cavern or mine shaft; places where a cache may be hidden. With an experienced operator, GPR can also detect a cache of gold bars at more than

Ancient arrastras can still be found in the secret places of the Uintas.

twenty-feet. Its search head can even be lowered down a shaft or drill hole to even greater depths. Newly developed super-sensitive models can be mounted in low flying aircraft to search the earth's surface. GPR does have one small drawback; a unit costs about $25,000! Of course, you can always rent one for about $1,000 per day. Hopefully, that isn't more than the treasure is worth!

In reading old Spanish documents, one finds still another way Spanish miners found mineral veins; even though it may sound a little spooky. In the dark of the night, strange flashes of blue-colored light are said to hover over places where silver veins lay just below the surface. Other colors of light flashes are said to mark ore bodies of gold, copper or cinnabar. Even today some prospectors claim to have seen light flashes or colored vapors rising above hidden caches. But there are those who say those vapors also hover over the graves of murdered miners; so it might pay you to be careful where you dig!

Without doubt, the Spanish miner's most reliable means of locating veins or outcrops of precious metals was simply to show a piece of ore to Indians met along the trail and ask where such rocks might be found. The Ute Indian lived in harmony with nature, and he knew where every kind of animal, plant or rock could be found. There was very little the Indian didn't know about his world, and although he might wonder why the white man prized colored stones, he had no reason not to show him where he might locate them. The famed western explorer, John Wesley Powell, noted how familiar the Indian was with his surroundings: "It is curious to observe the knowledge our Indians have. There is not a trail they do not know; every gulch, canyon and draw is familiar to them. They retain the features of the land like a map engraved in their mind." When Aztec Indians asked the Spanish conquistador, Hernan Cortez, why he placed such great value on the yellow stones Indians gave to him, Cortez replied that the Spaniards suffered from a malady which only gold could cure.

Many of the ore deposits shown to the Spaniards by native Indians were low grade which might be thought of little value today, but plenty of free labor every ounce of gold recovered was profit. After the Indians learned they could be easily enslaved by the well armed Spanish miner, they became reluctant to tell of places where precious metals could be found. There was a name for Spaniards who made a business of capturing Indian slaves to work in the mines, "lazadores." A tale is still told of a place where Indians were forced to work in a mine high above Vernal Valley, where there is a vein of nearly pure silver. Long after the Spaniards were gone and American settlers claimed the valleys for farms, one of the old Indians would go to that place, where he would cut enough native silver from a ledge with his knife to trade for needed food and supplies. He would not be seen again until he needed more silver. Although many tried, none could follow his trail into those mountains; neither could he be persuaded to tell where that ledge was located. When more white men came and claimed even the hills and mountains, that old Indian left, to where no one knew. But he didn't take his silver ledge with him; it is still there, somewhere above the Vernal Valley.

One of the old Spanish padres who trod the Uinta Mountains seeking for Indian souls to save, recorded that several Indians he met along the way wore twisted strands of copper metal around their necks. Others gave him small pieces of that same metal as a sign of peace. Those early Catholic missionaries kept a close watch for "tepustete," a heavy hematite iron ore they called "the guide to gold." They knew that if a miner dug where tepustete outcropped on the surface, they had a very good chance of discovering a gold bearing vein. The padre's name for a Spanish prospector was "cateador" while a Mexican prospector was a "gambucino." Today's prospector can still find countless prospect holes all across the Uintas where those cateadors and gambucinos dug into tepustete outcrops with the hope of finding gold. In that same manner they explored copper outcrops in search of silver.

Measurements and weights used by the Spanish miner are also of interest. His pack of gold ore might weigh one "quintal," or one-hundred pounds. A deerskin bag of silver might equal one "arroba" or twenty-five pounds. A "marc" or "marco" of gold was equal to eight ounces. Many old documents refer to a "carga" or "jackload" of gold or silver bars. The weight of a carga varied greatly from place to place and time to time. It was usually meant to be the weight a pack animal could carry, commonly from one-hundred and fifty pounds to as much as three-hundred pounds, depending upon the distance to be traveled and the roughness of the trail.

A term of Spanish measure purposely meant to confuse those following old maps or waybills was the number "forty." Like most Spanish mine measurements, that numeral came from the Catholic Bible, and is really a non-number. It could mean any number the maker of that waybill wanted it to be. An example from the Bible is the reference to the great flood, when it rained for forty days and forty nights; while another is the reference to wandering in the wilderness for forty days. In that same manner, a Spanish waybill might describe a cache of forty jack-loads or gold bars; when in fact the maker might have meant that figure to represent two jack-loads, or perhaps a dozen, the number forty being symbolic only. When used in distance, such as a tree sign indicating the treasure hunter should go forty varas to the east, continue eastward being alert for the next sign or symbol. Similarly, the number seven on a map or waybill represented gold, while a five referred to silver. The number seven along a trail might tell you that trail leads to gold, while a waybill which tells of five cargas hidden in a cave probably indicates that cache is of silver, not necessarily that five jack loads are cached there.

A Spanish furnace or "castellano," built of stone and adobe clay to smelt precious metals.

Spanish miners crushed their ore in an arrastra of the type used by Rhoads and Hathenbruck at their Granddaddy Lake Basin mine. After Mexican miners replaced Spaniards, the arrastra became known as a "molino," but by any name it was frustratingly slow in its operation, although it did grind quartz rock to a fine powder; what the miner called "cabuzuela," literally rich concentrates. There is an old saying which aptly describes the arrastra: "The mills of the Gods grind slowly, but they grind exceedingly fine." After the ore was crushed to powder, one of two processes was used to recover the metallic content. Silver was recovered by use of the "porpie" or patio process, an ancient method whose design is often credited to Don Pedro de Velasco in Mexico in the year 1566, even though that same

Fine this brush-covered cave along Dry Fork and you'll be close to the Mine of Lost Souls.

process was described in Agricola's De Re Metallica, published ten years earlier. In that process, finely ground ore was mixed with salt and a small quantity of water. If it was available, copper sulfate was added to hasten the process; the latter being obtained by roasting copper-iron ore, which is usually found wherever silver ores are mined. The mix was spread thinly over a closely fit stone floor, or patio, where it was left to bake in the sun while it was constantly being mixed by hooves of mules or horses being walked over it. Over the years since the 1500s, all sorts of mechanical means have been tried to agitate the mixture more thoroughly, but nothing has ever worked as well as animal hooves.

Quicksilver, "azoque," was added to the mix as it was turned and baked in the sun, usually for approximately thirty days. The mix was then washed in a large wooded vat, with the fine sandy waste rock being carried away, leaving a mixture of silver and quicksilver which is called amalgam. That amalgam was then roasted in a pan or retort over a hot fire, a process Spaniard's called "quema," which removed the quicksilver by evaporation or vaporization, leaving nearly pure silver which was then poured into molds to form bars or ingots. Spanish miners classified three grades of silver ore. "Molinque" was first-class ore containing as much as 1,000 ounces of silver to the ton; "calichal" was second-class ore, containing about 150 ounces to the ton while "fierrios" was low-grade ore. The patio process works best with low-grade or second-class ores, giving the best result with simple silver sulphides. High-grade or chloride ores do not respond well to the process. It took a long time for American mill and smelter men to learn that simple fact; something Spanish miners knew four-hundred years ago.

During Spanish mining days, gold was usually recovered by the chlorination process. As with silver, the ore was first crushed to powder in an arrastra, after which it was mixed with common salt and roasted in a "castellano" or furnace. The ruins which are sometimes found in the Uintas and called smelters are usually crude adobe or stone furnaces. A Mexican furnace was called a "galemar" and was little more than a shallow pit lined with stone, after which cedar bark and firewood would be laid in it, with a carga of approximately 200 pounds of gold and salt mixture placed on top of the firewood. More layers of ore and wood would be added until the furnace was full. At a permanent camp that furnace was well constructed of fitted stones, but at many small mountain mines the ore and firewood would be encased in a thick cover of clay, which would bake and harden as the fire burned. Since a high temperature was required, charcoal or cedar and pinion pine was preferred over softer woods. A strong flow of oxygen would be forced into the base of the furnace with a bellows made from animal hides. The work of nature has so eroded many of those old adobe clay furnaces that little remains of them today.

If the Spanish miner had oil of vitriol, a form of sulfuric acid, it would be added to the ore mixture before roasting. Vitriol could be obtained by dissolving copper ore in water. The furnace heat would consume or drive off most unwanted minerals, such as sulfur and arsenic, allowing the gold to react with the salt and vitriol, and become chlorinated. The chlorination process was used in western mining until well into the twentieth century. Using that process it was possible to obtain bullion with as much as ninety percent gold content. Most bars found over the years have been considerably less pure, their gold content depending upon the skill of the miner who made them.

The Field Museum at Chicago has described a process of smelting and casting gold objects of the type used in remote Catholic missions, as was researched by the late Henry Nichols. A brief paraphrase of his report is as follows:

> The exact method Spaniards used in smelting is unknown, but charcoal was probably used and a bellows of some sort to create a forced draft. One technique for creating objects such as a golden figure of Christ or a Virgin Mary was casting around a core of clay, which may have been baked before casting began. Decorations and ornamentation could be added to the casting by soldering or hammering wire gold after the cast was heated to the point of fusion.

With plenty of time, an unlimited supply of free labor and their knowledge of the patio and chlorination processes, Spanish miners accumulated large quantities of gold and silver bullion. They called their bullion bars "oroche" and it is significant that none of those bars found to date have been pure metal; even their most perfectly refined bars contain approximately ten percent base metals, usually copper. Still, that means those bars were mostly gold, or gold and silver mixed. As noted earlier, it was no coincidence that the gold coins made at the Mormon Mint were stamped from that same bullion mix. John Kay could only mint coins from the bullion which was brought to him; he had no way to further refine that metal. For more than three centuries countless tons of that same bullion was shipped to Spain from mines all across the northern mountains, and a lot of it came from Spanish mines in the Uinta Mountains.

Caleb Rhoads hired Matt Warner to stop Knight's miners. Warner stopped them permanently!

It is difficult to imagine such wealth today. Silver was so common-place that common iron was more valuable. Almost all of the iron used in New Spain was brought across the Atlantic on sailing ships. Pack animals were often shod with silver horseshoes because iron was more difficult to obtain. So much silver was shipped to Spain that it overwhelmed the monetary system of Europe. Two-thirds of the world's silver came from the mines of New Spain. From records still available, we know that more than four-billion dollars in silver bullion arrived at the port of Seville on antiquated and over-loaded galleons. It is believed that at least that much more went into Davy Jones' locker at the bottom of the Caribbean. The value of gold sent to Spain is even more staggering. Spanish records reveal that during the three-hundred years from 1500 to 1800, more than ten-million pounds of gold was shipped from New World mines. That half of what was shipped ever reached Spain was probably a miracle, considering the design of the galleons it was shipped on.

Spanish sailing ships were nothing more than large wooden tubs. They rode high in the water so that they had to be loaded with heavy ballast just to keep them from tipping over. It was almost impossible to keep them headed into the wind, so that only a moderate cross-wind would blow them off course. The entire Plate Fleet of treasure ships in 1622 was lost in hurricane winds off the coast of Florida, where gold and silver pieces of eight still wash ashore after every tropical storm. The bullion cargo of only one of those ships, the Atocha, now being salvaged by Mel Fisher, is estimated at four-hundred million dollars. Ten treasure ships of the 1715 Plate Fleet were lost during a violent storm

You can see this well-hidden shaft from the cave on Dry Fork; that is if you know just where to look!

which drove them onto coral reefs, ripping open their worm-eaten hulls and spilling countless millions in gold and silver into the depths. One single ship lost with the 1756 galleons, the Maravilla, is expected to surrender one-billion dollars in bullion!

Those poorly designed top-heavy tubs carrying tons of treasure were also easy prey for sleeker and faster pirate ships built in England. No one knows how many Spanish galleons lost their cargo's and were then sent to the bottom by free booters flying the Jolly Roger Flag, or how many sunk under cannon fire before their cargoes could be seized. Many of those galleons were so rotted and worm-eaten that they actually sunk from their own weight. Still others were blown far off course, to be dashed to pieces on some rocky shore where their ruins and treasure cargo will probably never be found. In June, 1683, piles of gold and silver bars were stacked like cordwood at the port of Veracruz, waiting arrival of the annual Plate Fleet. Thousands of animal hide bags held pieces of eight, fresh from the mint at Mexico City. During the eighteen months since the last Plate Fleet had docked, countless mule trains had brought treasure from the mines of northern Mexico, New Mexico and the Spanish borderlands even further north. On June 17, hundreds of pirates under command of Lorenzo de Graff rushed ashore, killing Spanish soldiers and sacking the seaport. Uncounted millions fell as booty to the pirates. One can't help but wonder how much of that precious cargo came from mines in the Uinta Mountains.

Because of the isolation of Spanish mines in the Uintas, as well as the difficulty of packing heavy bullion from them to Santa Fe and the gulf coast seaports beyond, it was not uncommon that bars of silver and gold would be stored at or near those mines and smelters until enough had accumulated to warrant sending a pack train on that long and dangerous journey. Those pack trains were still carrying treasure southward from those mines until well after arrival of the Mormon pioneers; as witness the one which camped at Mormon Selman's homestead at the mouth of Provo Canyon in 1852. For many different reasons, not all of those bars of bullion sent along the Old Spanish Trail ever reached Santa Fe. When the old Catholic mission and real of mines at Rock Creek Canyon were attacked by Ute Indians and the miners there either killed or driven away, many mines in the surrounding mountains were abandoned forever.

At many of those mines, caches of treasure were left behind, waiting for a pack train which never came. Massacres like the one at Lightning Ridge left still other treasures for today's prospector to search for. When Utes burned Fort Robidoux in 1844 and drove French fur trappers and Mexican miners from the mountains, the same thing happened again, with mines abandoned in the Yellowstone, Whiterocks and Mosby Mountain country. Nor does anyone know how many small parties of miners were killed at other isolated mines, from Kamas Valley to the Green River, their bodies and treasure thrown into some now forgotten shaft. Who can doubt that Caleb Rhoads brought treasure from just such a place?

Recently treasure was recovered from one of those old sites, but through ignorance one of the finders forfeited his share. Two treasure hunters from Colorado were guided into an area not far from Smelter Springs, located along a fork of Crow Creek north of LaPoint. Using a waybill they brought with them and taking advantage of their guide's knowledge of that area, they weren't long in finding what they were looking for; a cache of heavy gray-colored metal bars. But the guide mistakenly assumed those bars were lead because of their color and declined his share, choosing instead to take a few hundred dollars bonus for his service. Only after his two partners cached in their find of silver bars did he learn how close he had come to being rich. But close only counts at horseshoes!

Names which the old-timers used to describe places or landmarks are not always the names those same places are known by today. A case in point is one where gold nuggets can be picked off the ground. Caleb Rhoads, Pick Murdock and Happy Jack all brought gold from such a place. They even staked their claims on placers and a mine there, but there is a problem locating exactly where those claims are, for they weren't recorded at the county seat. Instead, each man staked his claim by carving his initials on several trees to identify his property. Those trees have the initials CBR, PM and HJ cut into their bark; all you have to do is find them. But that shouldn't be too difficult, for Happy Jack sketched a map of that area; as a matter of fact over the years he sketched three maps, no two of them exactly alike. The problem with interpreting those maps is that none of the landmarks are known today. (See Map S)

The Happy Jack Map shows a trail coming through the mountains to Three Forks Creek. One problem with that is, there are a lot of streams in those mountains which have three forks. The trail follows the creek, circles around Wade's Knoll and then continues past an area marked "Spanish Trees." No one today knows where Wades Knoll is located, or even why it was called by that name, and there are many places which have Spanish sign trees. Some of those signs are shown on the Happy Jack Map, including several Catholic crosses and a cannon. On the far side of Wade's Knoll there is a place marked "Deep Shaft," and close to it is a partially buried cannon, with part of its barrel protruding above the ground. Some of the old Indians say that cannon barrel is filled with gold nuggets!

Beyond Wade's Knoll, the trail enters the head of "Jagged Canyon," where six claims are marked on trees, with initials for Rhoads, Murdock and Happy Jack. In that Jagged Canyon just beyond the claims there is a grove marked "Big Trees," and near them a place marked "yellow colored ground which looks like sulfur." A Spanish smelter is shown close to that yellow ground. All you have to do to find those placers and mine is to follow Three Forks to Wade's Knoll. From there it is only a short way to Jagged Canyon, the Big Trees and the old smelter where the ground looks like sulfur. Sounds easy to me!

The flower stem shows the trail, the petals tell the direction, along the old trail to Carre-Shin-Ob.

Happy Jack kept detailed notes in which he recorded a lot of mundane things, such as trading horses or buying supplies. He also noted going into the mountains with other prospectors and described the places they went. Many of his notes tell of distances traveled and landmarks passed, as well as where he camped and the quantity of gold he and his partners found. Over the years he filled four notebooks with entries from the commonplace to the exciting. They contain everything from livestock prices to hand sketched treasure maps. As a group they are often referred to as the Happy Jack Journals, but they are really only leather-bound pocket notebooks, not journals of the type kept by many Mormon families. They are well worn, dog-eared and soiled with sweat and grease. There are missing pages and places where pages have been added, and some are almost illegible, but they are probably the nearest thing to a lost mine waybill we are likely to see.

I first heard of the Mine Of Lost Souls from Arlo McDaniel back in the early 1950s. McDaniel was a traveling man, playing guitar with a western band. While in New Mexico he learned of the mine from an elderly Mexican gentleman who said that as a small boy he had gone into the northern mountains with a mining party which included both his father and grandfather. They went to a rich silver mine located in a wild canyon somewhere in Ute Indian country. Work at the mine went routinely for most of the summer and the miners smelted and cast many heavy bars of nearly pure silver. Each day more of those bars would be cached in a pit dug beneath the floor of a small cabin they built. As summer turned to fall, Ute Indians began to become troublesome, running off their pack animals and attacking anyone who wandered far from camp.

One day while all the men were working at the mine and smelter, the young boy and his grandfather climbed high above their camp into a steep canyon where they picked ripe chokecherries. As they returned to camp near dusk they heard screams and yelling and saw below them the end of a terrible battle in which all of their comrades were being slaughtered. Bodies of miners lay scattered on the ground and their cabin was afire. In fear that the Indians might find them, the old man and his grandson hid until long after dark. At the first light of dawn they dragged the boy's father into the mine tunnel and closed its entry by pulling down the portal timbers to keep wild animals away. The old man wept that no priest was on hand to administer the last rites of the church, to save the souls of his son and the others who had been killed. Ever afterwards that rich silver mine in the Uinta Mountains was known as the Mine Of Lost Souls.

The old man who told his tale to McDaniel recalled the terrible journey he and his grandfather made after they fled from the mountains. They survived on wild berries and seeds during their long flight to Santa Fe. The grandfather died not long after they reached safety and the boy was left alone, but he always hoped that one day he might return to the place where his father's body lay in a forgotten mine tunnel. He hoped to have a priest say the words of the dead for his father. He also remembered the cache of heavy silver bars which still lay hidden where the cabin was burned. But Indian hostilities only became worse as the boy grew into a man and in time he had a family of his own to care for. The years passed until the boy became an old man, the old man who told his tale to McDaniel. Although nearly a century had passed, he still remembered the quiet mountain canyon and his father's grave, and the old mine and the cache of silver bars still hidden there. The old Mexican told McDaniel that the place where the mine was located was marked by a high promontory of land, where two small creeks came together. One of those creeks sank from sight in a swirling sink hole, only to surface again a little further down canyon. Close to that sink hole there was a cavern where he had lived with his father and grandfather, and just downstream from that cave there was a natural stone bridge or arch spanning a steep side gulch. Although he was too old and infirm to make the long journey north to that hallowed place, the old man who as a boy had played atop that stone arch gave McDaniel a map which had been made by his grandfather. (See Map T) It was his hope that his father's grave might be found so that a priest could say the Holy Words for him, and for the other miners who had died at the Mine Of Lost Souls.

When McDaniel's western band moved on to Nevada, he passed the old Mexican's map to me. The following summer I found all the landmarks which had been described to him; the creek forks, the sink hole, the cavern and the stone arch, but I found no trace of the burned cabin or the mine tunnel hidden under the rocks which covered it. If you would like to try your luck, go to the little town of Maeser, just west of Vernal, and from there follow the Dry Fork of Ashley Creek into the mountains for seventeen miles, to where a high promontory of land divides the canyon. Two small creeks come together just below that point of land. The creek coming from the left disappears into a sink hole about a mile further upstream. Only a short way up that left fork there is a natural stone arch across a side canyon, and about a quarter of a mile beyond it there is a cave, its entry nearly hidden behind a thick growth of chokecherry bushes. The mine tunnel and the burned cabin can't be far away.

Those sinks and caverns along Dry Fork should be of interest to the prospector who wants to understand the geology of that canyon, and where minerals might be found. Those sink hole ponds are shallow, most less than ten feet in depth, but they are often large in surface area. The water level in those sinks is often well below the surrounding ground surface, with a swirling whirlpool which sucks driftwood, sand and gravel into its vortex. Ancient alluvial deposits well above the present level of the creek indicates that the creek bed was once much higher than it is now. Those ancient creek bed benches along the dry hillsides are excellent places to look for water-worn gold nuggets.

The stream which disappears into those sink holes is at its greatest flow during spring months, but the flow of the stream which reappears further downstream is not greater in the spring, which indicates that much of the flow does not come to the surface anywhere nearby. Colored dye poured into those sink holes reveals that part of that underground water course is the source of Ashley Springs and the Brush Creek Springs, but even they do not account for all of the water which is lost in those sinks. A logical conclusion might be that there are unknown caverns and passageways somewhere beneath the surface, most likely in the Park City limestone formation which breaks the surface there. If that is true then it may also be true that the Mine Of Lost Souls is also in that mineralized formation. That could narrow the search area considerably.

Just downstream from that natural arch and the sink hole, a branch of Dry Fork Canyon turns right or northward to the Red Cloud Loop, a scenic drive. If you watch carefully where that road begins to climb steeply, you will see a heart carved into a very old pinion pine. There is an old Spanish mine about fifty yards from that tree, but it has been filled with rocks and brush. During the 1930s an old Indian named Hackford with several helpers cut dozens of trees and carried thousands of rocks to fill it. One of them said it took a month of hard work to do the job, but when he was asked why they went to such labor, he said it was just an old hole with nothing of value in it. It seems like an awful lot of hard work to cover nothing!

Caleb Rhoads knew of an old mine in Dry Fork Canyon. On one occasion he mentioned that along the trail to one of his mines he would ride his horse up a stream to hide his tracks. The bed of that stream was either dry or filled with water, as the stream sank into sink holes and then re-surfaced again. He said he could hear the echo of his horse's hooves ringing from hollow places beneath that stone stream bed. He also mentioned making camp in a cave along that creek to get out of a snow storm. Rhoads told of a mine not very far from that cave, in what he called the "Sawtooth Range." He said that mine was on a rocky ridge which outcropped along a mountain which was "jagged as a dinosaur's back." That "dinosaur" ridge was not far from the cave and natural arch.

Frank Murdock, the blacksmith from Whiterocks, also found a mine which sounds very much like the one Rhoads talked about. Murdock said that mine was on the "rooster-comb." There is a place along Dry Fork which even today one might describe as a dinosaur's back or a rooster-comb. It is only a short way downstream from the Dry Fork sink hole. Murdock said that inside that mine there is a vein of pure silver, but the Indians were afraid to go in it, calling it the Snake Hole. Murdock said it was a dark and dangerous place, where miners had left a lot of black gun-powder; that explosive being very old and unstable now. If you can find Murdock's "rooster-comb," you might be real close to Caleb Rhoads mine on the "dinosaur" ridge, along the "Sawtooth Range."

There was another and even better reason for Caleb Rhoads to know and remember Dry Fork Canyon. There are several versions of the story, but the account by Charles Kelly, Utah's premiere historian, probably told it best and most nearly correct. With the discovery of the Murdock-Dyer Mine, there was a prospecting rush to the east end of the Uintas, to Brush Creek and Dry Fork. When a sheepherder reported finding some copper-stained green-colored rock high in Dry Fork Canyon, prospectors from all along the range hurried to be the first there. At that time Rhoads was prohibited from entering the Uinta Mountains, so he hired a part-time prospector and geologist named E.B. Coleman to keep an eye on that mining rush for him. Coleman and his camp tender, David Swift, were to stake claims in Rhoads' name if anything of value was found. They were making their camp at Taylor's Hollow high on Dry Fork when they were approached by Dick and Ike Staunton and Dave Milton, three local toughs who were posing as prospectors. It was later learned that they had been hired by Jesse Knight to follow Coleman and report on any claims located for Rhoads. More about Jesse Knight and his connection with the Rhoads Mine will soon be explained. Both groups got into a heated argument over who was claiming what and when it looked as if a fight was imminent, Coleman slipped out of camp and hurried back to Vernal for help.

Knight's men made their camp close to Coleman's, so Swift, the camp tender, could keep a close watch on them, but still from a safe distance. As soon as Staunton's camp was made, he and his men began digging a tunnel into the side of Taylor Hollow. It was later said that Knight had selected that site so that his tunnel would be located outside the Ute Reservation boundary, but it would be driven under reservation land, to where he expected to uncover a rich ore body at depth. Some believed that Knight hoped to run that tunnel right into the Carre Shin-Ob cache! As soon as Rhoads learned from Coleman what Knight's men were doing, he sent Aaron Daniels to help Coleman. Daniels was then seventy-four years of age, so he would be no match for Staunton's crew. Daniels did note in his journal that he did go to Taylor Hollow, and upon his return from Dry Fork he reported to Rhoads and asked him if he thought Knight might strike ore in the tunnel his men were digging. Daniels recorded that Rhoads replied: "Jesse Knight may be the smartest mining man in the West, but he will lose on this one!" The short-lived mining boom in Dry Fork Canyon would soon add considerably to the toll taken by the Killer Mountains.

When Coleman arrived at Vernal, he paid $500 to Matt Warner and Bill Wall, both notorious outlaws, to return to Taylor Hollow with him as his body guards. There is good evidence that Warner and Wall were at Vernal specifically to go to Coleman's rescue and protect Rhoads interests. Among the Rhoads-Hathenbruck papers are records which show that Rhoads entered into a legal contract with Willard E. Christiansen in which Rhoads conveyed to Christiansen a one-twentieth interest in the Rhoads-Hathenbruck Indian land lease, and also agreed to pay Christiansen "certain expenses paid in hand as they are merited." In return Christiansen was expected to "serve in such capacity as is needed to complete the consummation of the lease between Rhoads and the Ute Indians." Christiansen was to act in Rhoads' behalf and protect his interests, including those at Dry Fork. That legal contract is of special interest, since Willard E. Christiansen was the real name of Matt Warner, a top gun-hand with Butch Cassidy's Wild Bunch! That was the same Matt Warner who took part in the gun-fight at the Lost Cabin Mine.

At daybreak, May 7, 1896, Coleman, Warner and Wall arrived at Taylor Hollow and were riding past the Staunton camp when for some unknown reason Ike Staunton suddenly and unexpectedly opened fire on them, shooting Warner's horse from under him. But Knight's men were no match for hired gunmen like Warner, who while jumping from his horse returned fire, killing both Dick Staunton and Dave Milton. Ike Staunton was badly shot-up and later lost a leg from his wounds. Warner fired so fast that the battle was over in seconds; Wall never had time to open fire. To Warner it seemed to be a simple case of self defense, so he and Wall surrendered to Sheriff Pope at Vernal, anticipating that no charges would be filed. But feelings at Vernal against the two gun-men were so strong that Sheriff Pope had to move both men to Ogden City to escape a lynch mob. Because both men had long criminal records, each was sentenced to five years in prison, but both were pardoned after serving just over three years. There's no telling what Jesse Knight's Taylor Hollow tunnel might have uncovered if his men had continued digging it, but just as Caleb Rhoads predicted, Knight lost that one. But the Killer Mountains claimed two more as its own.

Aaron Daniels knew Dry Fork well. He once told how he came upon an ancient looking mine tunnel there, badly caved in but still with a small hole he could crawl back into. He said there were skeletons of dead men inside that tunnel, and Spanish marks on its walls. He returned later to dig out that caved portal, but by then tons of loose rock had slid down the mountain to cover it completely. But on a ledge close by that tunnel, he found a name cut into ledge rock: Alvarez de Leon, 1669!

Another who may have stumbled onto an old mine near the Dry Fork sink hole was a prospector from Oklahoma. Every year for ten years or more he made the long trip from his home to prospect that area. Sometimes he worked with a Ute Indian who was then a member of the tribal council. Those who met him in the mountains said he had an old map given to him by his Indian partner. They said that map was almost exactly the same as the Mine Of Lost Souls Map given to the author by Arlo McDaniel. That Oklahombre found many interesting places along Dry Fork, including old stone trail markers and Spanish signs on trees and rocks. During his last trip to Dry Fork, he climbed atop a large rock so that he might get a better view of the forested mountain side he was on. To his surprise he saw what looked like a hand-built ladder sticking out from under a giant pine which had fallen by that rock. Investigating that ladder, he discovered a near-vertical shaft under that log.

It didn't take long to ascertain that sometime long ago that huge pine had been purposely felled over the shaft. When it fell, the tree trunk broke into several sections, so that he was able to move the piece which covered the shaft. The old ladder of poles with cross-pieces tied to them with rawhide was so old it fell apart when he began to put weight on it, but he was able to see that the shaft led down into a square-cut room-like chamber, with a drift or side tunnel leading directly back under Lake Mountain. Being alone at the time, he couldn't get down into that room, but when he returned to Oklahoma, he sent photos and directions how to find that shaft.

I relayed the prospector's story and photos to another prospector who lives near Dry Fork, and he examined those old diggings. He quickly found evidence that the shaft and underground chamber were man-made, and he even crawled back into that caved tunnel for more than two-hundred feet, to where it was closed off at a cave-in. He examined a large stope or room where he found parts of very old crude tools, and he pocketed samples of ore which were undoubtedly silver. An assay verified his findings, but because of the shaft's remote location he decided against trying to reopen those caved workings. But if you check a topo map of Lake Mountain and Dry Fork you will see something of interest. That tunnel was being driven straight towards Smelter Springs on the opposite side of the ridge, a place many believe is close to Carre-Shin-Ob.

A few years ago I was given a "genuine" Spanish map of the Dry Fork area, one which shows the Mine Of Lost Souls and several un-named diggings. That map bears the date 1777 and an official looking Mexican Seal, but if you take a real close look at that map, you might decide that it is just too good to be real. If you compare it with a modern forestry map of Dry Fork Canyon, you will note how perfectly it corresponds to actual landmarks and topography. Even Marsh Peak is placed exactly as it should be. In the first edition of Faded Footprints, I gave an opinion that map was a phony. For those who may not have seen it, it is included in this volume as Map "U". Since that time I have been contacted by several people concerning that map, a few of them very upset that I had published it, since they said it was their very own secret map; something quite strange since the original came from an Indian named Jimmie Redhorse, who gave or sold copies of that map to just about everyone. Several readers spotted the obvious, that the map was dated 1777, but it bore a Mexican Seal which never came into being until well after the revolution of 1821 brought independence to the Mexicans. One reader from Central America questioned the map legend, "Aqui lo tiene..." He correctly noted that no Spanish speaking person would write that legend as it was, or as another critic wrote, "It is poor gringo, not Spanish!" The moral of the story is, not all "genuine" Spanish maps are genuine!

Many old Spanish mines have been found on both sides of the Dry Fork ridge east of Lake Mountain. Aaron Daniels left a waybill and map to one of them before he died. (See Map V)

> Near Mosby Mountain there is a place where Spanish miners marked large pine trees with strange signs. There are many different ones, but ignore all except the Catholic crosses. They lead in a straight line to the edge of the ridge, where you can look across the valley to a canyon at the base of Lake Mountain. That canyon is about two or three miles north-east of those Spanish signs. A small draw or gulch runs back into the mountain, where there are other trees which have the same markings on them as I saw at Carre-Shin-Ob. On

the mountain front above, there is a giant pine with an arrow carved on its bark, which points right at the mine. That mine lays in a very hard to find place, and you can walk right by it if you don't watch closely. You can see where the tailings which came from the mine were strewn across the mountainside. I have seen gold from that mine, and I still have some pieces of it. It is very rich ore. Senator Tom Kearns wanted me to show him the location, but Pick Murdock told me that Kearns had cheated him out of a mine, so I never followed the lead. I could have made it one of the best mines in the western country.

Some say that the ghosts of murdered miners stand guard over the hoard of silver bars cached under that old burned cabin at the Mine Of Lost Souls in Dry Fork Canyon. They say that when the moon is dark and the night winds wail, that if you listen closely you can hear the war whoop of attacking Utes and the cries of souls lost in purgatory. On a stone along the creek there is an inscription which records the bloody legacy of the Killer Mountains: "¡Cuidado! ¡Maldicion Espanol! ¡Los Muertos No Habla!" Beware! The Spanish Curse! The Dead Do Not Talk!

CHAPTER 14

GHOST GOLD
The Search for Carre-Shin-Ob

Nearly everyone who has seriously researched the origin of the gold which Thomas Rhoads brought from the mountains for Brigham Young has come to the same conclusion: He never dug that gold in a mine; he simply never had time to do so. At some point in time that gold certainly came from a mine, and without doubt it came from a Spanish mine. We may never know the location of that mine or mines, but we do know that Rhoads didn't have to dig for gold at that place. That gold was already dug and waiting for him, in the form of bullion bars or ingots. Some call the place where he got that gold the Rhoads Mine, or the Church Mine, or the Brigham Young Mine; and some have called it Carre-Shin-Ob, the sacred mine of the Utes. But whatever you call it, it is the place both Thomas and Caleb Rhoads vowed they would never reveal the secret of. To avoid the confusion of calling that place by several different names, let's call it the Church Mine. If you have been reading the foregoing chapters carefully, you already know a lot about that place or mine, so it shouldn't be impossible to find it.

We know that gold was brought to the Mormon Mint by both Thomas and Caleb Rhoads. Research of early issues of the Deseret News and the Journal History of the Mormon Church describes both the departure and return dates as well as other details of their trips into the mountains. Individual journals and personal diaries, such as those of Aaron Daniels and W.W. Phelps confirm those findings and reveal even more information. Typical entries are those which tell of their return to Salt Lake City from journeys onto the Ute Reservation. Of interest is the time element involved in those trips. Careful scrutiny reveals that Thomas Rhoads was gone from the city from fourteen to seventeen days. One need only take into consideration the distance from the Salt Lake Valley to the Ute Reservation, from 150 to 180 miles depending upon the route followed, and the same distance back. When one considers the rough mountain trails then in use and the speed at which pack animals can be driven, it becomes obvious that all of the time consumed, fourteen to seventeen days, was used just traveling to the mine and back.

Rhoads' journey would have been made either through Emigration or Parley's Canyon and then over the mountains above either Kamas or Heber Valley; most likely via Kamas Valley since that is the place he chose for his land grant. His journey would have been made long before any pioneer trail or wagon road existed, each round trip totaling more than 300 miles, all of it through a rough, roadless terrain. To complete such a journey in little more than two weeks with pack animals was an accomplishment in itself, leaving no time to work in a mine digging for gold and then breaking, sorting and loading heavy ore into pack sacks. When considering the time element involved, it becomes obvious that Rhoads had only time enough to go to the sacred Church Mine, load gold bullion all ready to be loaded and return to Salt Lake City. He had no time to work in a mine.

Even today some old-time cowboys doubt that Rhoads could have made the described journey in the time allotted. One other thing is also apparent; Rhoads was not packing gold ore to the Mormon Mint. Gold in rock ore would have been far too bulky to carry so far with pack animals. Even a dozen horses couldn't carry enough gold ore to produce any large amount of gold metal after milling and refining. But even if he had packed gold ore, there wasn't a mill or smelter within a thousand-miles of Salt Lake City where that ore could have been reduced to metal. Rhoads packed bars of

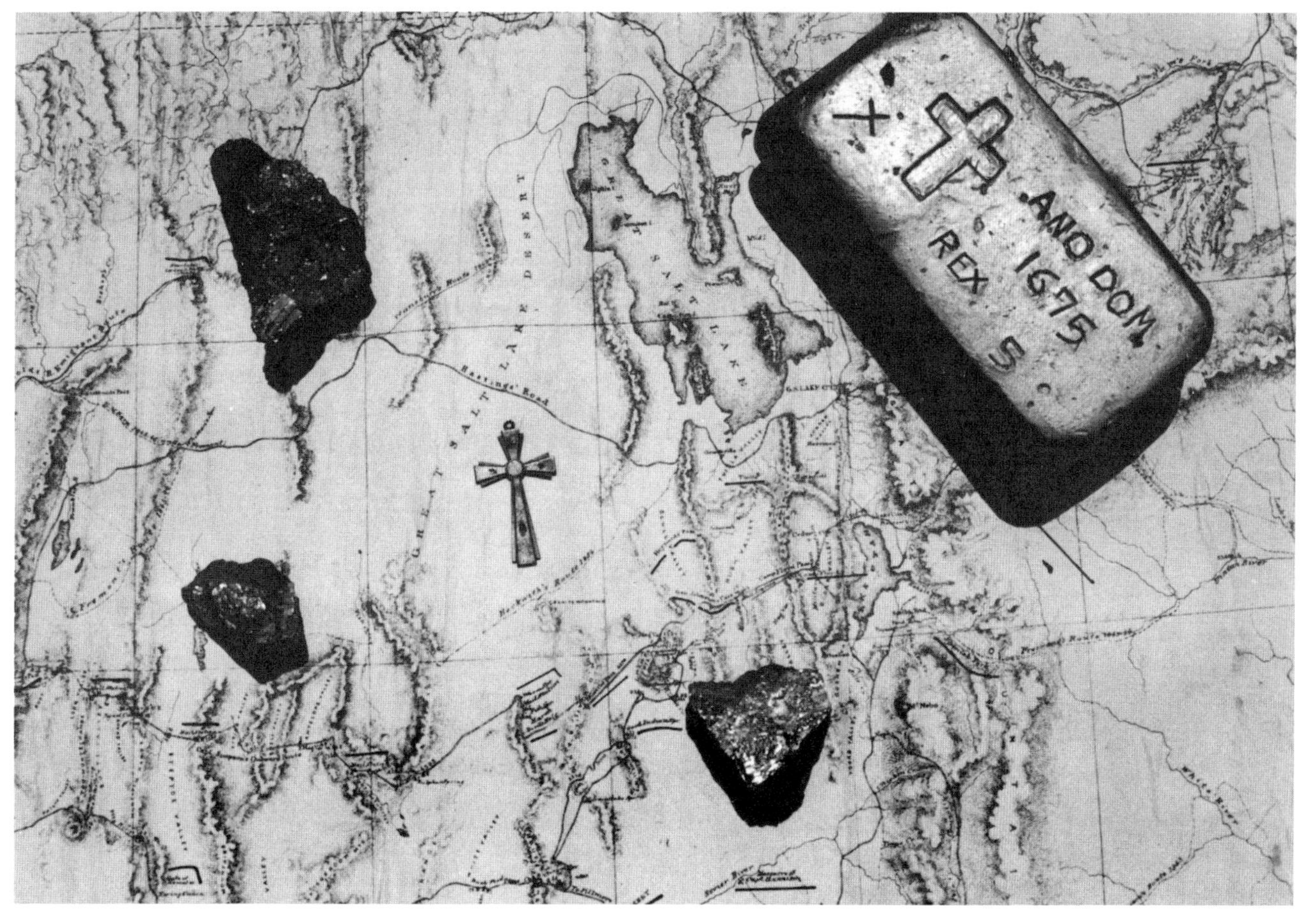

Lost Spanish Treasure! Somewhere in the Uinta Mountains.

gold bullion to the Mormon Mint. There is ample proof of that.

Several reliable pioneer journals describe the gold Rhoads brought from the mountains. Remember, when Ephraim Lambert weighed a sixty-pound sack of gold at Kamas, it was chunks and pieces of gold that he weighed, not gold ore in quartz. An article in the Deseret News issue of October 5, 1850, along with a similar entry in the Church's Journal History tells of gold, not gold ore, being rolled into long strips similar to a wagon tire; the iron band which is shrunk around a wagon wheel. That band is approximately two inches wide and ten feet or more in length, depending upon the size of the wheel. Mormon coins were cut from those strips of gold with a circular shaped cutter and were then stamped with a die in a hand operated press to inscribe the denomination of the coin. The words, "Holiness To The Lord," the initials "GSLC" for Great Salt Lake City and "PG" for pure gold was also on those coins. The Deseret News reported, "This is what makes trade brisk!"

We now know from examining Mormon coins owned by collectors and coin shops or in museums that the metal used to make those coins wasn't pure gold, but was in fact a bullion metal containing varying amounts of silver and copper. Mormon coins taken to California by emigrants and traders were discounted twenty percent from their face value because of that impurity; a ten-dollar coin being worth only eight-dollars. Army troops stationed at Camp Floyd near Utah Valley were directed not to accept Mormon coins, in General Order #4 issued by Brevt. Col. C.F. Smith, on January 10, 1860:

> The Commanding Officer has been informed that there is a large amount of gold coin — several thousands of dollars — purporting to be worth five dollars, commonly called "Mormon Coin," about to be put in circulation at Camp Floyd. As this coin is understood to be worth only four and a half dollars, soldiers are not to receive them for more than that sum, and better still, not to take them at all.

It is no coincidence that those Mormon gold coins were made of exactly the same bullion which Spanish miners produced at their crude mountain smelters. The

Ruins of an old smelting furnace along Mosby Creek.

metal found in Mormon coins and that found in bullion bars during recent years is almost identical!

John Kay and Thomas Bullock, minters at the Mormon Mint, had no way to refine bars of bullion brought to them by Thomas Rhoads; no way to remove the silver and copper content of those bars so the resultant metal would be pure gold. During that time period there were few places other than the United States Mint which could refine bullion into pure metal. In his book, The Marvelous Country, dated 1874, Samuel Cozzens wrote of Spanish smelting:

> Commenting on the vast quantity of treasure taken from those mines with only rude implements and the crude manner of assaying, fully one-half of the gold or silver was lost, which is verified by the richness of the waste rock and smelter slag left behind, thousands of tons of which can be found today near those old mines worked by the Spaniards.

So we know that Thomas Rhoads brought bullion metal to the Mormon Mint; the obvious question then becomes, what was the origin of that bullion? There are several theories, most of which can easily be discounted. That bullion didn't come from an Indian mine. Historical research has proven conclusively that until arrival of the Spaniards in the New World, native Indians had little interest in metals, except for small amounts of gold or native copper or silver which they hammered into tiny bells or ornaments. One early account relates that some Indians were capable of extracting wire or nugget gold from quartz, and could fashion it into articles of religious significance, but those trinkets had no value to them except as a sacred object or ornament. The Northern Utes did not have the necessary skills to mine gold from deep hard-rock mines or smelt the ore in a furnace to reduce it to bullion.

Some claim that after Hernan Cortez was driven out of Mexico City in 1520, the Aztec Chief Montezuma caused his national treasury to be carried far into the north, some say even beyond the Grand Canyon, to keep it from falling into the hands of the Spaniards. Some believe the cache on the Ute Reservation is Montezuma's gold. It's a good story, but a pretty hard one to swallow. That's an awful long way to carry tons of gold when there must have been lots of good hiding places a lot further south. Almost as far-fetched are fanciful tales that the Rhoads gold came from a great hoard cached by people known as the Jaredites just before they were all killed by the Nephites; as some interpret scripture from the Book of Mormon. There is only one logical explanation of where the gold from the Church Mine originated. It came from Spanish mines and was cached by Spanish miners, at the place Thomas Rhoads was shown by Ute Indians.

At places like Rock Creek, Whiterocks and the Yellowstone Canyon, places where old Spanish mines and smelters have been found, it is reasonable to believe that gold ore was taken to several centrally located smelter sites where it was reduced to bars of bullion. There were reals or groups of mines all across those mountains, from the Weber and Provo rivers at Kamas Valley to Lightning Ridge, Moon Lake, Pole Creek and the Dry Fork country further east. Spanish mines and smelters have been found at all those places. Gold and silver ore was brought to smelter sites like those found at Farm Creek, Crow Creek and Mosby Mountain. Pack trains such as those described by W.P. Mecham at Hoyt Peak and Mormon Selman at Provo Canyon carried that bullion to Santa Fe and the gulf coast. But there was always more bullion being smelted, so that those central repositories always contained some treasure. When the Utes rebelled during the 1700s and again in 1844, many of those pack trains never returned and those cache sites became known only to Chief Wakara and his Utes, that is until they were shown to Thomas and Caleb Rhoads.

Thomas Rhoads was taken to just such a place by Chief Wakara's Utes. He was not shown a place where he could dig for ore, he was taken to a place where bars of Spanish bullion were already stacked like cordwood, ready to be loaded onto his pack animals. There are good reasons to believe that is what happened. Remember the short time it took him to go to that place from Salt Lake City and return, plus the fact that we know he did not pack gold ore, but bars of bullion metal. There are convincing arguments that the Spanish cache Rhoads was taken to, the place which we will call the Church Mine, probably was not very far from Whiterocks. Several of Caleb Rhoads' closest friends later said the Church Mine was near Whiterocks.

Strange and intriguing stone ruins can still be found at forgotten places in the mountains.

Both Pick Murdock and Art Murdock said that Rhoads would be gone from Whiterocks only two or three days before returning with his pack animals loaded with gold. We know that Caleb Rhoads confided in Pick Murdock more than any other man, and Murdock once said that the trail to the Church Mine followed the Whiterocks River. That Murdock may have accompanied Rhoads to that sacred place would not have been breaking his vow to Chief Arropene, for Murdock himself was a Ute, a brother to those who first guided Rhoads to those sacred caches. Nor could those who knew Murdock best help but notice that the Whiterocks area was the place where he brought gold from after Rhoads' death.

If you study a good topographic map which shows the eastern boundary of the Ute Reservation as it was surveyed just prior to 1905, you will notice a peculiarity. While most of that boundary between reservation and non-Indian land carefully follows recognized township and range lines, that eastern boundary is a checkerboard of offset sections with irregular borders. The wording of a 1905 treaty signed by President Roosevelt and the Secretary of The Interior explains the reason for that crooked and uncertain boundary line. That treaty states that the Ute Indians were to be paid the unusual sum of $70,064.48 for "certain lands" located at the eastern end of the reservation, but there is no explanation why those "certain lands" were to be excluded. Could there be valuable minerals on those excluded lands?

At that time most geologists believed there were no valuable minerals in that eastern end of the Uinta Mountains, but as time passed they began to hedge their bets a little, saying there "probably" weren't any minerals there, "except" for some iron, or "perhaps" some copper, or "maybe" a little silver or "possibly" some lead or zinc, or some other exception. Now that geology is a more precise science there are logical explanations for all of those exceptions, and tacit admissions that valuable minerals might be found there, such as a statement found in the prestigious Intermountain Association of Geologists yearbook: "Then too, there are still those of light heart and good faith who scour the flanks of the Uintas for the legendary, yet well substantiated and documented, Lost Rhoads Mine; the same which supplied the gold for the Angel Moroni, atop the Mormon Temple at Salt Lake City." Note the words used: "well substantiated and documented," a wording which would not have been used only a few years ago.

Happy Jack, a Ute Indian, often accompanied Caleb Rhoads on his trips into the mountains beyond Whiterocks. He let slip several clues over the years about the location of the Rhoads Mine. On one occasion he said that he remained at their camp in the canyon below Whiterocks Lake while Rhoads went to the mine. Since he was a Ute who no doubt was knowledgeable of the location of Spanish mines and smelters just as Pick Murdock was, we can only speculate why he stayed behind at camp; perhaps to guard Rhoads' back-trail. He mentioned that Rhoads went up a side gulch which led off from the main canyon, and that he was gone a long time. If he was gone long enough, he could have crossed over into an adjoining drainage, which on the east is the Mosby Mountain-Crow Creek area, or to the west to the upper Yellowstone country. Caleb Rhoads himself once said that he often went up the Left Fork of the Whiterocks River. On another occasion he said that the entry to the sacred or Church Mine was so small that a person could ride right by it and never know it was there.

As already noted, there are several old mines on Mosby Mountain, and the ruins of several smelters in the foothills below those mines. If you check your topo map carefully, you will see a spring near one of those mines, a spring which Utes call Shin-Ob Spring! Art Murdock once recalled Happy Jack saying that the sacred or Church Mine was in the mountains above Whiterocks: "Up there, not very high, are caves, and by a little meadow is the mine." At another time he said: "Over on Mosby Creek, near bottom of mountain, are

Aaron Daniels discovered strange treasure signs along the ancient trail to Carre-Shin-Ob.

many trees with strange signs on them, not far from sacred mine."

Having narrowed the search area considerably, only one thing more is needed; an eye-witness account, and we have that. In 1885, Aaron Daniels married an Indian woman he called Rose. Some accounts say she was a Navajo girl captured by the Utes, while other records indicate she was a Shoshoni; a child survivor of the Bear River Massacre, and that she was raised by the Daniels family. In any event, Daniels and Rose had four children, but Rose helped raise nineteen other children by an earlier marriage. No matter what her tribal origin, Rose was accepted as a member of the Ute Tribe and was given an allotment of 160 acres near Whiterocks, where she and Daniels made their home. Daniels was related to the Utes through his marriage to Rose, even though he had long been friends with many of the tribal elders. As noted earlier, Daniels often went prospecting with Pick Murdock, Happy Jack and Caleb Rhoads. As a confident of theirs, he learned the location of many Spanish mines, but revealed their location to no one in his family.

Aaron Daniels kept a detailed daily journal, and in an entry dated during July, 1895, only a few months before his death, he recorded the following:

> While prospecting with Kale Rhoads I learned the location of a sacred Indian mine. I have often been asked why I haven't profited from that knowledge. I took an oath not to reveal its location, so I am not at liberty to profit from it. This I owe in loyalty to the Mormon Church, who I have heard Rhoads say the gold belongs to. I have also discovered other old mines worked by the Spaniards, which will yet prove to be as valuable as the sacred mine. Now being old in years and infirm, yet sound of mind an of good memory, I leave a record of my life, including the location and description of some of the mines I have found or have been shown. I do this so that my family might profit from my experience. I also attach my maps and charts showing the location of those mines to my wife, Rose Daniels, with instructions to develop them for the benefit of my family after my death.

Rose Daniels had little interest in her husband's prospecting and made no attempt to locate any of the places described by his maps and documents. She lived a simple Indian life, caring for her home and garden. She lived for nearly fifty years after Daniels' death, dying on July 6, 1943, at the age of one-hundred and three years. Daniels left many accounts of Spanish mine locations, among which is a map and description of a trip he made to the sacred or Carre-Shin-Ob Mine. (See Map R) His description of that trip and mine is too lengthy to quote in its entirety, but all of the pertinent landmarks and the trail he followed are given exactly as he recorded them. Those who have prospected at the places he tells of will easily recognize them and will appreciate the accuracy of his narrative.

> I first saw Carre-Shin-Ob, the sacred Indian mine, in the year 1889, when I was told by my wife's people that I might see it up close, but only if I didn't know where it was located. With Happy Jack, David Copperfield, Cessapoonch and some others, I was taken blindfolded to an old Indian trail which was near the white chalk cliffs, just below the old boundary of the reservation. I was not allowed to see the route by which we climbed the mountain. At one place we climbed so steeply that I had to cling to my horse's neck to keep from falling off. After some time the Indians removed my blindfold and showed me many old Spanish signs cut on trees. Happy Jack tried to explain their meaning to me, saying that many years before, in his grandfather's time, the Spaniards had forced Utes to dig for gold in their mines. Just down the slope of the hill from those marked trees there was a well hidden mine tunnel, which could only be seen by going up a short draw to where some mine timbers were sticking out of the ground. It was on a dark-colored vein, and right behind that vein there were some long white cliffs of caleche clay. I was blindfolded again

before leaving those marked trees. We came down from the mountains and watered our horses at a place I suspected was Indian Spring. When they next uncovered my eyes, I was on a bluff near the edge of a steeply sloping ridge. To the north were three large knolls, two to the west close together and a third more to the east. On a hill below there were two long ridges running parallel to each other. Happy Jack pointed out an area to my right, where the vein was exposed in a crumbling gray colored rock in yellow colored soil. On the ground there were thousands of small crumpled looking gold nuggets, which looked like kernels of corn thrown to chickens. We left the horses on top of that hill and went down the first ridge on foot. As we descended to the bottom between the two ridges, the Indians became very nervous. Happy Jack said they wouldn't take me any closer because Towats guarded the mine and he would kill us all if we went there.

I was disappointed, so finally Happy Jack took me to a place where I could look back towards the north ridge, and said, "Tunnel is there, at bottom of ridge. You go by yourself, up the slope to some thick brush. When you get close, you will see it." I climbed up where he had pointed and could see a pile of loose rock back in some brush which didn't look natural. I suspected they concealed a tunnel portal, so I rolled several large rocks away to open the entry. I crawled in to where I could stand up. The tunnel had been run in the same gray colored vein I had seen on top of the ridge. I could see a vein of yellow gold running up one wall and across the top of the tunnel. As I moved around my foot struck something, which I discovered was a pile of leather bags with straps securing them. I tried to open one but found they had all been sewn shut with stiff rawhide. I didn't dare cut one open with my knife for I had been warned not to touch anything and I knew those Indians meant it. I removed a few rocks to make it easier to crawl out, and the light coming in the portal revealed some bones and a skull from a human skeleton, and beneath them were pieces of Spanish armor. I replaced the rocks exactly as I found them and slid down the slope to the bottom f the ridge. Again I was blindfolded and was taken away as fast as I had been taken there. Not one of those Indians said a word as we rode out of the mountains, so I knew they were very scared. I tried to return to that mine several times later on only to discover that each time I was followed. I am bound by an oath not to reveal what I know of the sacred mine, but my oath like my flesh ends with my death. It is my hope that my family may some day profit from my experience.

Aaron Daniels saw the Carre-Shin-Ob Mine in 1889. One-hundred years later another white man on the reservation was shown where that mine is located. For reasons which are obvious, his identity must remain undisclosed. That man, who we will call Bob, was in a position of trust and some influence with the Ute Tribe, which allowed him to become close friends with several high-ranking Indians. It wasn't unusual that on occasion he would be invited to go on hunting trips with his Ute friends. One day he accompanied several Utes on a trip which took them past the Chalk Cliffs and Indian Springs, almost exactly the same route Aaron Daniels had followed a century before. Knowing of his interest in early Spanish times in the Uintas, one of the Indians pointed to a place of rough breaks low on the mountain and said, "There is Carre-Shin-Ob!" Immediately there was a strained silence among the other Indians and the one who had spoken realized he had made a grave error. He warned Bob: "You must never repeat what I have told you, and you must never come here again!"

Take Warning! No one looks on the sacred gold of Carre-Shin-Ob and lives to tell of it!

Several months passed and since it was deer hunting season, Bob returned to the place he had been to with his Indian friends, slowly driving his truck along an old road which skirted the edge of the mountains. He thought it strange there were no other hunters, nor were there any vehicle tracks on that dirt road. When he was near the place where he had been told, "Here is Carre-Shin-Ob," he got out of his truck and lifted his binoculars to scan the mountain front. He heard the crack of a rifle and instantly felt a bullet strike the side of his vehicle. As he jumped back into his truck, another bullet struck the spot where he had just been standing.

Knowing that was a place where he wasn't supposed to be, he quickly turned his truck and drove directly back to town. He parked his truck at the curb-side and had just stepped onto the sidewalk when he saw an old Indian walking towards him. As they passed the old Indian spoke to him: "You plenty lucky Bobby! If anyone else, he would be dead!" To this day he doesn't know how that old Indian could have known where he had been, or of the shots fired at him in the hills many miles from town. Bob hasn't been back to that place where he was told, "There is Carre Shin-Ob!" What's more, he says he isn't going back!

As described earlier, Caleb Rhoads took Frederick Hathenbruck as a partner, and they had been trying to lease several sections of Ute Reservation land before those lands were opened to settlement. Hathenbruck worked hard and even lived among the Indians for more than a year before he was able to get them to agree to the lease. In December 1897, the Salt Lake City Herald-Republican printed a dispatch sent to the editor by Rhoads: "My land lease has been signed by all twenty-one Ute Chiefs. It grants me 480,000 acres for ten years, with a renewal option. I am to give the Indians ten percent of the net proceeds of all mines located under the lease." The Rhoads Mining Company was formally organized on December 18, 1897.

The Herald-Republican noted that government officials at Washington had yet to approve that lease before Rhoads and Hathenbruck could begin to develop their mines. Hathenbruck explained that the actual acreage he wanted to lease was much smaller than the area agreed on, but since there were no survey markers to measure from, the lease was of necessity much larger than actually required. He also pointed out the fact that Rhoads had actually located five mining claims on July 10, 1859, prior to authorization of the Ute Reservation in 1861; an important clue which treasure hunters today might well ponder over.

The Editor of the Herald-Republican reported that he had personally examined ore specimens from the Rhoads Mine, and had also seen an assay report on that ore, noting that it carried values of 1,800 ounces of silver and 1,000 ounces of gold to the ton. In an editorial he wrote: "The Rhoads Mine is accounted as being one of the richest propositions in the state. There is no doubt but that it is a bonanza!" But in spite of the editor's glowing endorsement, Hathenbruck did not fare as well among Washington politicians as he did with Ute Indian chiefs.

Kerry Ross Boren, researcher for and co-author of *Footprints In The Wilderness,* has uncovered a conspiracy of fraud, perfidy and intrigue at the highest levels of government regarding the fabulous Carre-Shin-Ob or Church Mine, only one of the treasure sites believed to have been on the Rhoads land lease. If Boren's research was not so well documented, it would be unbelievable, for it proves that one of Utah's best known mining millionaires not only played a secret part in organizing the Raven and Florence mining companies in order to keep Rhoads and Hathenbruck off Indian land, that man also knew the location of and exploited the riches of Carre-Shin-Ob. That man was Jesse Knight!

Jesse Knight, mining millionaire. Caleb Rhoads claimed he used sacred gold to save the credit of the Mormon Church.

Provisions of the congressional acts which authorized the Ute Reservation also prohibited prospectors like Caleb Rhoads from working or developing mines within its boundaries. But due to public pressure to allow mining and homesteading on the reservation, the congress passed amendments to the several Ute treaties in 1888 and 1894; amendments which provided loopholes which allowed prospectors to lease Indian lands for mining purposes. When those amendments became law, Caleb Rhoads moved quickly to take advantage of them, but he realized that negotiating with congressmen and government attorneys was beyond his talents; that being the reason he took in a partner who was well educated and capable of working with government agencies. That man was Frederick Hathenbruck.

At first, Hathenbruck was reluctant to join Rhoads, for like many he doubted the claims Rhoads made regarding his mines; for example that he would pay the national debt. But when Rhoads took him to several of those mines, he quickly recognized their great value and he too became enthusiastic to obtain a lease on Indian land. But Rhoads and Hathenbruck weren't the only ones to appreciate the financial rewards to be gained from leasing Indian land; Jesse Knight also saw his chance to strike it rich. Knight became a shadowy and often mysterious background figure in the Rhoads Mine story.

Like Caleb Rhoads, Jesse Knight had been born into a pioneer Mormon family; at Nauvoo, Illinois, on September 6, 1845, at the time when angry mobs were driving the Mormons out of Illinois and Thomas Rhoads was readying the first Mormon wagon train to head west. Knight's father died in 1847, and when Jesse was only five years old he walked barefoot across the plains and mountains to Utah, with his widowed mother and seven small brothers and sisters. Even then he was the man of the family, herding cows in the Salt Lake Valley. He was working at a sawmill while still only a boy and was driving freight wagons to Nevada mining camps as a teen-ager. It was then that Knight decided his financial future would be in mining. He spent every spare minute prospecting, sometimes as far eastward as the Ute Reservation and Uinta Mountains, but more often near his Utah Valley home, in the Tintic District near Eureka. It was at Eureka that he first became acquainted with Caleb Rhoads, who then owned a boarding house at Diamond City. The two men worked closely together, first at Tintic and later at coal properties in Carbon County. For many years they remained close friends.

In time Jesse Knight became a multi-millionaire, controlling more than eighty corporations; including railroads, power plants, banks, farms and ranches, mines, mills and smelters, as well as his own town, called Knightsville. After he became wealthy and when the Ute Reservation was being settled, he built an elaborate system of canals and flumes to divert the waters of Rock Creek to irrigate thousands of acres of sage-covered wasteland on the Blue Bench north of Duchesne City and turn it into productive farmland. That huge project had to be abandoned when his flumes filled with sand, but the ruins of his flumes can still be seen on the rimrock above the ghost town of Utahn. But irrigation wasn't Jesse Knight's first interest on the Ute Reservation.

In 1885 Jesse Knight was still a very poor man, a sometimes prospector trying to support his aged mother and put his brothers and sisters through school. Even as a youth Knight was known for his vision and foresight, something which would in time bring him great wealth. It was during that year that Knight had a remarkable vision. He told his family that while on a prospecting trip into the back-reaches of the Ute Reservation, he had a dream in which he saw an ancient mine in that mountain fastness near where he was camped. He saw himself standing on a hilltop from where he could look down at two ridges running parallel, and in a brushy draw below one of those ridges he could see an old mine. He pulled loose some heavy stones so he could enter that tunnel, and inside that cavern he saw piles of gold and silver bars stacked high, and rotted leather bags of gold nuggets spilled across the floor. A vein of gold crossed the end of that chamber while artifacts of ancient design were piled against its walls. Knight said the walls of that chamber were made of gold-gilt, engraved with strange hieroglyphics, similar to those he had seen in the Book of Mormon. When Knight told his family of his dream, he made a statement so strange that they couldn't help but wonder about it. Knight said that while at that ancient treasure repository he was visited by the Angel Moroni. The angel showed Knight many precious artifacts of purest gold, including records on plates of gold, the same kind of plates Joseph Smith translated the Book of Mormon from. The angel then told him that in days yet to come, the Mormon Church would face financial ruin and when that time came, Knight would be allowed to return to that place and take enough gold so that he would be the means of saving the credit of the church.

Florence Mining Company geologists failed to find Caleb Rhoads' gold mine along Rock Creek.

Knight later acknowledged that he had made several trips to that place in secret and upon return from one of those journeys he told his family that he had seen "marvelous things, wonderful things, like those which the angels had shown to Joseph Smith." Few people today know that Caleb Rhoads also described that place to some of his family, telling them that he had seen things there which "were of worked gold, all of fine workmanship." Rhoads said that Indians guarded that place, "not because of the gold which is there, but because there are artifacts of great religious significance to them." Isaac Morley, who had baptized Chief Wakara and who had been present during meetings between the Ute Chief, Brigham Young and Thomas Rhoads kept a very detailed daily journal. In that journal he

Government engineers led the Raven Mining company's quest for Rhoads gold, but in vain.

wrote of what he heard during those meetings, writing that the sacred mine contained "hand-crafted gold objects of great beauty, and that the walls were of gold-leaf with ancient hieroglyphics on them."

Knight continued prospecting near Utah Valley whenever time permitted. One day during the early spring of 1896 he climbed the barren slopes of Mount Godiva, located near Eureka, where he had still another vision. Knight heard a heavenly voice say to him: "Jesse, this country is here for the Mormons." He interpreted that heaven-sent message as a sign that he should stake a mining claim on that exact spot. There was no sign of ore on Mount Godiva, so when he offered his friend, Jared Roundy, an interest in his claim if he would help dig a discovery hole, Roundy replied that he wanted no part of such a "damned humbug." Knight countered by telling Roundy that he had just named the claim: "I will call it the Humbug!" Weeks turned into months as Knight toiled digging a tunnel deep into the mountain, finding nothing but worthless rock. Then in August, 1896, Knight's Humbug tunnel broke through into a huge body of ore which assayed four ounces of gold, one-hundred and seventy-five ounces of silver and thirty-four percent lead to the ton. It was a bonanza! In coming years Knight would have still other visions which would lead him to other bonanzas; including the Godiva, Tintic Central, Dragon and other equally rich finds, which would make him fabulously wealthy. On his discovery at the Humbug, Knight again repeated his earlier statement that in time he would be the means of saving the credit of the Mormon Church.

Not long afterwards Caleb Rhoads read an account of the treasure chamber Knight had uncovered on Mount Godiva, and he likened it to the same kind of treasure he had found in one of his mines, perhaps the Carre-Shin-Ob. That account described Knight's treasure chamber as follows:

> The shaft broke into a huge cavern, the top of which was made of beautiful crystals of every shape and color. The floor of the cavern was a solid layer of carbonate silver, so soft it could be kicked loose with the toe of a boot. The width of that silver was fully forty feet, while its depth was from forty to fifty feet. One could walk in that cavern for a thousand feet or more, on that clean silver ore, completely free of waste rock, just like wheat in a bin ready for market.

Knight's first paying shipment of ore from the Humbug Mine was made in November, 1896. It is important to remember that date. That first ore shipment returned $11,189, and from that payment he had to pay all of the costs of operating his mine which had accumulated before he struck ore, and all of his miner's wages and other costs after the strike was made, leaving very little profit for him. Even though he had sold his first ore, Jesse Knight was still a very poor man. To quote from the Mormon Church's Treasures Of Pioneer History, "Knight had great difficulty finding financing, but he finally secured a loan of fifteen-hundred dollars at 12% interest, giving a mortgage on his mother's home. Considerable time elapsed before he could obtain adequate means to begin developing his mine."

During the 1890s the Mormon Church went into bankruptcy as a result of the Edmonds-Tucker Act of 1887, which authorized seizure of all church property because church leaders refused to renounce polygamy. The First Presidency of the church was desperate for cash, and in order to save the credit of the church, in 1896 they appealed to Jesse Knight for help. In retrospect, it is difficult to understand why those leaders came to Knight, for at that time he was poor as Job's ox; that is unless those leaders knew something about Knight's finances which others didn't know. Church President Wilford Woodruff authorized Bishop Joseph B. Keeler to seek funds from Knight to pay church debts, then said to total more than two million dollars. On November 22, 1896, Knight gave Bishop Keeler his personal check for $10,000; a generous gesture, but only a pittance compared to the millions church leaders needed. It was hardly an amount which would have saved the church's credit. There has to be some question of where even that small amount came from, for as related, Knight had very little money left from his first ore shipment and even less from the fifteen-hundred dollar loan he had just received by mortgaging his mother's home.

Giving that money to the church president was typical of Knight, for he never hesitated to give anyone in need his last cent. He was known to grubstake others and make bad loans until he was penniless. For a long time until the Humbug Mine began showing a profit, Knight was still very poor and very much in debt. Two years after his strike at the Humbug, he had to borrow money to purchase the Uncle Sam claim. Knight had a favorite saying: "The earth is the Lord's bank!" In time Knight became very rich by making with-drawls from the "Lord's bank," but that time was still years in the future. There is good reason to believe the gold Knight gave to save the credit of the church did in fact come from the "Lord's bank," but that "bank" was located on the Ute Reservation!

In Mormon Church archives there are letters of thanks to Jesse Knight from church officials for saving

Caleb and Sidsie Rhoads. Their mines and treasures were real, but who can find them today?

the credit of the church and the honor of the church presidency. One of those letters, written on the same day Knight gave his personal check to Bishop Keeler states in part:

> Dear Brother Knight:
>
> I feel that this kindly act is an answer to my prayers, to open some doors of relief whereby we may meet pressing demands upon us. May God bless and prosper you!
>
> /s/ President Wilford Woodruff

Somewhat of a mystery is a telegram of thanks sent to Knight by the church's First Presidency, also for saving the credit of the church; but that telegram wasn't sent until two years later! That telegram, which was a priority means of communication before the turn of the century, reads as follows:

> September 3, 1898
>
> Brother Jesse Knight:
>
> God bless you and yours forever. May you have a great abundance of peace and prosperity, and an eternity of bliss in the life to come; this is the prayer of your brethren in the Gospel.
>
> /s/ Joseph F. Smith
Lorenzo Snow
Heber J. Grant

Note that Knight gave his personal check to President Woodruff on November 22, 1896, and received his letter of thanks dated that same day. But the telegram of thanks from the First Presidency wasn't sent to Knight until September 3, 1898, two years later. Surely they wouldn't have sent such a priority message as a telegram to give thanks for something President Woodruff had given his personal thanks for two years earlier. It seems logical to believe the First Presidency was giving Knight thanks for an entirely different and probably much more significant gift. By that time it had become known that the $10,000 check given by Knight had nothing to do with saving the church's credit, but had been used to stop a financial run on the church owned Utah Trust & Loan Company. It seems unlikely that the First Presidency would have sent such a fawning, wheedling telegram of thanks for such a relatively insignificant gift, especially one which had been given two years earlier.

It would be odd if the First Presidency had been thanking Knight for saving the church's credit two years earlier in 1896, for $10,000 wouldn't have paid the interest on the church's debt. It seems to apparent that they were thanking him for a much more substantial gift, but if so, where would such a large sum of money have come from? It couldn't have come from Knight's Humbug Mine, for he never made his first shipment of ore from that mine until November, 1896, which shipment hardly paid for his expenses. The answer to where Knight could have obtained so much money before the Humbug Mine began paying dividends can be found in correspondence between Caleb Rhoads and Utah's senior Senator, Reed Smoot, some of which correspondence can be found in Mormon Church archives.

At the same time Jesse Knight was saving the credit of the church, Rhoads and Hathenbruck were trying to obtain a lease on Indian land, but unknown persons and unexplained events kept blocking their every effort. In order to promote their case, Hathenbruck had entered into secret arrangements with Utah Senator Reed Smoot; who incidentally, was one of the Twelve Apostles of the Mormon Church. Senator Smoot promised to help Rhoads and Hathenbruck, but unknown to them, he was also working closely with Jesse Knight to gain him access to those same lands; or if that could not be done, to keep Rhoads and Hathenbruck from obtaining a lease on them. Rhoads had been aware for some time that men posing as prospectors had tried to follow him during his covert and probably illegal trips into the Uinta Mountains, but he never suspected they were in the employ of Knight. But then Rhoads made a startling discovery. Jesse Knight had been to the sacred Carre-Shin-Ob Mine and had removed large quantities of gold from it!

For reasons which are now obscure, but which were no doubt associated with his pledge to Brigham Young that he would not reveal the location of that fabulous hoard, Rhoads seldom visited Carre-Shin-Ob other than on the infrequent occasions when he brought gold to Salt Lake City. On one of those rare occasions he discovered that Knight had been there, which fact he made very clear in a letter to Senator Smoot. That letter, with his underlining and capital letters, is in church archives and reads in part as follows:

October 23, 1901

Honorable Reed Smoot
Washington, D.C.

I enclose information regarding my strong indignation at the actions of Brother Jesse Knight, and his interference in my efforts to secure a lease on the Indian Reservation. I need not inform you that Brother Knight brought out significant amounts of gold from THAT MINE, against all my warnings that he should not do so. You will remember that those were the same amounts which he gave to President Woodruff, to save the credit of the church. President Heber J. Grant knew about those things, and could testify to them if he would, but the church is not about to do so until this Indian lease thing is settled with the government. But the church IS involved, and between them and Brother Knight's efforts against me, I am left in a hard place.

I do not worry that Brother Knight will ever reveal the location of THAT MINE, or that he will ever use it for any purpose other than which his conscience dictates. I will not go near THAT SACRED MINE again, because I know he has agents everywhere in the government. I do not accuse you, because you have been a good friend to both of us, but instead to let you know my feelings about Brother Knight's bid to overcome my mining lease. I hope you are still a friend and in support of our efforts.

/s/ Caleb B. Rhoads

Note that in his letter to Senator Smoot, Rhoads stated that Knight took "significant amounts of gold from THAT MINE," the value of which was the same amount he gave to Mormon Church President Wilford Woodruff to save the credit of the church. That amount had to be significant, since the church's debts were then said to be several million dollars. Also, note that Rhoads twice referred to "THAT MINE" and "THAT SACRED MINE," and stated he would not go near it again; which of course indicates that he had in fact been to it recently, at least recently enough to know that Knight had been there. We do not know how many times Knight went to Carre Shin-Ob, but his comments to his family suggests that he went there at least two times without taking anything from it. How many times he went there to obtain gold to pay church debts is unknown. Caleb Rhoads was soon to learn that Senator Smoot was not the friend he had thought he was, and he would also learn that Jesse Knight had far more influence at Washington than either he or Hathenbruck had.

Rhoads and Hathenbruck had known for some time that someone high in government had been blocking their efforts to have their Indian lease approved, and it soon became apparent that those persons included Jesse Knight and Senator Smoot. In correspondence between H.C. Henderson, an attorney representing Rhoads and Hathenbruck, and who incidentally was also a congressman from New York State, there is a startling statement and accusing finger pointed directly at Senator Smoot. On July 21, 1902, Congressman Henderson wrote to Hathenbruck, telling him that certain persons at Washington were defaming the character of both he and Rhoads, and added: Upon receipt of this letter, let me know definitely just how much money it is going to take to satisfy Reed Smoot. Also, state whether you will leave this matter entirely in my hands, giving me your best knowledge and experience in these matters." It became obvious that Senator Smoot's favor would go to the highest bidder, and Rhoads knew that Jesse Knight's pockets were much deeper than his. At that time Knight was fast becoming a millionaire as the largest owner of patented mining ground in the West, as well as the largest tithe payer to the Mormon Church.

Because of the rapidly growing conspiracy involving Senator Smoot, Jesse Knight and others to keep Rhoads and Hathenbruck from obtaining a lease on Indian land, Hathenbruck went to Washington to meet with Secretary of The Interior Cornelius N. Bliss. There he learned that H.P. Myton, Indian Agent for the Ute Reservation, expected a share of the Rhoads Mine in return for his support of the Indian lease. On April 25, 1902, Hathenbruck made a sworn and notarized statement that Myton told him that if he was brought in "on the ground floor or was given an interest in said lease, he could secure approval of that lease, but if he was not given a share, the lease would not be approved." Hathenbruck presented Myton's demand to Caleb Rhoads, who refused to share the lease with Myton. Several days later Hathenbruck informed Myton of Rhoads' refusal, to which Myton replied: "You will never get your lease approved."

At Washington, Secretary of The Interior C.N. Bliss still appeared to be ready to give his stamp of approval to the Rhoads lease, when suddenly without explanation he resigned his position. Secretary Bliss had found himself in a tight spot when he received a letter from George Q. Cannon, Utah's junior Senator, and like Smoot, one of the Twelve Apostles in the Mormon Church. A portion of that letter from Cannon to Bliss stated: "Rhoads and Hathenbruck are poor, ignorant prospectors, unable to swing a proposition worth $100,000,000." After the unexpected resignation of Secretary Bliss, the Rhoads application for a lease on Indian lands was pigeon holed and forgotten. Shortly thereafter, Congress chartered the Raven and Florence mining companies, awarding them the mineral rights to the lands Rhoads and Hathenbruck had tried for so long to get.

At the last minute had Hathenbruck somehow been caught up in the web of intrigue between Jesse Knight and Senator Smoot? Had he somehow been forced to sell his soul to the devil? On April 24, 1901, Hathenbruck signed a release "of all my rights, title and interest in that certain lease on the Ute Indian Reservation, as well as all of my rights, title, interest and capital stock of the Rhoads Mining Company for value received." Amazing as it might appear, those rights were transferred to Senator Reed Smoot! When questions arose over the roll he played in that intrigue, Hathenbruck told the editor of the Salt Lake City Herald-Republican, "It is not easy for me to recognize the justice of this arrangement. If this matter is ever opened for investigation, there will be some very disgraceful revelations!" Needless to say, there was never an investigation.

The identities of the men behind the Raven and Florence mining companies revealed how badly Caleb Rhoads had been duped. H.C. Henderson, his attorney, had actually been a pawn of Knight and Smoot, and it was he along with James Sherman, Warren Hooker and William Ward who organized those companies and had congress grant them an exclusive charter. All four were congressmen from New York. Contrary to what some have thought, the Raven and Florence companies were two separate entities. On May 27, 1902, the Raven Company gained exclusive rights to all gold, silver and precious metals on Indian lands which were granted to it, while the Florence Company had rights to all coal, oil, gilsonite and hydro-carbons. Both companies were granted the right to prospect the entire reservation before those lands were opened to others; while Rhoads and Hathenbruck were specifically prohibited from entering those same lands. Some specifics of the land grant given to the Raven and Florence companies were as follows:

> Up to thirty days before said lands are restored to the public domain, the Florence Company shall have the preferential right to locate not to exceed 640 acres of mineral land. The Raven Company may locate one-hundred mining claims, twenty-five of which shall be for precious metals.

Apparently, Senator Smoot had promised Raven and Florence officials that when Rhoads and Hathenbruck learned that they would not be granted a lease or be allowed on Indian lands, they would quickly capitulate and reveal the locations of their mines and sell or lease them to the New York firms, probably at a bargain price. This is proven by a letter sent to Rhoads by Senator Smoot which states in part:

January 15, 1904

Caleb B. Rhoads
Frederick W. Hathenbruck

I certainly hope that you will be successful in obtaining a good large sum of money from the Florence Mining Company people, for you certainly have worked very hard for a long time. I thank you for the interest you have shown in my behalf.

/s/ Reed Smoot

But if Senator Smoot thought that Rhoads and Hathenbruck would surrender their claims or reveal their locations to the Raven and Florence people, he was very much mistaken. When their efforts to obtain the Indian lease failed, Hathenbruck may have been forced to assign all of his rights in those lands to Senator Smoot, for reasons which are now unknown, but that may have been a hollow victory, for Hathenbruck did not reveal the location of those lands. Of course, he knew only a few of Rhoads' mining locations. Caleb Rhoads signed nothing over to the Senator who he had believed was his friend; instead he became even more secretive about the location of those mines.

Executives of both the Raven and Florence companies were well aware that Rhoads and Hathenbruck had bargained with Senator Smoot to obtain certain Indian lands, and that Rhoads had promised to pay the national debt in exchange for that lease. The government's position had been that Rhoads must identify the exact plots of land he wanted before a lease could be granted, but Rhoads was too wise to reveal those exact locations and in doing so lose his ace in the hole. Anyone who has hiked that country knows the futility of trying to find a well concealed mine portal in 480,000 acres of rugged mountain wilderness. When Rhoads refused to identify those lands, he and Hathenbruck were forbidden to enter Indian lands, on threat of arrest if they did so. Hathenbruck wrote to newly elected Utah Senator Thomas Kearns appealing his restriction from the reservation, in part as follows:
August 27, 1904

Honorable Thomas Kearns:

I have been employed as a topographer and guide for F.M. Lyman & Co. in their survey of the Ute Reservation, but I have been treated as a criminal and prevented from entering those lands. For many years I have treated Indians when they were sick, clothed them when they were naked and fed them when they were hungry. I have paid to old Chief Tabby $500, which on account of his age he wants now. I would deem it an extreme favor to enlist you in my behalf for justice. Respectfully,

/s/ Frederick W. Hathenbruck

While Rhoads and Hathenbruck were barred from entering the Ute Reservation, high-priced geologists employed by the Raven and Florence companies as well as government engineers hired at public expense were given free access to those same lands. During that summer teams of expertly trained men trampled back and forth across the Uinta Mountains, from Kamas Valley to the Green River. They examined nearly every canyon and gulch, walked across every hillside and mountain front and panned for color in every river and stream; all to no avail. When reporters asked Rhoads if he wasn't concerned that one of those experts might find his mines, he replied: "My mines are in a place where a geologist would be least likely to look."

Accompanying this text are two never before seen photos of Raven and Florence geologists being led by Col. George F. Timms, prospecting along Rock Creek. But without guidance from Rhoads, their search was in vain. With all of their expertise and the unlimited financial backing of Congress, not one of Rhoads' mines or caches was ever found. But the very fact that both of Utah's Senators, the entire Congressional delegation from New York State and the most highly qualified experts the federal government could find all believed those old Spanish mines and caches existed somewhere in the Uinta Mountains is a fact which cannot be overlooked. Of course, that was something that Chief Wakara, Brigham Young and Thomas and Caleb Rhoads knew all along!

There will still be some who will say that Carre-Shin-Ob, the sacred Church Mine, never existed. They will say that only those who hoped to benefit from it ever claimed it existed. To determine whether such a place ever existed, we should have the testimony of someone who could not hope to benefit from it, someone not personally touched by it as Caleb Rhoads was or as Aaron Daniels hoped his family might be. It would be desirable if that person was also recognized as a historian or authority on the Ute Indians, and the Spanish mines they guarded. Mormon Selman was just such a person.

Selman was a missionary to the Utes for more than forty years, a man who had no aspirations of ever benefiting from his close relationship with the Indians. He was also a man of unquestioned integrity, a man who did not lie. Selman was content to live among the Utes and teach them how to farm and irrigate their crops. He wanted nothing more than to help them. He attended their most sacred council sessions and later compiled a Ute-English dictionary. He was trusted by the Utes as few white men ever were, and he became as close to becoming a Ute as any white man ever became. In a place well removed from the problems of reservation life and at a time when the sacred mine was still relatively unknown and not controversial, Ute elders told him of that mine. Their story, paraphrased from his journal is as follows:

> There is a place which the Utes consider to be sacred; only a Chief may go there. It is where the gold used in our ceremonies comes from, the place we call Carre-Shin-Ob, or Here Dwells The Great Spirit. Once in the long ago, white men riding horses, the first horses our grandfathers ever saw, found the place where the sacred gold came from. The Utes drove them from the mountains, following them beyond Timpanogos Lake, where they caught and killed them in a canyon. All of the gold they had stolen was taken back to the mountains, where it came from. The leather bags of gold which they stole from Mother Earth were hidden in the sacred place. Those Spaniards were not allowed to take even one grain of the sacred gold.
>
> After the Mormons came to the mountains, the Ute Chief Wakara became their adopted son, and allowed one of them to take some of the gold to his father, the Mormon Chief. All of them are dead now; Wakara, the Mormon Chief and the tall bearded one who was given the gold. Now only a few of the Old Ones know where the sacred gold is hidden, and they too will soon be dead. Our friend Selman has asked that the sacred place be shown to one of the young braves before the Old Ones die, so that its secret will not be lost. But we are afraid that a young brave might reveal it to a white man, so that the sacred place would not longer be fit for the Great Spirit to dwell in. It is better to let the secret die with the Old Ones.

One of the present generation of Ute elders said almost the same thing recently; that when he was young only a few of the Old Ones knew where the Spanish mines and caches were hidden. If they chose, they could pass that knowledge on to an eldest son, but they were not allowed to pass it on to a daughter or to a son who married a white woman. If the Old Ones died without a son to pass the secret to, it would be lost forever. Thus in time the secret of Carre-Shin-Ob was lost even to the Utes, so that now few if any known where the sacred gold is hidden. But they know that it did exist, and that it still does exist. The trail to the sacred cache or Church Mine is still there, the landmarks described have not changed. The Chalk Cliffs, the Spanish sign trees, the three knolls and the two ridges are just as they always were. They are not hard to find, not if you follow the true waybill. I have found those places, and so can you; but finding Carre-Shin-Ob, well, that is another matter. That's hard to do when you have to keep looking back over your shoulder! Don't forget what Happy Jack told Aaron Daniels: "Towats guards the sacred gold, and he will kill you if you go near it!"

THE WAYBILL

In the preceding chapters you should have discovered your own waybill to the lost treasures of the Uinta Mountains. I've given you all of the pieces to the puzzle; it's up to you to fit them together. Those pieces are the clues to where Coronado's children found gold and silver in those Killer Mountains, and where they cached a lot of it. The maps are clues also, most because they lead the way to treasure, but some because they don't!

When you follow your own waybill to treasure, you may find an empty money pit where someone has been there ahead of you. Some Spanish mines and caches have already been located; but a lot more are still out there to be found. You have the waybill and the maps necessary to find them. When you are in those mountains, please don't deface any of the old signs and symbols which you may find. Take only photographs and leave nothing but footprints.

Be warned: There are strange happenings associated with those lost mines and treasures, with real dangers involved, including very real disappearances and some very real deaths. Once while prospecting at Currant Creek, the late Gale Rhoades said that he believed that one day he too would die in those mountains. He did, and so have a lot of others. Look for treasure, but don't forget to keep a close watch over your shoulder!

MAP INDEX

MAPS

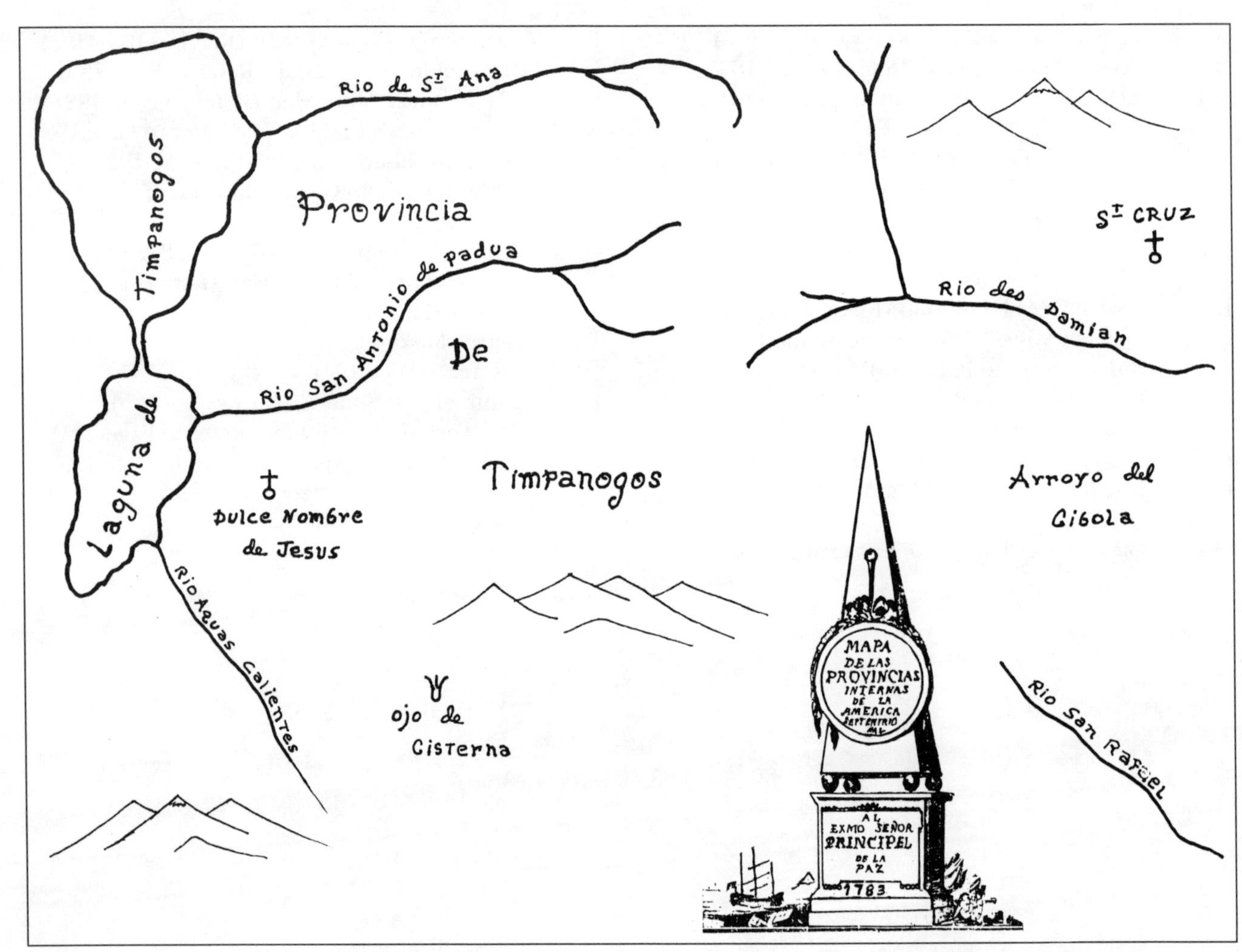

A The La Paz Map of 1783: Mapa de los Provincias de la America Septemtrioal, the Province of Timpanogos.

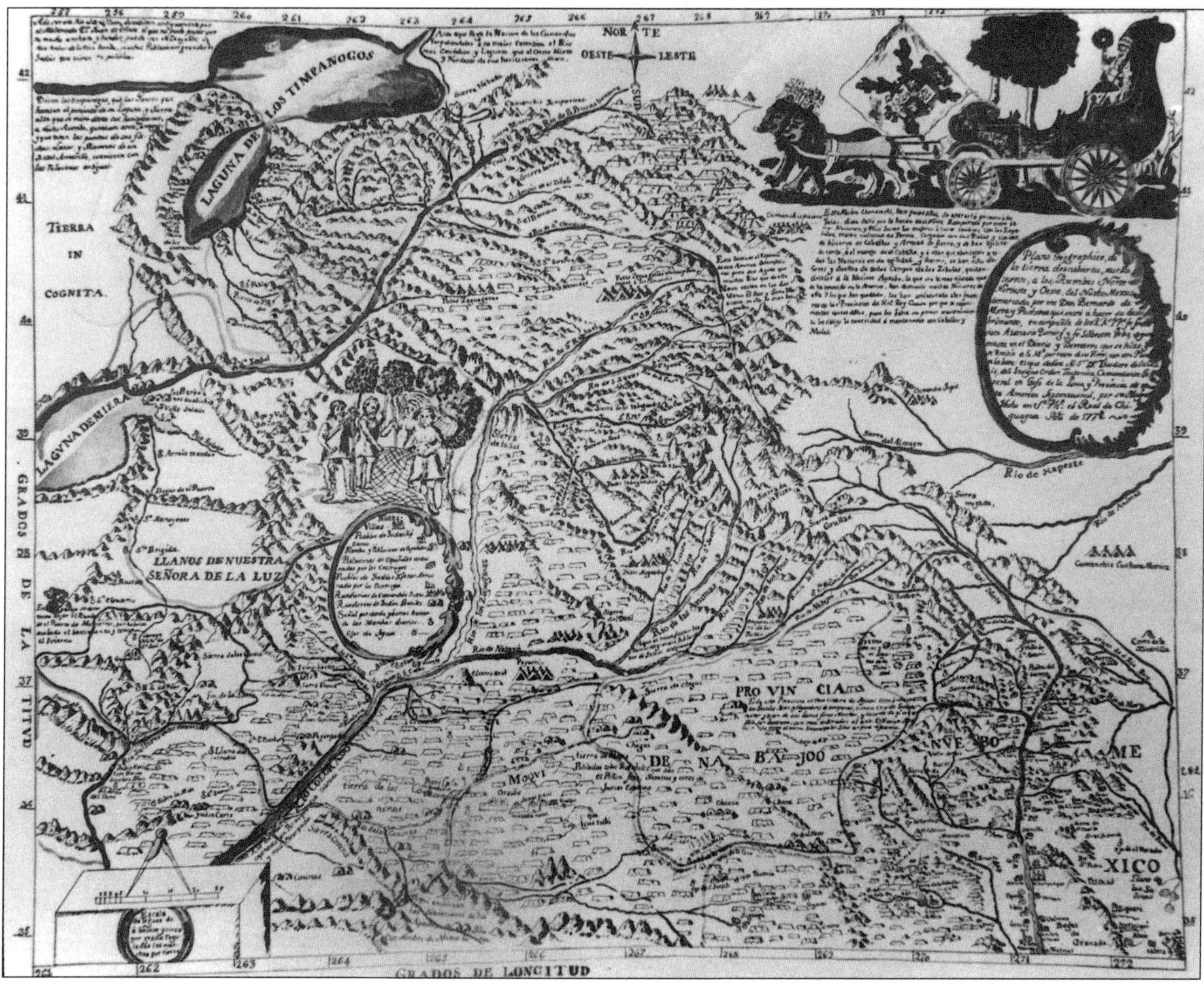

B The Don Bernardo Mierra Map of 1776. Laguna De Los Timpanogos at the upper left.

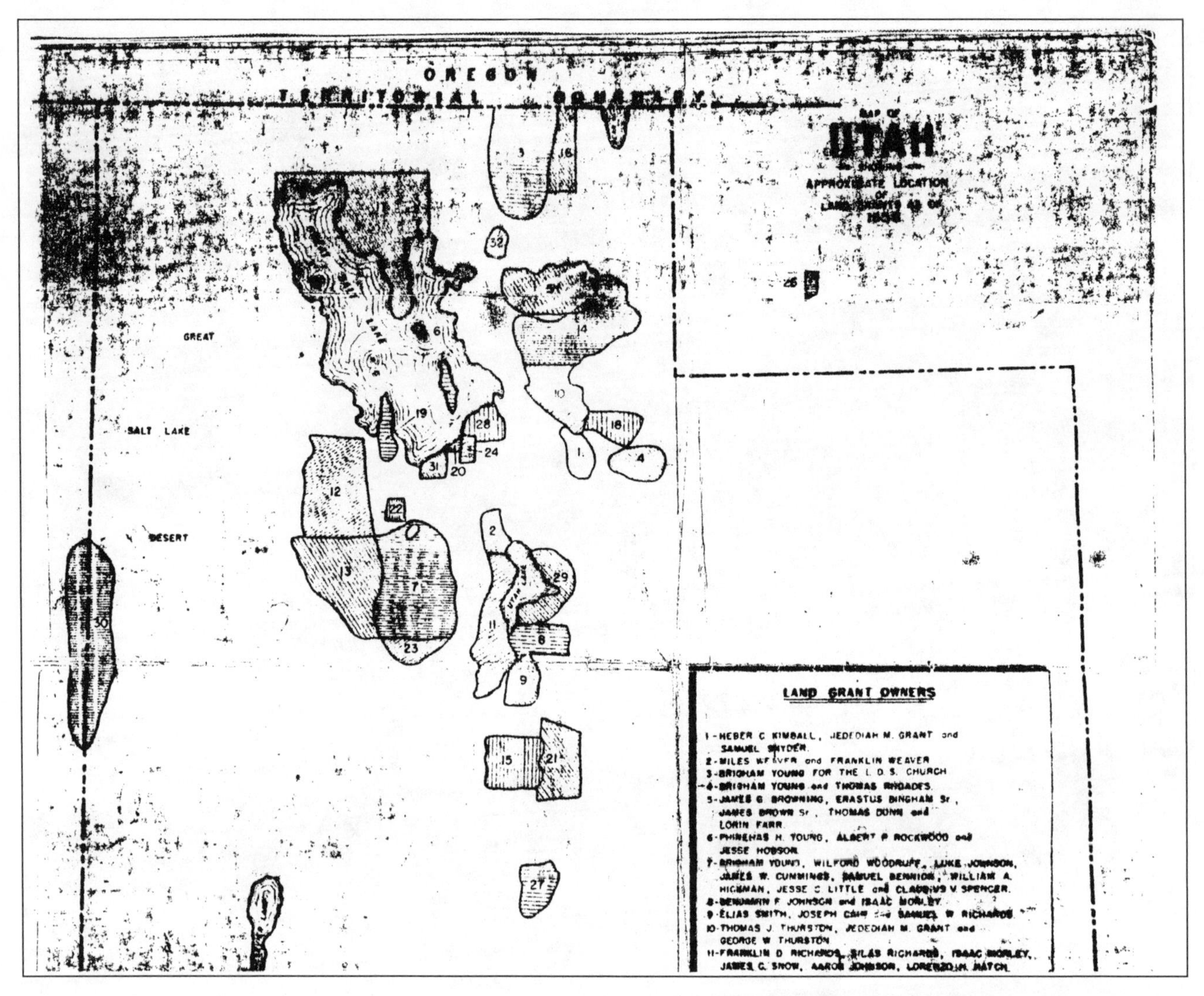

C-1 The LDS (Mormon) Church Land Grant Map, dated 1858. Grant #4, right center, is the Thomas Rhoads grant.

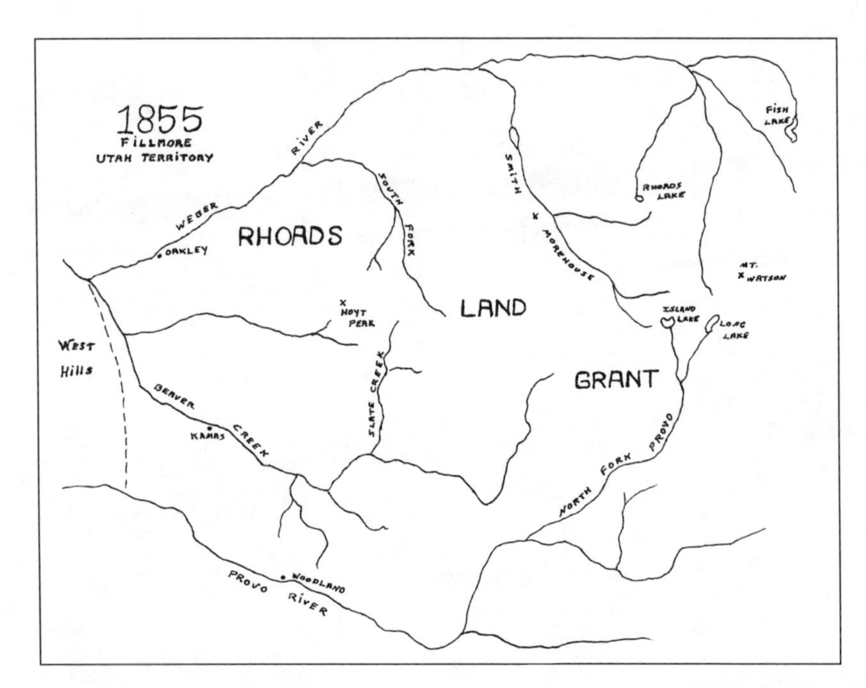

C-2 The Rhoads Land Grant of 1855: All of the lands between the Provo and Weber rivers, to their headlands.

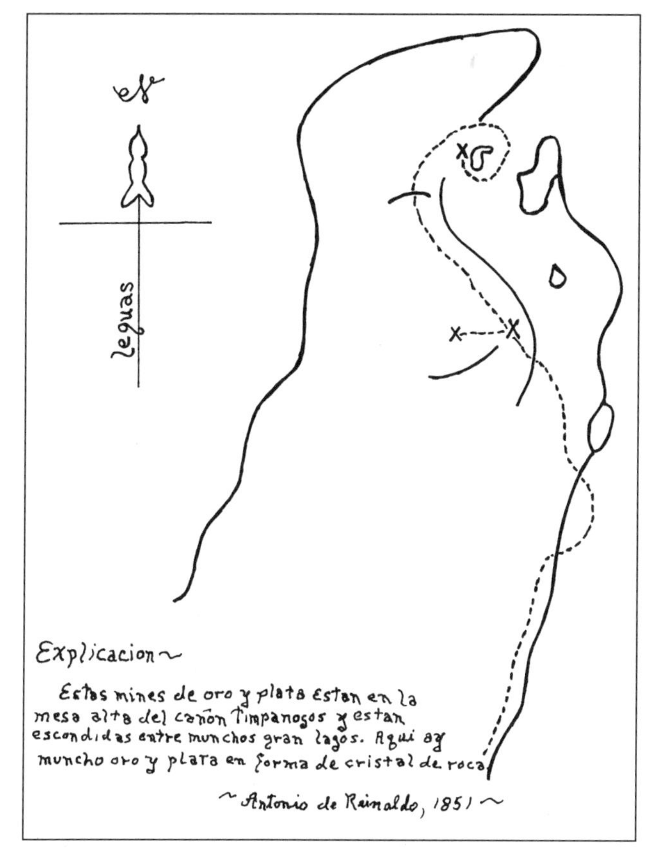

D-1 The Antonio de Reinaldo Map of 1851, showing mines of silver and gold at the headwaters of the Provo River.

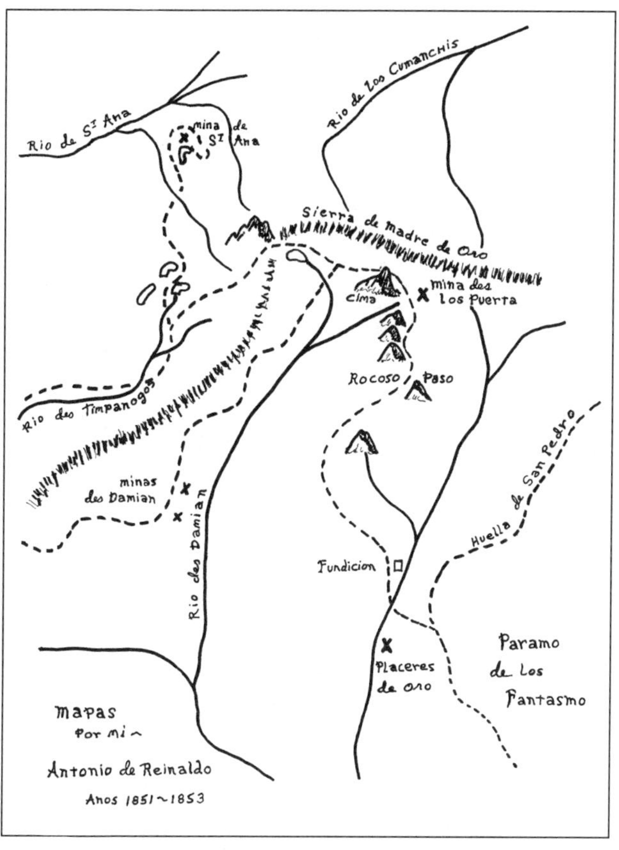

D-2 The Antonio de Reinaldo Map of 1851–1853. Mines of the Upper Mesa, headwaters of the Provo and Weber rivers, Uinta Mountains.

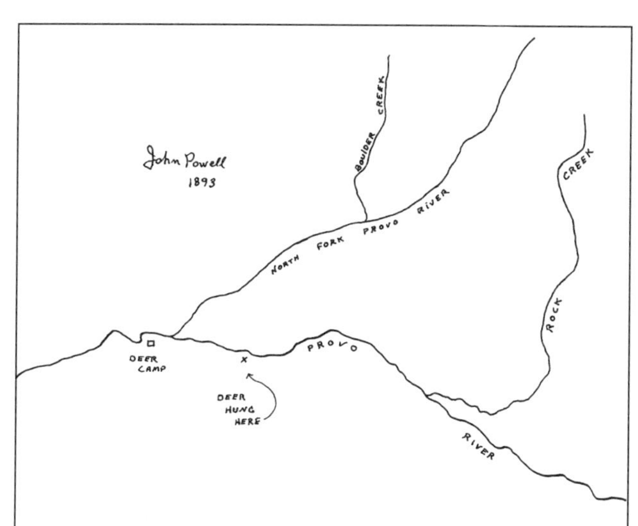

E The Powell "Dear Hunt" Map. Caleb Rhoads found gold "not far" from where he hung his deer. (Compare this with the Garcia Map.)

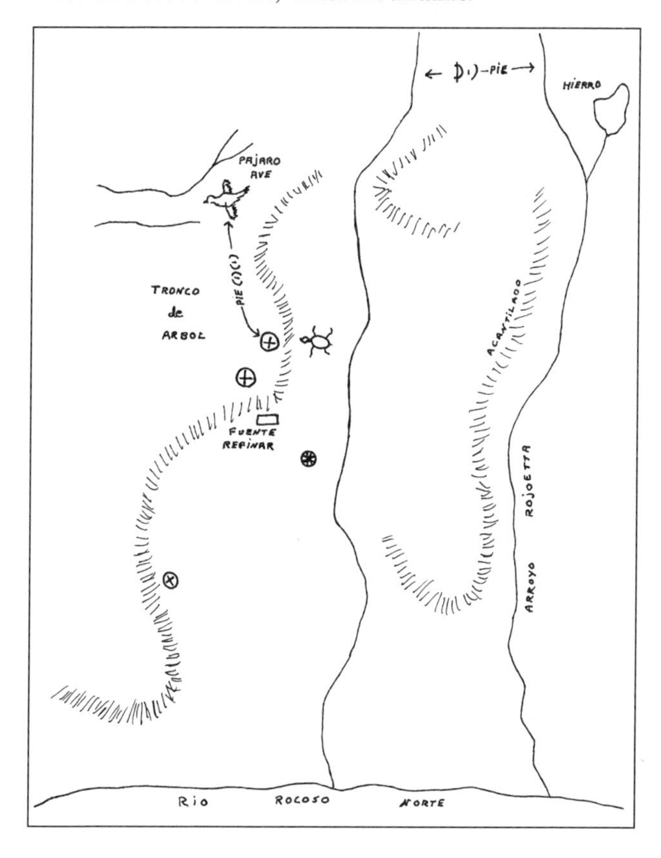

F (Right) Spanish mines of Soapstone Mountain and Lightning Ridge, area of the lost Stone Staircase Mine.

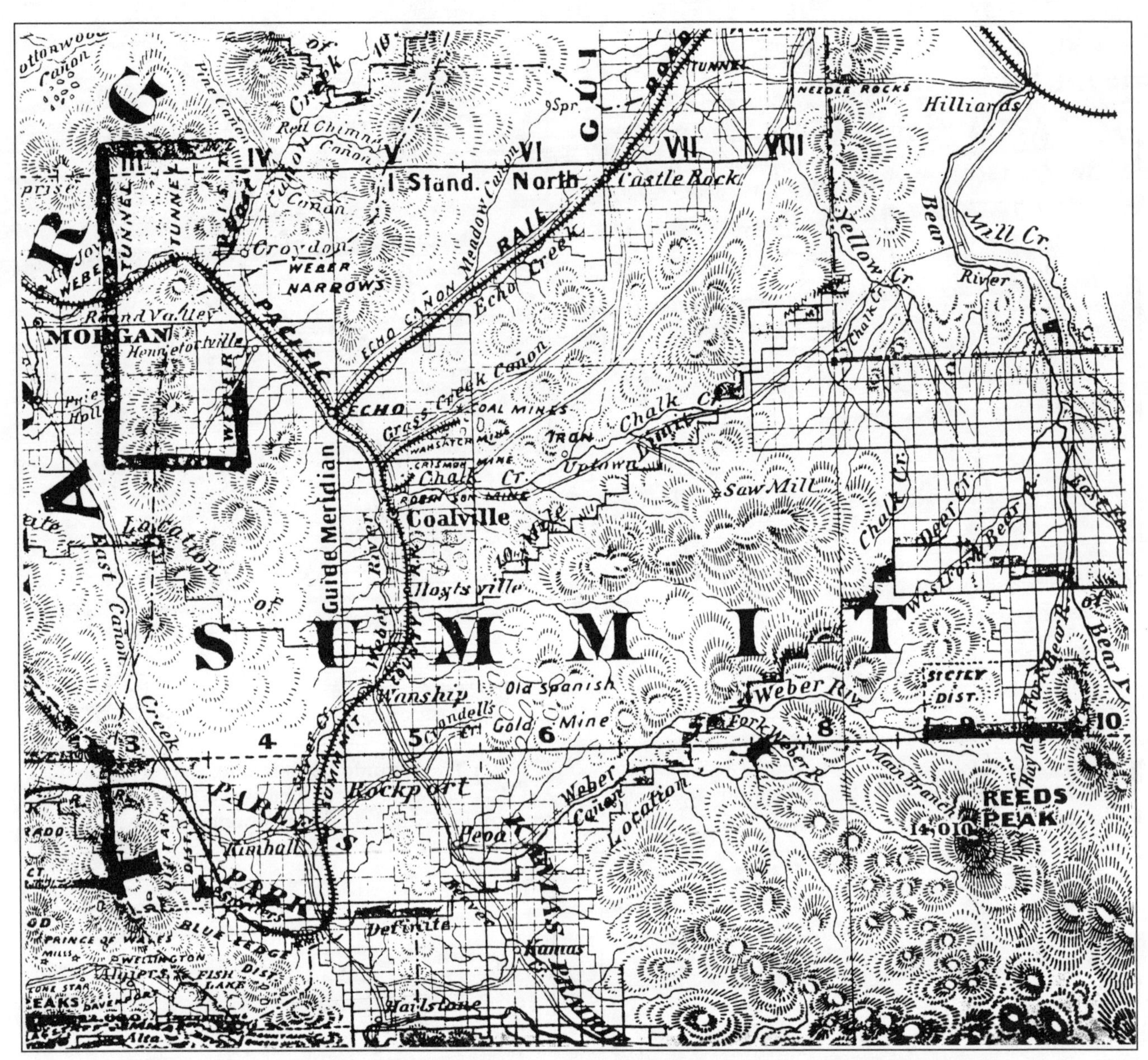

G Froiseth's New Section & Mineral Map of Utah, 1878. An old Spanish gold mine overlooks the Weber River.

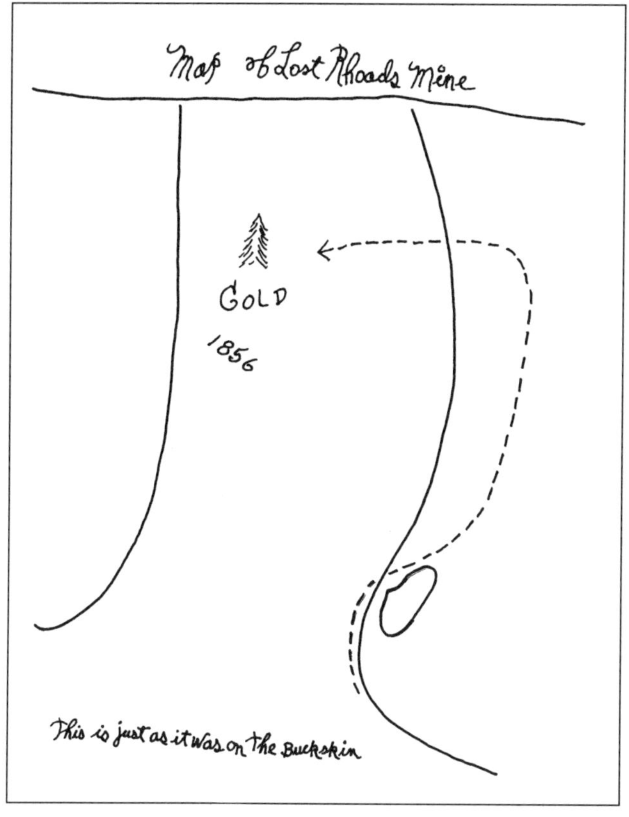

H The Thomas Rhoads 1856 Pine Mine Map. What area does it really portray? (Compare it with the 1851 Reinaldo Map.)

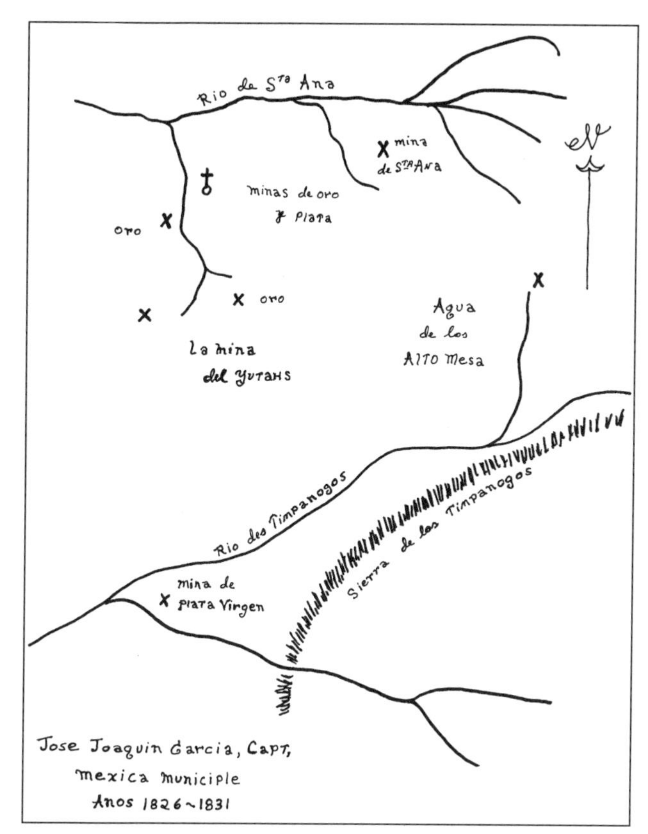

I The Jose Joaquin Garcia Map of 1826–1831. Spanish mines on Hoyt Peak and the upper Weber and Provo rivers. (Compare it with the Reinaldo maps.)

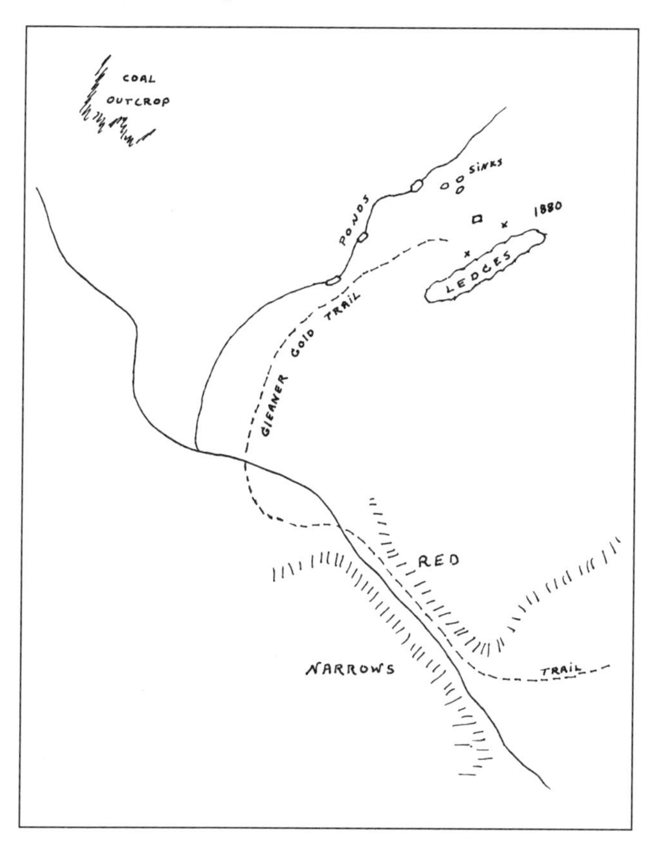

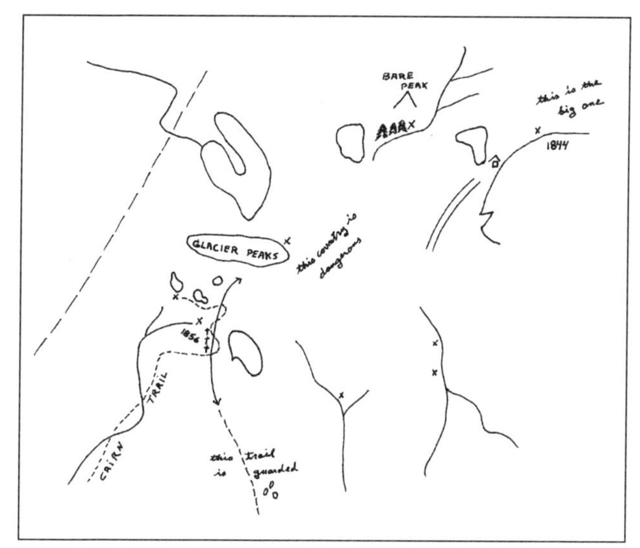

K Caleb Rhoads' "Dangerous Country" Map. Some call it a "Mystery" Map.

J (Left) The Gleaner Gold Map, lost mines of the Currant Creek country.

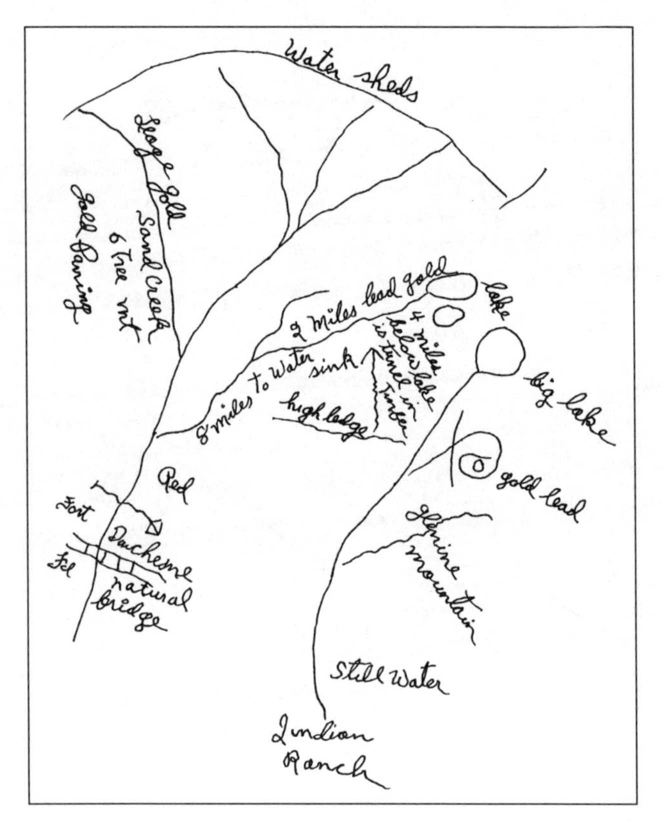

L The recently discovered original Tunnel In Timber Map, used by Louchious Miles and "Hap" Stevenson.

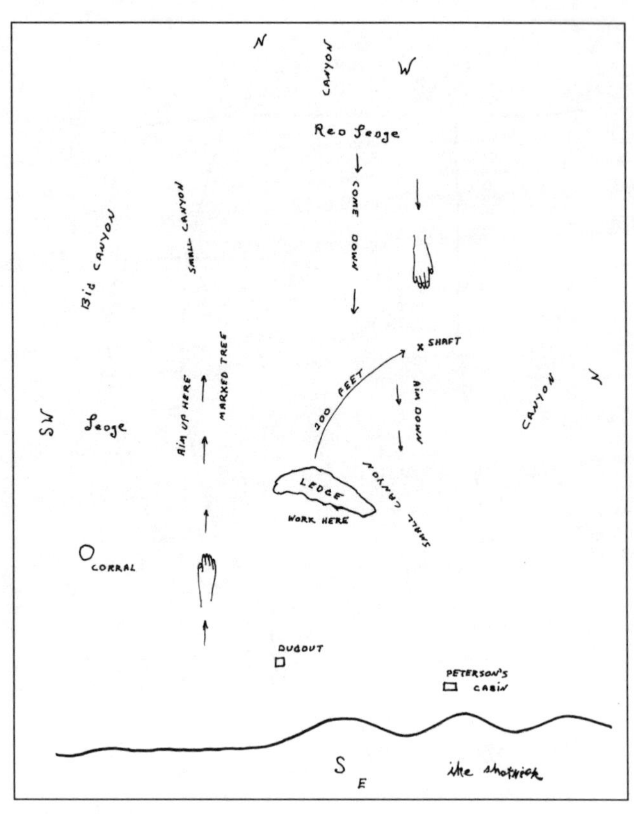

M The Ike Shotnick Mine Map. Incorrect directional markers only make it more confusing.

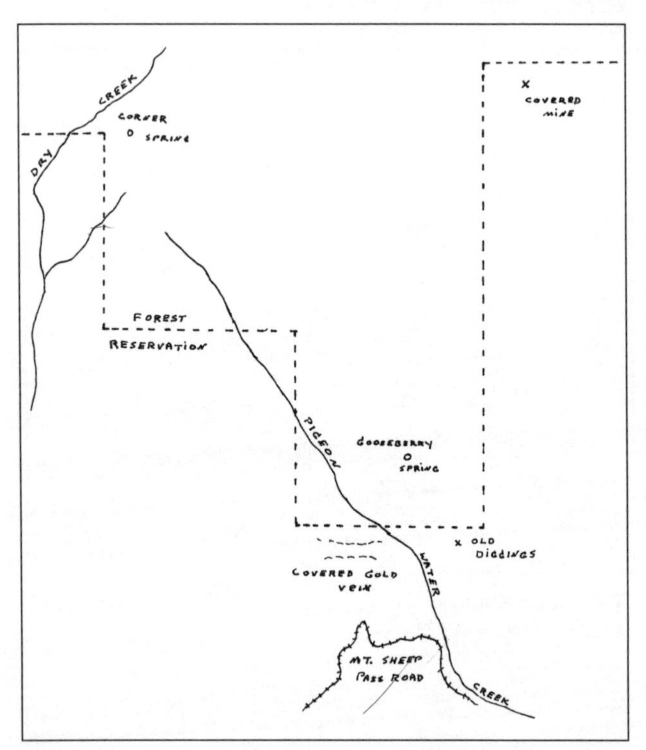

N The Gooseberry Springs - Pigeon Water Map. Hidden gold on Indian land.

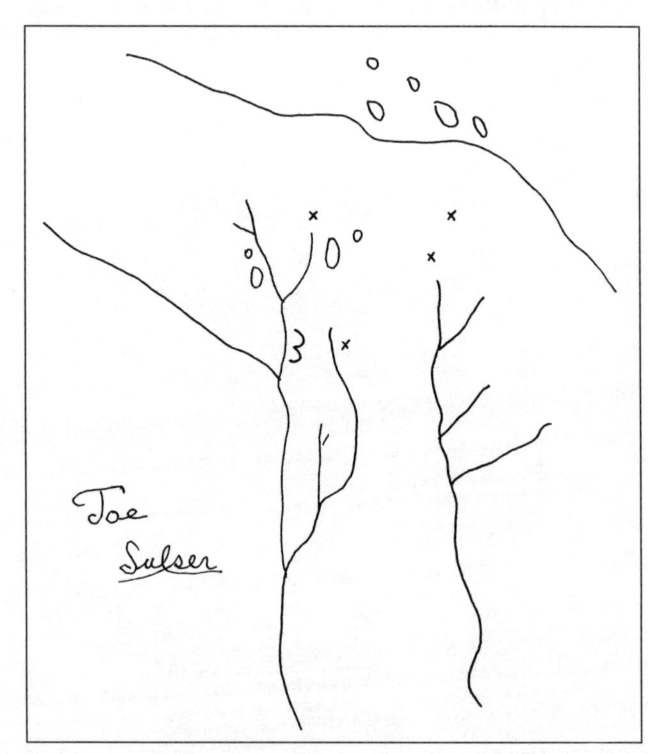

O The Sulser Map. What streams and lakes does it show? Did it lead Ben Bullock to treasure?

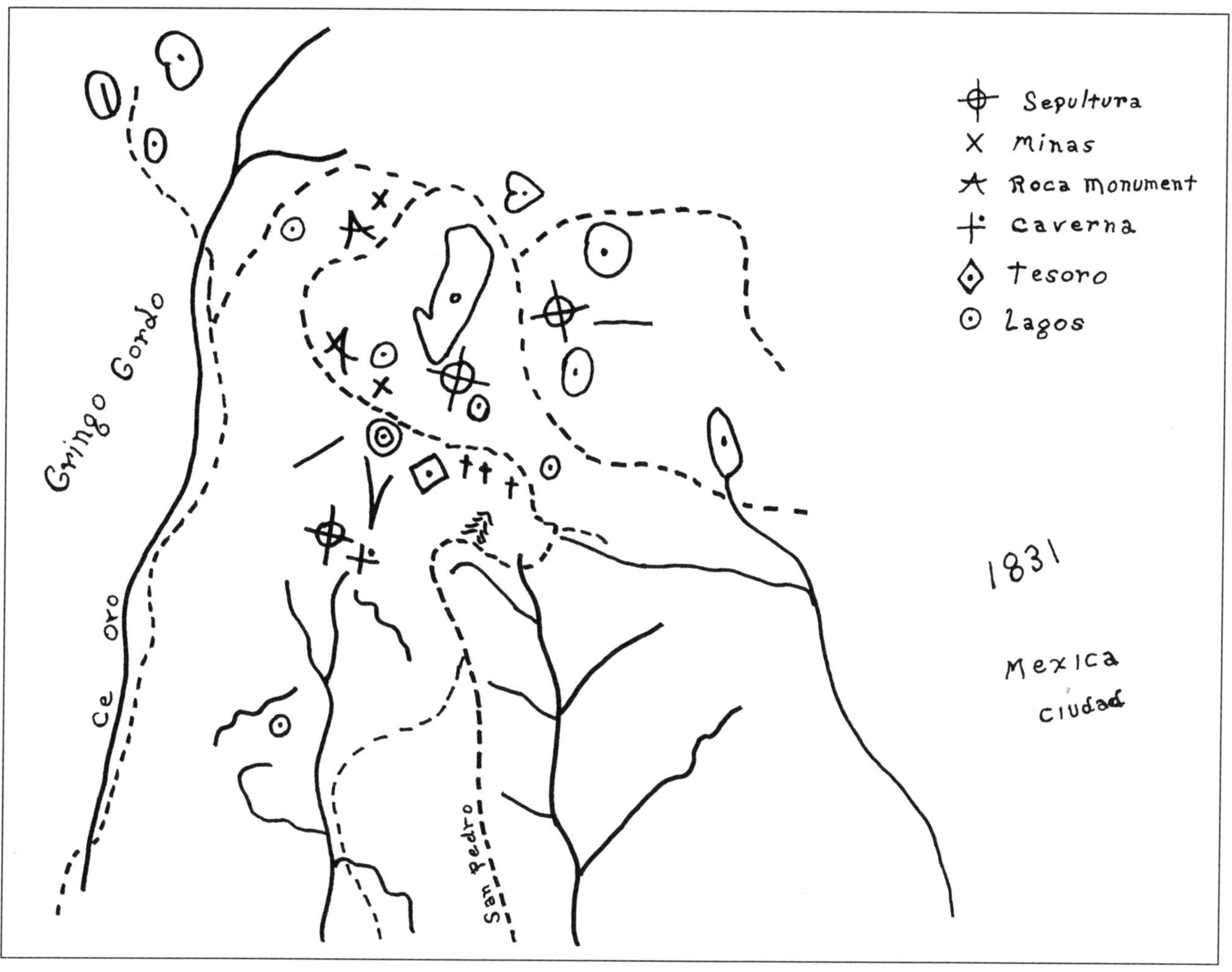

P The original Gringo Gordo Map. An enigma. Can you decipher its hidden meaning?

CBR
NAME TREE
OLD MINE
CREEK
OLD MINE
POLE CREEK
Pole Creek Sink + Mine
LEFT FORK
UNDERGROUND WATER COURSE
SMELTER
Pole Creek Cave
POLE
SMELTER
DARK CANYON
A. Daniels

Q Aaron Daniels map of the Pole Creek mines. Prospecting here could be hazardous to your health!

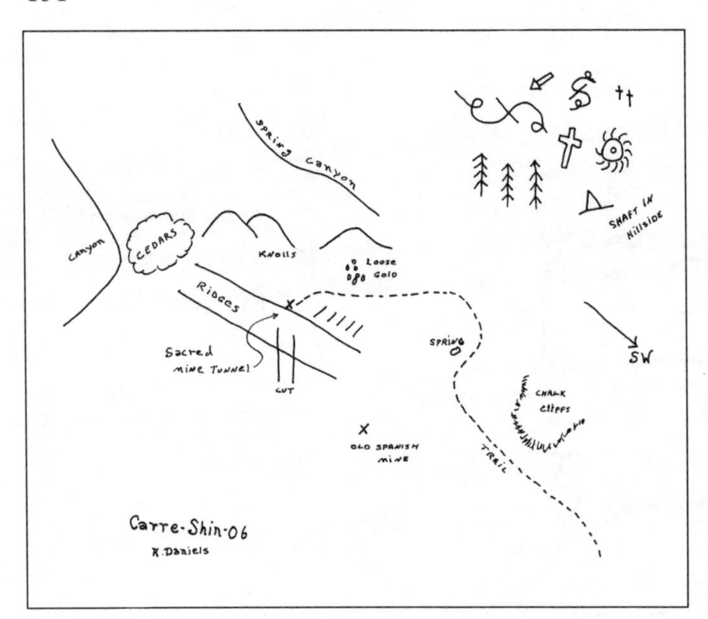

R The trail to Carre-Shin-Ob, an Aaron Daniels map of the Brigham Young or Church Mine. Gold from it was used to mint Mormon gold coins.

S (Below) The most detailed of Happy Jack's Three Forks Spanish mine map.

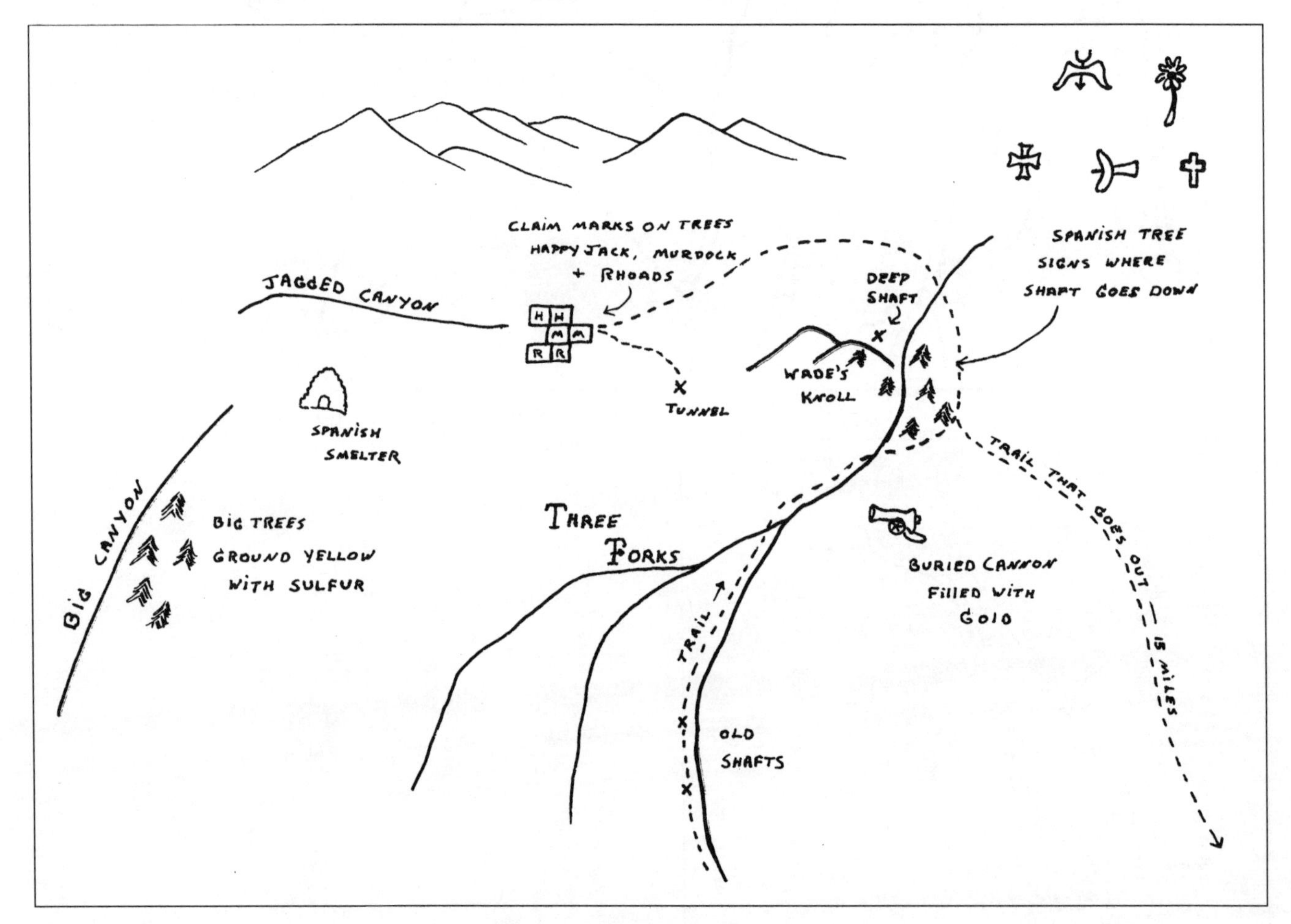

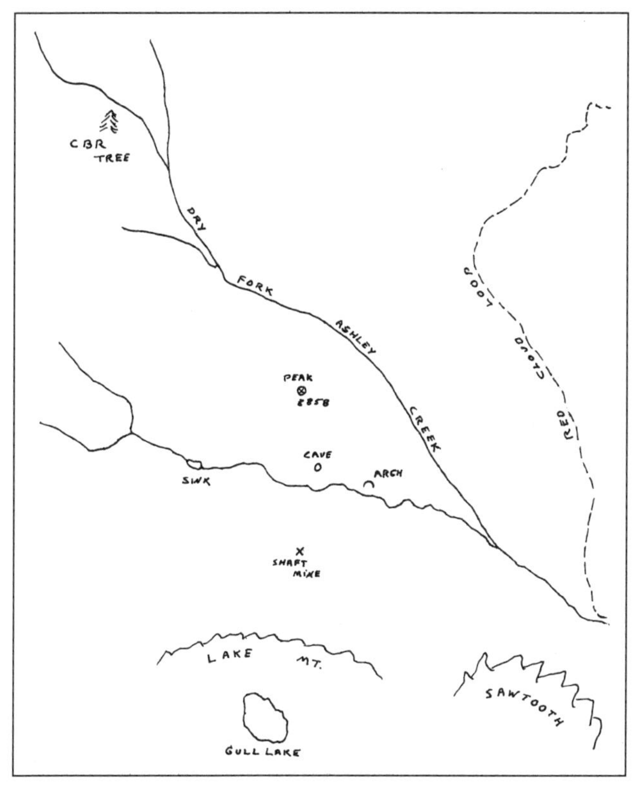

T The Arlo McDaniel Mine of Lost Souls Map. All of the landmarks are there, but where is the mine?

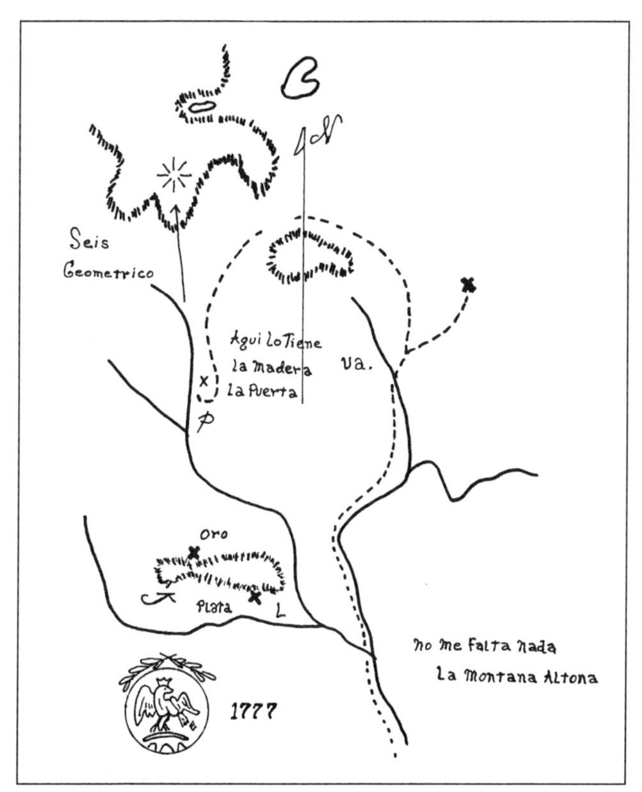

U The Dry Fork of Ashley Creek Map. Is it real? What do you think?

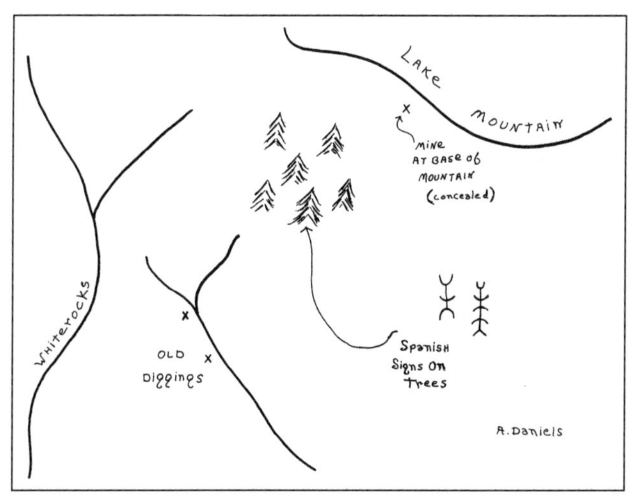

V The Spanish Mine at Lake Mountain. Aaron Daniels found gold here.

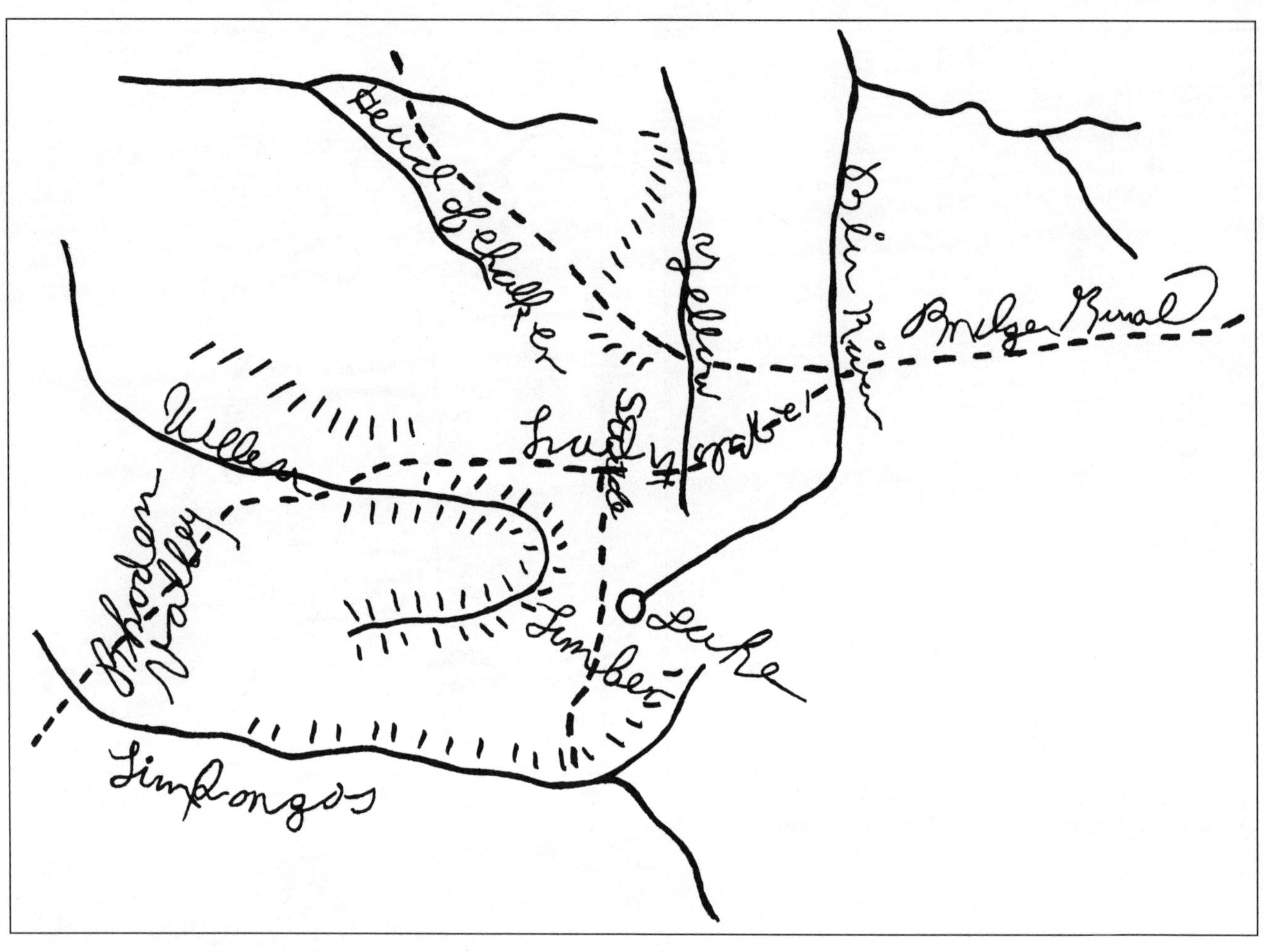

W The Weber River - Bridger Road Map. What does it really show?

CACHE INDEX

MINES INDEX

SPANISH GLOSSARY

Glossary of Spanish-Mexican mining and prospecting terms often encountered on old maps, charts, waybill and documents. Compiled from an early text published at Zacatecas, Mexico, of archaic words in common use since early Spanish colonial time.

Adarme: Weight for gold or silver, about 1.8 grammes
Ademe: Timber in mines, timbering in general
Afloramiento: Outcrop of an ore vein
Amalgama: Amalgam
Amojonar: To set claim monuments or markers
Apique: A vertical shaft
Arrastra: A crude mill for grinding ores
Arroba: Twenty-five pounds
Aurifero: Gold bearing rock
Azoque: Quicksilver, mercury
Azogues: Free milling ore, free gold visible
Bartolina: Mine entrance, portal
Barra: Bar or ingot, bar of gold or silver
Barrena: A hand-held stone drill
Boca: Tunnel portal
Bonito: First-class silver ore, 1,000 ounces per ton
Braguetilla: Crude smelter, hole in the ground
Buscones: Prospector, miner working abandoned mine
Cabezuela: Rich concentrates, both gold and silver
Calichal: Second-class silver ore, 150-500 ounces per ton
Capellina: Bell-shaped retort for silver amalgam
Cardenillo: Ruby silver
Carga: Mule load, up to 300 pounds
Caretta: Wood wheeled cart
Castellano: Small furnace, about four feet high
Cata: Prospect hole
Cateador: Prospector
Cebar: To melt rich ores
Cendradilla: Crude smelter for silver ores
Chispa: Nugget, ore containing visible gold
Chispeada: Rich silver ore, one-third silver
Choclo de Oro: Mass of native gold
Ciguena: Windlass, hand crank hoist
Cpbre: Copper
Conducta: Bullion pack train
Crisol: Melting pot, slag pot
Cuero: Leather ore bucket or sack
Cueva: Cave, cavern
Derrotero: Continuing diary, chart or set of directions
Descubridora: Initial mineral discovery, first mine
Dureza: Hardness, as of an ore
Ensayar: Assay, assay office
Escalera: Notched pole ladder, chicken ladder
Escondita: Hidden
Escoria: Furnace slag or cinders
Faiscador: Placer miner, gold washer
Fierros: Low grade silver ore, 20-30 ounces per ton
Frente: Working face of tunnel or drift
Fundicion: Smelting plant, furnace
Galemar: Reduce ore in a Mexican furnace
Gambucino: Mexican prospector
Grasero: Slag from Mexican smelter
Hacienda: Smelting works, custom mill
Hierro: Iron, iron ore
Huaira: Ancient Indian smelting furnace
Impuesto Minero: Tax paid on mining claims
Jales: Rich tailings from concentration or amalgamation
Lazadores: Men who capture Indians for slave labor
Legua: League, 2.52 English miles
Magistral: Roasted copper pyrites, used in amalgamation
Malacate: Windlass, horse operated
Mena: Mineral vein, ores
Mina: Mine
Molino: Mexican arrastra, grinding mill
Molonque: Rich specimen, more than half silver
Muesca: Notched square post ladder
Negrillos: Black sulphide of silver
Oeste: West
Oro: Gold
Oroche: Bullion containing both gold and silver

Patio: Stone floor where silver ore is amalgamated
Pertenencia: Mining claim
Placeras: Placers, gold or silver
Planchera: Bar or ingot of bullion, sand and earth
Plata: Silver
Plata Mixta: Gold and silver mix
Plata Pella: Silver amalgam
Plata Pina: Silver bullion
Platino: Platinum
Plomo: Lead, lead ore
Pozo: Winze, inclined shaft
Puerto: Mountain pass
Quema: Retorting amalgam, roasting ore
Quintal: Weight equal to four arrobas, 100 pounds
Quinto: Mining tribute or tax paid to the Spanish King
Real de Minas: Group or cluster of mines
Reliz de Alto: Hanging wall of mine formation
Reliz del Bajo: Foot wall of mine formation
Rico: Rich
Rosicler: Ruby Silver
Socavon: Adit, mine tunnel portal
Sur: South
Tanatero: Mine laborer, ore-bag carrier
Tejo: Gold or silver ingot
Tentadura: A crude assay made by panning
Tepustete: Iron ore found in gold veins, placers
Tercio: Sack of ore weighing 150 pounds
Terrero: Mine dump
Tiro: Mine shaft, usually of two compartments
Vara: Measure equal to approximately 33 inches
Veta: Fissure vein, ore bearing
Zurron: Rawhide ore sack holding about 150 pounds

GENERAL INDEX